A Guide to Fluid Mechanics

This book is written for the learner's point of view, with the purpose of helping readers understand the principles of flow. The theory is explained using ordinary and accessible language, where fluid mechanics is presented in analogy to solid mechanics to emphasize that they are all the application of Newtonian mechanics and thermodynamics. All the informative and helpful illustrations are drawn by the author, uniting the science and the art with figures that complement the text and provide clear understanding. Another unique feature is that one of the chapters is wholly dedicated to providing 25 selected interesting and controversial flow examples, with the purpose of linking theory with practice. The book will be useful to both beginners in the field and experts in other fields, and is ideal for college students, graduate students, engineers, and technicians.

Hongwei Wang graduated from Beihang University with a PhD major in turbomachinery and has been teaching fluid mechanics for 20 years. His key publication is the Chinese edition textbook *Fluid Mechanics as I Understand It*, published in December 2014, followed by the second edition published in March 2019. This book is no.1 best seller in fluid mechanics at China's biggest online retailer.

A Guide to Fluid Mechanics

HONGWEI WANG
Beihang University, Beijing

Translated by

YAN ZHANG
School of Computer and Software Engineering, Nanyang Institute of Technology

Shaftesbury Road, Cambridge CB2 8EA, United Kingdom

One Liberty Plaza, 20th Floor, New York, NY 10006, USA

477 Williamstown Road, Port Melbourne, VIC 3207, Australia

314–321, 3rd Floor, Plot 3, Splendor Forum, Jasola District Centre, New Delhi – 110025, India

103 Penang Road, #05–06/07, Visioncrest Commercial, Singapore 238467

Cambridge University Press is part of Cambridge University Press & Assessment,
a department of the University of Cambridge.

We share the University's mission to contribute to society through the pursuit of
education, learning and research at the highest international levels of excellence.

www.cambridge.org
Information on this title: www.cambridge.org/9781108498838

DOI: 10.1017/9781108671149

Original Title: 我所理解的流体力学（第2版）

English edition translated by Yan Zhang

First published 2023

A catalogue record for this publication is available from the British Library.

A Cataloging-in-Publication data record for this book is available from the Library of Congress.

ISBN 978-1-108-49883-8 Hardback
ISBN 978-1-108-71278-1 Paperback

Contents

Foreword

This book is written from the perspective of learners. Its aim is to elucidate the physical principles of flow, rather than be oriented toward engineering calculations. There are no example solutions or exercises in this book, so readers can understand the principles of fluid mechanics and enjoy the beauty of fluid motion with a relatively easy and interesting reading experience.

Socrates said: "Education is the kindling of a flame." Learning is a very personal thing. Only learners themselves can determine the success of learning. It is quite common for teachers to be enthused and excited on the podium while their students sleep soundly below. Regardless of how extensive the content of a textbook may be, how in-depth its discussions, and how rigorous its logic, if no one is willing to read it, its value will not be appreciated.

Science books should not be strictly divided into textbooks and popular science books. It is very important to get readers interested and to understand the so-called profound theories in an accessible way. Textbooks addressed to students should analyze problems from the learner's perspective, so that more students can enjoy the beauty of science through them, rather than developing a love of science by reading popular accounts of it. Rigor and popularity need not exclude each other. Through our efforts, we strive not only to maintain the scientific level of the discussion, but also to make it easier for readers to understand.

It should be the responsibility and obligation of teachers to deeply understand the subject and then to present it in an even more understandable way. This is a creative process that may be called the "reprocessing" of knowledge. In fact, the knowledge we have acquired is more or less written after "reprocessing." As authors of science books, teachers should strive to conduct a deep "reprocessing" of knowledge. There is no need to write another book if it only repeats what has already been said in previous books. While original discoveries and inventions are certainly important, the "reprocessing" and dissemination of knowledge are the keys to wider application. Euclid's *Elements of Geometry* and Newton's *Mathematical Principles of Natural Philosophy* are classic theoretical books. Nowadays, however, the teaching materials for college students are not these abstruse works, but more understandable versions written by Euclid's and Newton's successors who have mastered that ancient erudition.

A good textbook is not simply a restatement of facts, but a creative process rich in contributions. When I was a student, in addition to learning in class, I also studied, as a hobby, the teachers' lecturing styles. By comparing the way my teachers and I understood a topic, I figured out why students understood when teachers lectured in a certain way, and why it wasn't easy for them to grasp a concept when it was taught in a different way. I finally became a teacher myself. I naturally love teaching and I am appreciated by my students. At the beginning my focus was on preparing lectures and teaching methods. Later, I paid attention to knowledge understanding and student responses, thus achieving a reverse transformation from educator to learner. Now I regard every class as a new learning opportunity. During classroom sessions new questions constantly pop into my head, and I can often deepen my knowledge and gain new insights, which is what every student should do in class. As I have my own unique understanding of what is taught, I believed that it should be written down for more people to see it and benefit from it. That's why I wrote this book.

However, there are risks in publishing my own understanding of fluid mechanics. One's interpretations could be faulty, or not rigorous enough. Will these shortcomings mislead students? I think this is why, although many teachers can make their teaching lively in class, the textbooks they write are obscure or difficult to understand. If we faithfully follow classical works and take rigor as the highest priority, it is not necessary to write another introductory book on some well-established area such as fluid mechanics. Therefore, I decided to take the risk and write a book based on my personal understanding of the subject, which I believe will be helpful to junior readers of fluid mechanics.

Now, let me introduce the contents and characteristics of this book. It is not a popular science book, but can be used as a textbook. For this, it only needs to be supplemented with examples and exercises. There are numerous formulas and derivations in the book, even more than in many undergraduate textbooks. It is said that each additional formula will scare one reader away. I admit that this claim may be right. However, scaring your readers away does not necessarily need formulas. There are actually very few formulas in Newton's *Mathematical Principles of Natural Philosophy*, but it is not any easier to read than modern textbooks that contain plenty of formulas. After all, mathematics is the language of science, and I have no intention of weakening its role. On the contrary, I even hope that readers will have a deeper understanding of some mathematical concepts through the application of them to mechanics.

Compared with existing teaching materials and books on the same topic, this one has some distinctive characteristics, among which the many exquisite color pictures are the most intuitive. All of these pictures have been hand-drawn by myself. Of course, some drawings refer to relevant books, but I tried to strike a balance between scientific accuracy and aesthetics. I can guarantee that all curve graphics can be directly used as a reference for engineering applications, and all flow images are in line with the actual conditions.

In the final chapter I included 25 interesting and useful flow examples for in-depth analysis, so that readers can enjoy the experience of learning and applying their knowledge. For example: What is the shape of falling raindrops? Why will outlet velocity increase if you squeeze the outlet of a watering hose? As long as their thinking is inquisitive, anyone who has learned the basics of fluid mechanics should be able to explain these everyday phenomena.

This book is suitable as a supplementary textbook for students, as well as for self-study material for engineering and technical personnel. Readers who are studying fluid mechanics for the first time, and using this book as a textbook or as self-study material, will find that a large number of concepts in classical physics, theoretical mechanics, and solid mechanics are used. Therefore, they do not need to regard fluid mechanics as a completely separate discipline, which will make their learning easier. By placing understanding at its core, this book is also highly suitable as a textbook for those who have studied fluid mechanics before and seek to refresh their knowledge of it.

I hope that this translated version of the book brings a new experience to English-language readers, and I would be very happy if it could also provide them with a deeper understanding of some facts or concepts.

I am indebted to my alumni Dr. Yan Zhang, who translated the entire book from Chinese into English, for his elaborate work. Also, the extensive efforts and excellent work of Prof. Arturo Sangalli are truly appreciated, for the intensive grammar checking and creative text polishing.

Nomenclature

Notation

$\vec{f}$	Vector quantities
$\overline{f}$	Average of f
f'	(1) Derivative of f
	(2) Perturbation of f
f^*	Dimensionless value of f
f_{cr}	Critical value of f
f_∞	Value of f far away from the point of interest
Δf	Change of f
δf	Infinitesimal change of f
$\mathrm{d}f$	(1) Differential of f
	(2) Infinitesimal change of f
$\mathrm{D}f/\mathrm{D}t$	Material derivative of f

Letters

a	(1) Speed of sound
	(2) Acceleration
A	Area or surface
AR	Diffuser or nozzle area ratio (exit area/inlet area)
B	Volume
const	A constant
c_{p}	Specific heat at constant pressure
c_{v}	Specific heat at constant volume
C_{f}	Skin friction parameter
C_{p}	Pressure rise coefficient
C_{D}	Drag coefficient
d	Diameter
D	(1) Diameter
	(2) Drag force
e	Total energy per unit mass

E	Total energy
Eu	Euler number
f	(1) Force per unit mass
	(2) Friction factor
f_b	Body force per unit mass
F	Force
Fr	Froude number
g	Gravitational acceleration
G	Gravitational force
h	(1) Enthalpy per unit mass
	(2) Height
h_t	Stagnation enthalpy per unit mass
H	Boundary layer shape factor
i	Imaginary root
k	(1) Thermal conductivity
	(2) Specific heat ratio
L	Length
m	Mass
$\dot{m}$	Mass flow rate
M	Molar mass
Ma	Mach number
n	Normal unit vector
p	Pressure
p_0	Atmospheric pressure
p_t	Stagnation pressure
$\dot{q}$	Rate of heat per unit mass
$\dot{q}_x$	Rate of heat per unit mass per unit area
$q(\lambda)$, $q(Ma)$	Mass flow function
$\dot{Q}$	Rate of heat
r, R	Radius
$\vec{r}$	Position vector
R	Gas constant of air
R_0	Universal gas constant
Re	Reynolds number
s	(1) Entropy per unit mass
	(2) Streamwise unit vector
S	Entropy
St	Strouhal number
t	Time
T	Temperature
T_t	Stagnation temperature
u, v, w	Velocity components in Cartesian coordinates
u_i	Velocity components in Tensor form

$\hat{u}$	Internal energy
U	Reference velocity or characteristic velocity
v	(1) Velocity component in y direction
	(2) Specific volume (volume per unit mass)
V	Velocity magnitude
w	(1) Velocity component in z direction
	(2) Work per unit mass
w_s	Shaft work per unit mass
W	Work
We	Weber number
x, y, z	Cartesian coordinates

Symbols

α	Planar diffuser half-angle
β	Shock angle
Γ	(1) Stress
	(2) Circulation
δ	(1) Deflection angle
	(2) Boundary layer thickness
δ_{ij}	Kronecker delta
δ^*	Boundary layer displacement thickness
Δ	Difference of change
ε	Strain rate
η	Dimensionless distance from wall
θ	(1) Boundary layer momentum thickness
	(2) Circumferential coordinate
λ	(1) Coefficient of thermal conductivity
	(2) Coefficient of velocity
μ	Dynamic viscosity
ν	Kinematic viscosity
ρ	Density
ψ	Stream function
τ	Shear stress
ϕ	(1) Velocity potential
	(2) Some mechanical property per unit mass
Φ	Some mechanical property
Φ_v	Dissipation function
ω	Vorticity
Ω	Angular velocity

Subscripts

b	Body (as in body force)
c	Center or core
cr	Critical condition
cv	Control volume
D	(1) Diameter (as in Reynolds number)
	(2) Drag (as in drag force)
e	Exit station
i	Inlet station
i, j, k	Indices of Tensor
n	Normal direction or component
r	Radial direction or component
s	Streamwise direction
sys	System
t	Stagnation condition
x, y, z	Components in x, y, z directions

1 Fluids and Fluid Mechanics

The wind blows into the sails, the ship sails through the huge waves, and the seabirds fly in the sky. The forces exerted by fluids are everywhere in our daily lives.

1.1 Fluids: Basic Concepts

Fluids are substances that deform more easily than solids, and includes liquids and gases. However, an object deforms only under forces, and whether it deforms easily depends on the intensity and type of forces acting on it. As a solid, nylon rope has high tensile strength but offers practically no resistance when pressing on it, and can also be easily cut by scissors. Therefore, we need a more rigorous definition to distinguish between fluids and solids.

The key difference between solids, liquids, and gases lies in the microstructure of the material. Figure 1.1 shows the microstructure in the three states of water, with oxygen and hydrogen atoms represented by small balls. The close-packed molecules of a solid strive to retain a fixed arrangement, while those of a liquid have no such

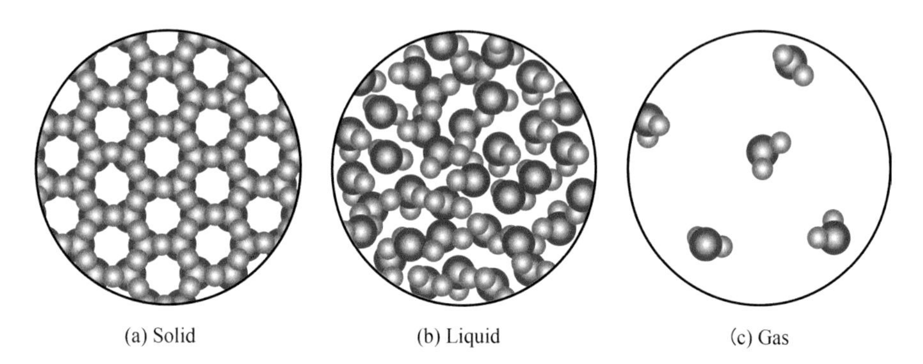

Figure 1.1 Microstructure in the three states of water.

intention. The wide-spaced molecules of a gas neither huddle together nor aim to retain a fixed arrangement.

Now we can try to understand the features of the three states of matter. A solid or liquid, being composed of closely packed molecules with strong cohesive forces, tends to retain its volume unless it is subjected to enormous pressure. On the other hand, under the action of external pressure the distance between the molecules of a gas will decrease, producing a change in overall volume.

Liquids share with gases the ability to flow because the atoms and molecules in both states have no tendency to be rigidly packed in a fixed pattern. Although tightly packed, the molecules in a liquid can be arranged randomly, causing the liquid to have a definite volume but no definite shape. Since the molecules in a gas move freely and independently of each other, the forces between them can be ignored, except when they collide with each other.

We understand that if the resultant force-torque acting on a rigid body equals zero, the object remains stationary or in uniform motion. However, this is not true for a real solid or fluid. Any solid material has a strength limit. Even if the resultant force-torque acting on it equals zero, the solid material inside may experience a tensile stress, a compressive stress, or a shear stress. A solid deforms under such stresses. Furthermore, stresses beyond the strength limit cause the solid material to yield and to fracture. At the microscopic level, the intermolecular (or interatomic) chemical bonds at the fracture point are broken, losing their interactive forces; this means the material cannot retain its original size and shape.

A fluid differs substantially from a solid in this respect. If a fluid is subjected to compressive stresses only, it will deform like a solid and remain stationary. When subjected to shear stresses, a fluid will deform continually and be unable to resist the external shear stresses in static equilibrium. This is somewhat similar to the case in which the forces acting on a solid far exceed its strength limit. But a fluid at rest has zero "shear strength" and will continually deform under the action of any small amount of shear force. Therefore, the essential difference between fluids and solids is that *a fluid cannot generate internal shear stresses by static deformation alone.*

To better understand the difference between fluids and solids, let us discuss the static friction between two solid surfaces. Place a solid block on a plane and gradually

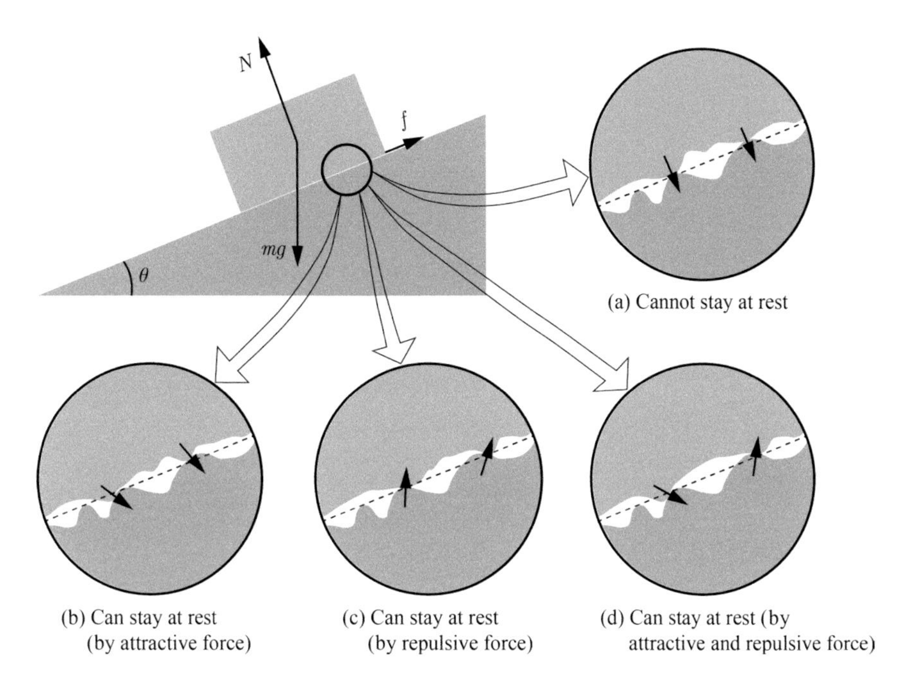

Figure 1.2 The forces exerted on a solid block on an incline and corresponding microscopic explanation of the static friction between two solid surfaces.

lift one end of the surface. Within a certain range of inclination angle the block will not move because the component of gravity acting along the inclined plane equals the static friction. Figure 1.2 shows the forces acting on the block (represented by arrows) and the corresponding microscopic explanation of the static friction between the two surfaces. As is well known, static friction results from the electromagnetic forces between molecules (or, similarly, atoms) that are in contact. In order to generate shear forces, the electromagnetic attractive and repulsive forces should not be perpendicular to the contact surface but have one component acting along it. This component may be caused by one of two factors, corresponding to rough and flat surfaces, respectively.

In general, as shown in Figure 1.2, neither the bottom of the block nor the inclined plane is a perfectly flat surface. There are some attractive and repulsive forces only between the molecules on the very small contact area between the two surfaces. When the plane starts to tilt from horizontal, the block will slide down an imperceptible distance. At the microscopic level, when the block slides down, as shown in Figure 1.2(b)–(d), the attractive and repulsive forces between the bulges of the block and the plane offer an upward component acting along the contact surface. This generates a shear force that keeps the block in static equilibrium.

If two perfectly flat planes at the molecular level are placed in contact with each other, is there any friction between them? Not only friction is present in this case, but it can be very high. This is because the contact area is quite large, and a certain form of molecular bond forms between the molecules of the two objects, both of which are similar to the situation inside a single object, resulting in a remarkable shear stress.

Figure 1.3 Water droplets at rest on a glass pane.

The contact between a liquid and solid interface is somewhat like that of two perfectly flat solid surfaces – that is to say, a full intermolecular contact. However, unlike the static frictional force occurring between two solid surfaces, there is no such force between a liquid and a solid interface. A typical example is a boat floating on water, which can be moved by an arbitrarily small amount of thrust. The liquid molecules adjacent to the solid surface will stick to it through the intermolecular attractive forces, forming a relatively fixed structure. However, these stuck molecules cannot apply a shear force to the adjacent molecules further inside the liquid. As a result, the solid has no way of exerting a shear force on the fluid through static displacement alone.

Even so, we notice water may flow at a uniform velocity along an inclined riverbed, which means there is a frictional force between the water and the inclined ground to balance the component of gravity along the riverbed. That is to say, *a frictional force does indeed occur in a moving fluid*. This property of a fluid is known as *viscosity*, and will be discussed in detail in the next section.

Here's an interesting counterexample: Water can remain stationary on inclines, even on vertical walls or glass. Consider water droplets clinging to window panes, for example, as shown in Figure 1.3. The water droplet is like a type of solid cantilever structure inside which internal shear stress occurs. Consequently, shouldn't there be shear stress in these water droplets?

Conclusion first: In this case there is still no shear stress. To further analyze this situation, we need to take into account the surface tension of a liquid, a concept that will be discussed later.

1.2 Some Properties of Fluids

The definitions of many properties of fluids, including density, pressure, and temperature, are consistent with those that apply to solids. However, liquids possess other properties as well, such as viscosity and surface tension. Viscosity is an important characteristic of fluids from the viewpoint of fluid mechanics. The surface tension arises due to cohesive interactions between the molecules on the liquid surface.

1.2.1 Viscosity of Fluids

Any shear force externally applied to a fluid, no matter how small, will cause a continuous deformation of that fluid. Viscosity is a property of a fluid that describes the shear stress caused by such a continuous shear deformation. Common liquids and gases are viscous, and only superfluids (such as liquid helium near absolute zero) can be assumed to be inviscid.

Before discussing the viscosity of a liquid, we need to clarify the difference between viscous and adhesive forces: these are in fact completely different concepts. The viscous force is the shear force exerted in a moving fluid, while the adhesive force is the attractive force between molecules of two different materials. It is the adhesive force between the molecules of the glue and those on the solid surface that holds them together, rather than the viscous force discussed here. They cannot be held together by the viscous force since it does not exist in a static fluid. When a solid object moves through a fluid it is subjected to the frictional force on its sides. This force results from the combined action of the adhesive force between the solid surface and its adjacent fluid molecules, and the viscous force within the fluid.

We know that honey has a much higher viscosity than water. Let us discuss the intrinsic factors that determine the magnitude of viscous forces in a fluid by analogy with the frictional force between two solid surfaces. For a block sliding down an incline, its dynamic friction is equal to the product of the coefficient of friction and the normal force between the two solids in contact. The coefficient of friction is a numerical representation of the magnitude of the intermolecular force between neighboring molecules of the solid and the contact surface roughness. This is easy to understand. But why is the dynamic friction proportional to the normal force between the two solids in contact? After all, the normal force is perpendicular to the motion and should not affect the magnitude of the friction force.

The reason is that the greater the normal force, the larger the contact area between the two solid surfaces. Since there is essentially a linear relationship between contact area and normal force, the frictional force is proportional to the normal force. If two solids come into gapless contact with each other at the molecular level, the friction force should be independent of the normal force.

For solids, the coefficient of dynamic friction does not in general equal that of static friction due to different mechanisms behind frictional force. Static friction is a consequence of the balance of forces exerted on a stationary solid, while dynamic friction also involves the conversion of kinetic energy into internal energy. When two adjacent solid surfaces slide relative to each other, not only do forces similar to static friction occur on the contact area, as in Figure 1.2, but also the bonds between the molecules of both objects are constantly broken and reformed, with molecules and molecular groups falling from the objects. Accompanied by the change in molecular vibrational

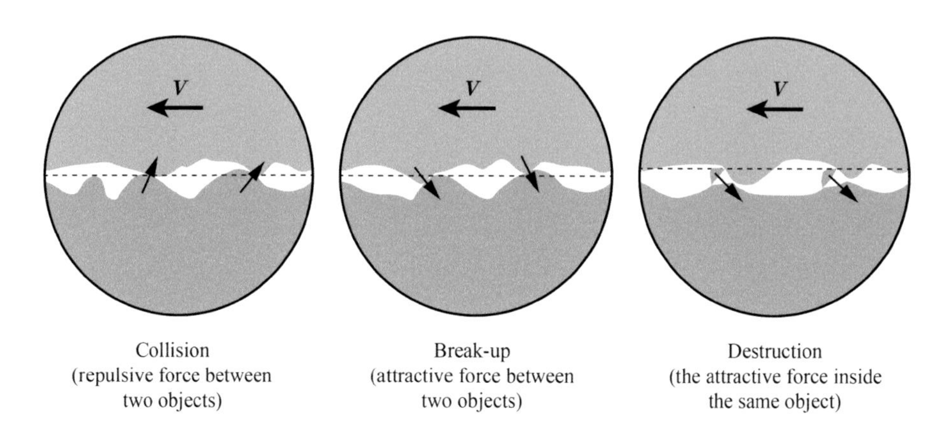

Figure 1.4 Microscopic explanation of the dynamic friction between two solid surfaces.

energy and the directional disorder of molecular motion, friction inevitably produces heat. Figure 1.4 shows three typical phenomena occurring at the contact area.

Whenever a fluid and a solid come into contact with each other, the molecules of the fluid right next to the solid surface will stick to and move with it. Thus, *the friction between a fluid and a solid surface is actually that between two fluid layers*. In other words, the friction between solid and fluid surfaces relates only to properties of fluids, independent of the properties of solid materials.

Since the liquid and the solid surface come into full contact at the molecular level, the frictional force should be independent of the normal force (i.e., the pressure). This explains why the magnitude of the viscous force in a liquid is basically independent of pressure. On the other hand, viscous force is closely related to the temperature of the liquid. *In general, for liquids, the higher the temperature, the lower the viscosity*. For example, a cold, thick syrup becomes less sticky when heated. Contrary to liquids, *the viscosity of gases increases with temperature*. The difference in viscosity between liquids and gases is directly related to the physical mechanism generating it.

If the velocities between two adjacent layers of liquid are different, the slower-moving layer is dragged along by the faster-moving one by intermolecular force. As the momentum of the molecules in the faster-moving layer decreases, that of the molecules in the slower-moving one increases. From a macroscopic point of view, the momentum transfer is caused by the frictional forces between the two fluid layers. Since friction is an internal force when we look at the two layers of liquid as a single system, the total momentum of the system is conserved. Figure 1.5 shows how the slower-moving liquid layer is dragged along by the adjacent faster-moving layer at the molecular scale.

The friction caused by the intermolecular forces in a liquid is somewhat similar to that acting between two solid surfaces in contact with one another. Unlike solid molecules, liquid molecules do not always flow in their respective layers, but instead move freely among different layers. In other words, the liquid molecules in each layer have random diffusion perpendicular to the direction of motion. Some molecules

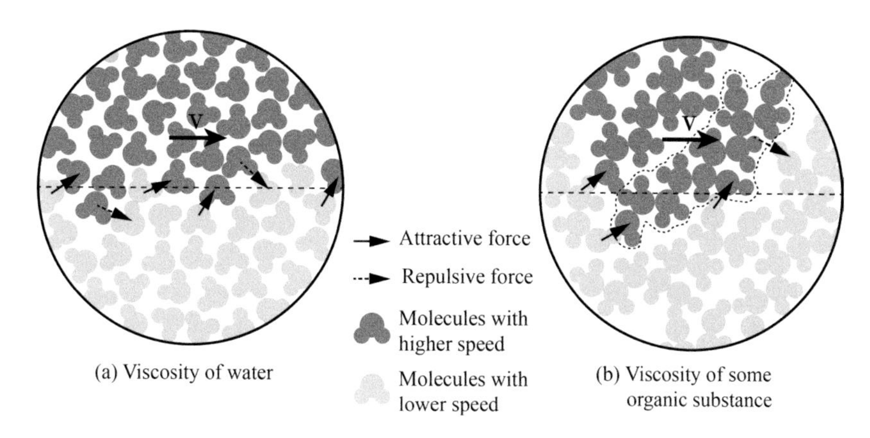

Figure 1.5 Microscopic explanation of the viscous forces in a liquid.

from the faster-moving layer may intrude into the slower-moving one, where after one or several collisions they are decelerated to move along with their neighbors. At the same time, some molecules from the slower-moving layer may intrude into the faster-moving one, where they are accelerated to move along with their neighbors after undergoing one or multiple collisions. The transfer of momentum taking place through the interaction between two adjacent layers is reflected as frictional forces along the direction of motion. For liquid molecules in laminar flow, the frictional forces generated by momentum transfer, being much smaller than their intermolecular attractive and repulsive forces, are usually ignored. Consequently, the viscosity of a liquid is considered to be caused by intermolecular forces, among which the attractive forces dominate.

Daily life experiences show that liquids of organic substances, such as paint, honey, and plasma, generally exhibit high viscosity. The large, irregular, and polar molecules of these substances are difficult to move but get easily tangled with each other. For example, glycerin molecules are not very large, but they are noncircular and polar. Glycerol molecules are more likely to get stuck when they pass over rough surfaces, and the molecules get tangled easily, so glycerol is highly viscous. On the other hand, mercury molecules are larger than glycerol molecules, but they are spherical and nonpolar. Therefore, mercury, though much more viscous than water, is less viscous than liquid organic substances of the same molecular weight. Figure 1.5(b) shows the shearing motion in a liquid organic substance, where the region enclosed by dashed lines represents a large organic molecule. As can be intuitively seen, the molecules of this organic substance, compared to the water molecules in Figure 1.5(a), are large and irregular in shape, inevitably experiencing substantial tangle and therefore having higher viscosity.

As the temperature of a liquid increases, the vibrations of individual molecules intensify and the tangle between them becomes more easily loosened, leading to a decrease in viscosity. This can be explained by analogy to the following example: When carefully positioned on a wet and muddy surface, a cellphone can be kept from

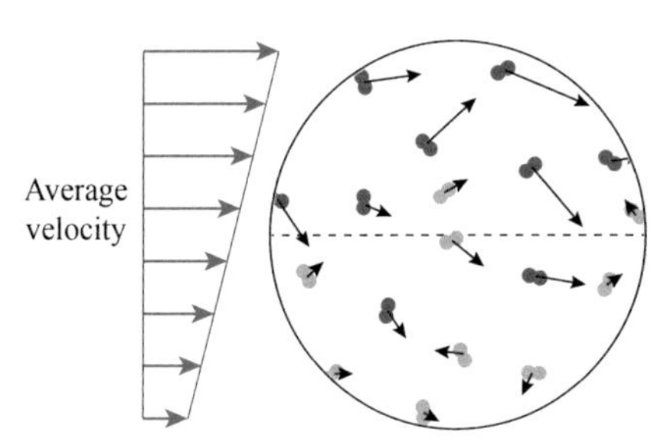

Figure 1.6 Microscopic explanation of the viscous force in a gas.

sinking for quite a long time; if someone calls, the cellphone will vibrate and quickly sink. Generally speaking, the more intense the tangle between molecules, the more sensitive viscosity is to temperature. Syrup, with elongated molecules, shows a rapid reduction in viscosity as temperature rises, but temperature has a much smaller impact on the viscosity of mercury, which has spherical molecules.

However, not all liquids show a reduction in viscosity with increasing temperature. For example, the viscosity of a single-component lubricant decreases rapidly with increases in temperature. This causes huge problems for the lubrication of mechanical parts, such as those in engines used in a wide operating temperature range. To overcome this shortcoming, a substance with large molecules whose shape changes from spherical to elongated as the temperature increases is added to modern lubricants in order to make their viscosity less sensitive to temperature.

Attractive and repulsive forces between gas molecules are negligible. The momentum transfer due to frequent collisions between molecules produces pressure force in a gas. The collisions between molecules can produce not only a normal stress such as pressure, but also a shear stress, such as viscous force. The latter may be considered as a frictional force acting between adjacent fluid layers traveling at different velocities. For instance, the molecules in the faster-moving layer constantly fly into the slower one due to random thermal molecular motion, dragging the slower neighboring molecules to accelerate; the molecules in the slower-moving layer also constantly fly into the faster one, pulling back the faster neighbors to decelerate. From a macroscopic point of view, the viscosity of a gas represents the drag forces caused by the momentum transfer between two adjacent layers of the gas. As remarked earlier, this kind of momentum transfer also exists between two adjacent layers of a liquid, but it is negligible compared to intermolecular forces. In a gas, however, molecules exert almost no forces on each other, but are themselves very active. It is the momentum transfer between the molecules that creates the viscosity of the gas. Figure 1.6 shows the mechanism responsible for gas viscosity. As slower molecules frequently collide with faster ones, the former accelerate while the latter decelerate, thereby resulting in viscosity.

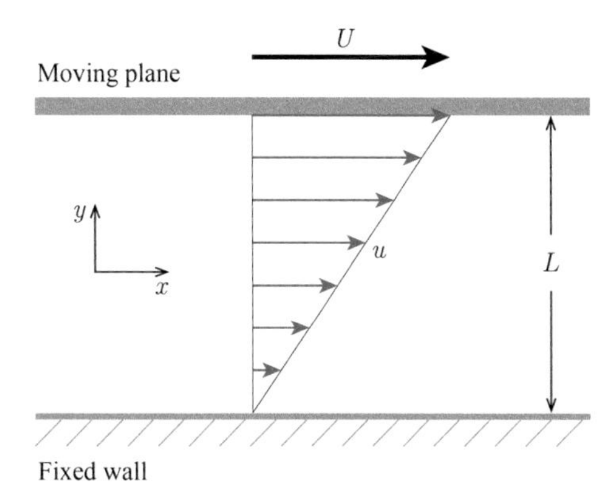

Figure 1.7 Newton's experiment for the viscous force in a liquid.

As the temperature of a gas increases, the thermal motion of the molecules intensifies, and the frequency of their collisions increases, generating more exchange of momentum between the individual gas layers. This is why the viscosity of gases increases with temperature. Just as in liquids, viscosity in gases is essentially pressure-independent. How can this phenomenon be explained?

For an ideal gas, pressure is determined by temperature and density. When the temperature is constant, a change in pressure results in a change in density and vice-versa. Therefore, we reach a conclusion that the viscosity of an ideal gas is independent of its density at constant temperature. It may seem odd that a denser gas does not generate a larger viscous force. When density increases, the collision frequency between molecules must also increase. Why is the viscosity of a gas independent of its density?

As shown in Figure 1.6. If the density of a gas is increased, more molecules in the faster-moving layer indeed fly into the slower layer in unit time. However, due to the decrease in mean free path, the two molecules in collision would have less velocity difference and therefore the momentum transfer caused by a single collision decreases. The percentage of decrease in the momentum transfer due to a single collision is equal to the percentage of increase in density, resulting in a constant momentum transfer. As a result, the change in density does not affect the magnitude of viscous force.

The above analysis, based on kinetic molecular theory, is qualitative. Quantitative derivation of viscosity can be found in some textbooks and will not be discussed here.

Classical mechanics is mainly concerned with the motion of macroscopic objects. Isaac Newton (1642–1727) was the first to attempt a quantitative study of the viscosity of liquids. In 1686, he measured the viscosity of liquids experimentally and established Newton's friction law to describe the frictional force in a fluid.

To make the flow model as simple as possible, Newton attempted an experiment in which the liquid in question was sandwiched between two large parallel horizontal plates, as shown in Figure 1.7. The bottom plate is fixed, and the top plate moves at a steady horizontal speed. When a fluid flows over a solid surface, the layer right next to the surface will attach to it. The fluid attached to the bottom plate remains stationary, while that attached to the top plate moves with a constant velocity of U. Newton's law of viscosity states that the force required to pull the upper plate is proportional to the velocity with which the top plate moves and inversely proportional to the distance between the plates – that is,

$$\frac{F}{A} \propto \frac{U}{L},$$

where F is the force required to pull the upper plate; A is the area of the contact surface between the fluid and the plate; U is the top plate velocity; and L is the distance between the two plates.

Each successive fluid layer from the top down exerts a force to the right on the one below it. Since the pressure is constant in the fluid, the shear forces between any adjacent fluid layers are the same, and relate to the pulling force to the upper plate. Newton found that the fluid velocity varies linearly with y. He stated that the shear stress between adjacent fluid layers is proportional to the velocity gradient between two layers, written as:

$$\tau = \mu \frac{\partial u}{\partial y}, \tag{1.1}$$

where τ is the shear stress; u is the horizontal velocity in the fluid; μ is a measure of the viscosity of a fluid, known as the coefficient of viscosity, or coefficient of dynamic viscosity. The coefficient of viscosity varies greatly from one fluid to another, while that of the same fluid changes with temperature. There exist tables listing the coefficients of viscosity that have been experimentally determined for various liquids and gases.

The fluids used in Newton's experiments satisfy Equation (1.1). All gases and many common liquids with small viscosity obey Newton's law of viscosity and are accordingly called Newtonian fluids. *A Newtonian fluid is a fluid in which the viscous stresses are linearly correlated to the local strain rate.*

But there are many fluids in nature that do not satisfy this linear relationship; these are called non-Newtonian fluids. Generally speaking, Newtonian fluids are less viscous, while most more viscous fluids, such as paint, honey, and plasma, are non-Newtonian. Unlike Newtonian fluids, most non-Newtonian fluids are composed of molecules or particles that are much larger than water molecules. Since these molecules are tangled up in the flow field where the velocity gradients exist, non-Newtonian fluids exhibit a more complex relationship between shear stress and velocity gradient than simple linearity. Furthermore, some non-Newtonian fluids have a shear stress, or viscosity, that depends not only on velocity gradient but also on time or deformation history.

1.2.2 Surface Tension of Liquids

Surface tension is an amazing force, often used in science shows to draw a crowd, such as through coins floating on water and oversized soap bubbles. Figure 1.8 shows one of the most common surface tension phenomena. Figure 1.8(a) shows an ellipsoidal droplet of water on a plane, and Figure 1.8(b) shows the diagram of forces on water molecules near the droplet's surface. As can be seen, any water molecules inside the

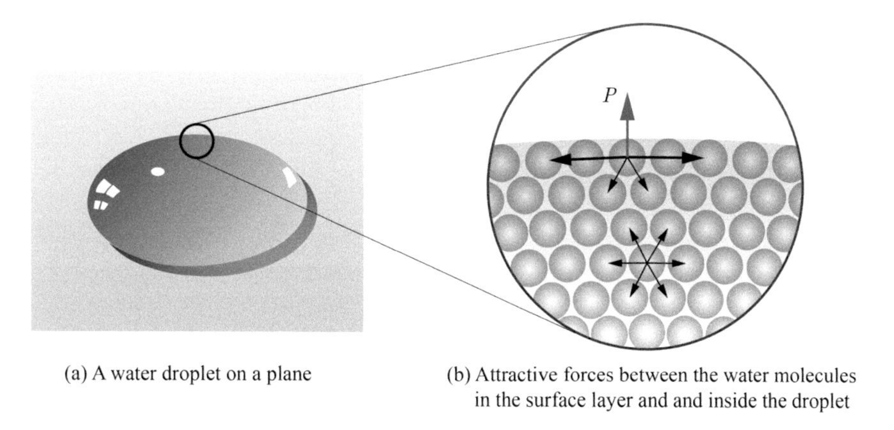

(a) A water droplet on a plane

(b) Attractive forces between the water molecules in the surface layer and and inside the droplet

Figure 1.8 An ellipsoidal droplet of water and the diagram of forces on water molecules inside it.

droplet do not experience a net force because the weak attractive forces exerted by the neighboring molecules all cancel out. On the other hand, since the water molecules in the surface layer are a bit farther apart than those inside the droplet, they are more attracted to each other and to the layers of water below. They are not pulled upwards, but only from the sides and downwards. Surface tension is viewed as the net force of attraction among the water molecules at the surface.

Surface tension seeks to minimize the surface area, so that all water droplets tend to be pulled into a spherical shape. Due to the phenomenon of surface tension, the pressure inside the droplet will always be higher than the atmospheric pressure on the outside. Therefore, in addition to bonding to each other, the water molecules at the surface are also subjected to a repulsive force exerted by the internal water molecules. The repulsive force P is directed outwards, away from and perpendicular to the surface, as shown in Figure 1.8.

Water is especially suitable for demonstrating surface tension since it has the greatest surface tension of any liquid other than mercury. The reason is that the attractive forces between water molecules include an unusually strong type of hydrogen bond between the hydrogen and oxygen atoms, which is stronger than the other intermolecular forces. Organic solvents have lower surface tension than water. For example, the surface tension coefficient of alcohol is only one-third of that of water, thereby forming much smaller droplets. Some surfactants, such as soaps and detergents, can significantly reduce the surface tension of water, playing an important role in human life and production. There is a common misconception that soapy water has greater surface tension than pure water. But in fact, the soap added to water decreases the pull of surface tension. The principle behind a soap bubble is that a water film does not quite have the same thickness all over. With soap added, where the soap film is thinner, there will be less soap and the surface tension will be higher, and thus the film is stronger; where it is thicker, there will be more soap and lower surface tension, and thus the film would be stretched thinner. In other words, soapy water forms a bubble with more uniformed thickness, allowing the soap bubble to last a longer time.

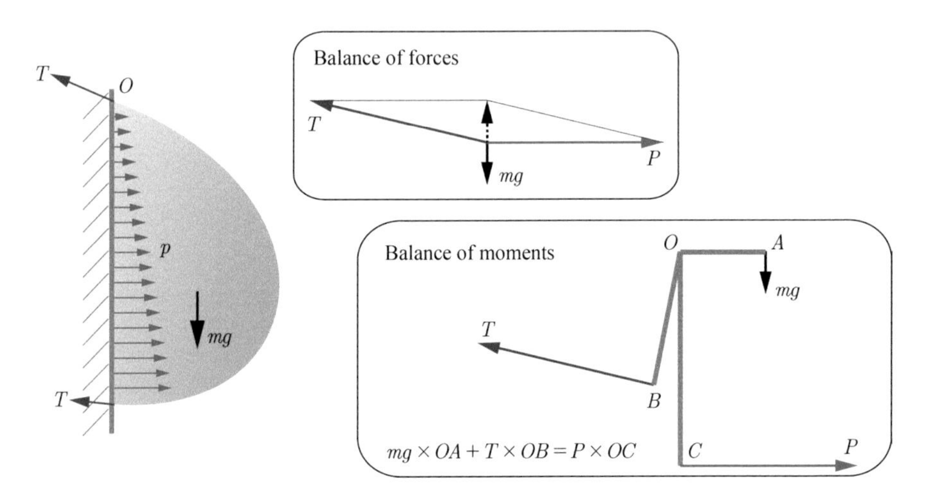

Figure 1.9 Force field analysis on the water droplet positioned on a vertical glass surface.

Take a look at the water droplets in Figure 1.3. Force analysis is carried out as shown in Figure 1.9. The water molecules in the surface layer play a key role in shaping the droplet. There won't be any attractive force, on average, on any single molecule inside the droplet from the surrounding molecules. Simply wrapped by a surface film of water, the droplet can remain at rest on the glass surface. Analogically, if you wrap a piece of cloth around some dry sand and pin all sides of the cloth to the wall, the sand bag will stay at rest. There can be no shear stress between the sand particles as long as the cloth is durable enough and pinned tightly to the wall.

For a water droplet to remain stable on a vertical glass surface, the surface film of water must be shaped as shown in Figure 1.9. Under gravity, the pressure in the lower part of the droplet is greater than that in the upper part. Thus, the curvature at the bottom of the surface film will be greater than that at the top to generate a higher pressure inside the droplet. The total tensile force between the surface film of water and the vertical glass surface, T, must be directed obliquely upward in order to balance the weight of the water droplet without the need for shear stress. The support force exerted upon the water droplet by the glass surface, P, indispensable to achieve balance, is directed horizontally. The most critical reasons why the water droplet can remain attached to the glass surface are the adhesive forces between the surface film of water and the vertical surface, and this surface film holding the droplet together. Through this example we seek to emphasize the highly important fact that *there is no shear force within a static fluid.*

1.2.3 Equation of State for Gases

The equation relating the values of pressure, volume, and temperature of a gas is known as the equation of state. A theoretical gas composed of many randomly moving point molecules that are not subjected to intermolecular forces is known as an ideal gas. The state of an ideal gas may be expressed by the following formula:

$$p = \rho \frac{R_0}{M} T,$$

where p is the pressure of the gas; ρ is its density; T is its absolute temperature; $R_0 = 8.314$ J/(mol·K) is the ideal gas constant; and M is its molar mass (the unit will be expressed in kilograms for use with the International System of Units).

The equation of state for an ideal gas is applicable to most real gases in which the mean free path of a molecule is relatively large at moderate pressure and temperature. The equation of state for air is usually abbreviated as

$$p = \rho RT, \tag{1.2}$$

where $R = 287.06$ J/(kg·K) is the gas constant of air.

Here it is necessary to discuss in depth the three macroscopic physical properties of a gas, namely, density, temperature, and pressure. Further discussion will inevitably involve gas kinetics and statistical physics on a micro-scale. In order to make it clear and concise, the discussion in this sector will be carried out from a qualitative perspective.

First, *the density of a gas is defined as its mass per unit volume*. Strictly speaking, since a gas is composed of molecules in thermal motion that are separated from each other by distances far greater than their own sizes, density cannot be defined at a point in space. The continuum hypothesis, however, considers fluids to be continuous, so the density of a gas at one point in space is actually the average mass of a small volume element at that point divided by its volume, and the element is large enough to contain many molecules. For a particular gas, the magnitude of density represents the number of molecules in a given volume.

Second, *the temperature of a gas is the measure of the average kinetic energy of all the gas molecules*. For a given gas, the magnitude of the average velocity of molecular thermal motion depends directly on temperature. For the gas in a specific confined space (i.e., density is constant), an increase in temperature obviously increases the collision probability between molecules, the number of collisions per second between molecules and walls, and the intensity of single collisions, which will be reflected as an increase in the pressure of the gas.

The pressure of a gas represents a tendency to expand itself, or its thrust against the adjacent solid–liquid interfaces or the vessel walls. The attractive and repulsive forces between gas molecules are negligible, and this expansion and thrust arise from the collisions of some gas molecules with each other and with the vessel walls. The momentum theorem states that the magnitude of pressure depends on the amount of momentum transferred from the gas molecules to the vessel walls per unit time. Two factors influencing the total amount of momentum transfer are the magnitude of the momentum transfer to the wall due to a single collision, and the number of molecules transferring the momentum. The former factor is determined by the temperature of the gas, and the latter depends on both its density and temperature. On the one hand, higher temperature increases the average kinetic energy and therefore the average velocity of the gas molecules. Thus, the molecules collide with the vessel walls more frequently and with greater momentum transfer in a single collision. On the other hand, higher density increases the number of molecules hitting a unit area of the wall per unit time. Therefore, the pressure of a gas is positively correlated to its density and temperature, reflected as the relation in Equation (1.2).

By the way, unlike gases, there's no unified ideal equation of state for liquids. Liquid density generally decreases with increasing temperature, with different rates of change for different liquids. It is difficult to describe the rates of change in density with a unified theory; these can be affected by many factors and are usually measured by experiments. However, liquid water is an exception to this rule: When its temperature rises from 0°C to 4°C, its density increases instead of decreasing as expected. Water reaches its maximum density at about 4°C. The pressure inside a liquid is independent of its temperature and density, but directly depends on the external forces exerted upon it.

1.2.4 Compressibility of Gases

Compressibility represents the relative change in the volume or density of a certain amount of substance as a response to a pressure change. The molecules in a solid or liquid are so tightly packed together that any attempt to bring them closer will be very hard due to significant intermolecular repulsive forces, so both solids and liquids are very difficult to compress. Consider water, for example. At room temperature (20°C), its density at 1 atm is 998.2 kg/m^3, while at 100 atm it has only a slight increase of less than 0.5% to 1002.7 kg/m^3.

Unlike solids and liquids, gases have larger intermolecular distances and negligible intermolecular forces, making them highly compressible. The equation of state for an ideal gas states that the density of a gas is affected by its pressure and temperature. For an isothermal (constant temperature) compression/expansion of a gas, its density is directly proportional to pressure, while for an isobaric (constant pressure) compression, its density is inversely proportional to temperature.

In fluid mechanics the so-called compressibility refers specifically to the compressibility of flow rather than the compressibility of a fluid. *Compressible flow is defined as a flow in which the volume of a fluid element undergoes a significant change.* Gases can be compressed much more easily than liquids. However, whether a gas would be compressed in a specific flow also depends on the external forces exerted on it. The gas enclosed in a cylinder can be easily compressed with a piston, but most of the time we deal with the flow of gases or air in open space. Being unconfined in all directions, the gas elements can move out of the way in time, thus avoiding being compressed. Air is a compressible fluid that displays as incompressible flow at normal speeds. For example, if a car is moving at a speed of 120 km/h, the air hit by the car is compressed very slightly (with a 0.5% or lower increase in density) and then disperses in all directions.

As we can see, the gas in open space won't be compressed as long as its escape velocity keeps up with the velocity of the moving object compressing the gas, so whether or not the gas is compressed depends on both the compression and escape velocities. In the previous example, if the compression velocity is that of the car, then what is the escape velocity of the gas? The escape velocity is related to the thermal velocity of gas molecules, macroscopically reflected in the speed of sound in the gas. *In general, when the velocity of a gas flow is below 0.3 times the speed of sound, the relative change in gas density remains negligible and the flow can generally be treated as incompressible.* Of course, this standard can be raised or lowered appropriately depending on the required accuracy. (The relationship between the relative change in gas density, the velocity of the gas flow, and the speed of sound in the gas will be discussed in detail in Chapter 7.)

The compression caused by air deceleration can also be understood as the effect of the inertial force of the gas behind on the gas in front during deceleration. The inertial force arises from a change in the magnitude or direction of the velocity of the gas flow. In a high-speed, rotating vessel, the pressure of the gas varies along the radial direction, correspondingly creating a radial density gradient. Even if the gas flows slowly from the center to the periphery of the vessel, there will be a significant change in its density. This type of compressible flow is caused by the effect of the centrifugal force.

Gravity can and does affect density, such as the apparent variations of atmospheric density with altitude. The gravitational effect is generally neglected when the height difference is small. For example, in some problems the air density at the top of a 10-story building is generally assumed to be the same as that at ground level, while there is in fact a slight difference of a few thousandths between them. For the gas rising from the ground to a cloud, the density change by gravity must be taken into account. The effect of temperature on the density of gases is also not negligible. If a gas element conducts heat to its environment while flowing, its temperature decreases. Then, the gas is compressed, and work is done on it. If a gas element receives heat from its environment while flowing, its temperature increases. In that case, the gas expands and does work on its environment.

If flow is incompressible, there is no volume work, then temperature has a negligible effect on the fluid flow. As a consequence, the flow problem can be significantly simplified. The incompressible flow model can be used to calculate the slow flow of a gas with small temperature gradient and weak centrifugal force, which is known as incompressible flow theory. Incompressible flow has been the most extensively studied and widely used flow theory in fluid mechanics.

1.2.5 Thermal Conductivity of Gases

Thermal conductivity is the property of a material that describes the ability to conduct heat from hot to cold places. The flow of heat through a unit area per unit time satisfies Fourier's law of heat conduction, that is:

$$\dot{q} = -\lambda \frac{\partial T}{\partial n}, \tag{1.3}$$

where $\dot{q}$ is the conductive heat flux; λ is the coefficient of thermal conductivity; and n is the direction of the temperature gradient.

The mechanisms of heat conduction in gases and solids are quite different. In solids, the molecules, on average, vibrate but stay in fixed positions, so the transfer of molecular kinetic energy is due to their vibrations (or electron flow in some substances). In gases, heat conduction also depends on diffusion of the molecules on account of thermal motion. Heat conduction caused by molecular diffusion still exists even if the macroscopic velocity of the gas is zero; thus, molecular diffusion is different from the mixing in a flow. The heat transfer by the macroscopic mixing of gases is known as convective heat transfer. Although facilitated by the molecular diffusion effect, the thermal conductivity of gases is still much smaller than that of solids and liquids due to the large distances between molecules, so that it can usually be ignored.

1.3 The Concept of Continuum

Matter is composed of elementary particles, and the concepts of molecules and atoms have been extensively used in the previous analysis. In classical solid and fluid mechanics it is not necessary to study the behavior of individual molecules or atoms. Even for gases composed of molecules that are separated from each other by large distances, macroscopic properties, such as pressure and temperature, can be used to describe their microscopic mechanical behavior. Therefore, *in classical mechanics, a matter is always considered to be continuously separable, which is known as the continuum hypothesis*. Both solid mechanics and fluid mechanics are branches of continuum mechanics.

When describing the properties of a fluid at a certain point, it actually refers to a small region of the fluid. The region should be small enough to be approximated as a point on the macro-scale, but still be large enough to contain a large number of molecules on the micro-scale, so that the macroscopic quantities of interest are not affected by the microscopic randomness of the molecular thermal motion.

The only criterion in determining whether the continuum hypothesis is applicable is the relationship between the length scale of the flow and the molecular size (solids and liquids) or mean free path (gases). For gases at room temperature and pressure, the molecular mean free path is of the order of 7×10^{-5} mm. As long as the investigated object is far larger than the molecular mean free path, the continuum hypothesis holds. Common flows satisfy the continuum hypothesis. However, tiny pollen grains suspended in still water move randomly due to the thermal motions of water molecules colliding with them. This so-called Brownian motion does not satisfy the continuum hypothesis. Furthermore, a rocket or spacecraft moving at an altitude of 120 km does not satisfy the continuum hypothesis either, because the local mean free path of air molecules becomes as large as 0.3 m.

1.4 Forces in a Fluid

In fluid mechanics, the forces acting on an object are traditionally classified into two main categories according to their type of action: (1) a force exerted on the fluid

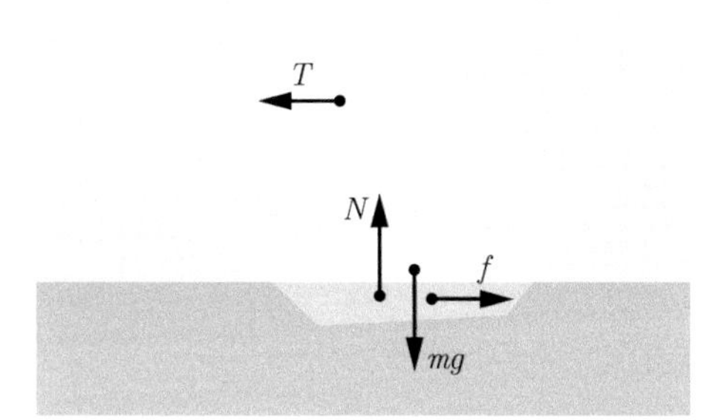

Figure 1.10 Forces exerted upon a sailboat.

without touching it, and therefore acting uniformly on the whole body of a fluid parcel, is called a body force; (2) a force appreciable only if fluid elements are in contact, and therefore existing only on the surface, is called a surface force.

Both gravitational force (universal gravitation) and magnetic force are body forces, and so is inertia force in non-inertial coordinate systems. Both pressure and viscous forces are surface forces. Pressure, as a normal stress, always acts inward and normal to any surface in a fluid. Shear stress inside a fluid is completely generated by viscous force, which also contributes to the normal stress. In most cases, the viscous normal stress is negligible compared to the pressure, so it is generally considered that viscosity only makes the fluid experience shear stress.

The pressure at any point in a stationary fluid or in a moving inviscid fluid is independent of direction. This property makes pressure a scalar quantity that can be regarded as a state parameter for fluids. Pressure is direction-independent because it is the only surface force in a stationary or moving inviscid fluid. Body forces (gravitational and inertial) at a point within a fluid tend to zero, and surface forces from all directions have to be identical in order to cancel out.

Figure 1.10 shows all the forces exerted upon a sailboat, among which the wind gives the sailboat a thrust of T, and f is the drag exerted mostly by the water. Because the sailboat is floating in static equilibrium, the magnitude of the buoyant force, N, must be equal to the magnitude of the weight of the boat, mg. Among these forces, gravity is the only body force, while buoyancy, thrust, and drag are the resultant forces on surfaces. A sudden increase in wind strength increases the thrust, which is not balanced by the drag, resulting in the forward acceleration of the sailboat. Then, the thrust needs to overcome both the drag and the inertia of the vessel. Taking the moving sailboat as a reference, the inertia is reflected as the inertial force, which is a body force.

Expanded Knowledge

States of Matter

In fluid mechanics a fluid is distinguished from a solid by its mechanical properties, while in general physics books the differences between solid, liquid, and gas are due to the arrangement of molecules or atoms. In fact, in modern scientific classification there are more than three states of matter, and the differences between them are not entirely clear. Solids can also be divided into two categories: crystalline and amorphous solids. Crystalline solids, such as ice, salt, and solid metals, have an ordered long-range arrangement of atoms or molecules within the structure. Amorphous solids, such as glass, rosin, and plastic, lack such an arrangement. Due to the similar configuration of atoms or molecules as a liquid, an amorphous solid is sometimes said to be a liquid with fairly low fluidity. That is to say, amorphous solids have the physical properties of liquids. Started in 1930, the tar drop experiment in Australia is one of the longest continuously running laboratory experiments to measure the flow of a piece of tar over many years. A "solid" piece of tar placed in a funnel at room temperature flows at a very low rate, taking several years to form a single drop. The experiment is conducted by more than one scientific organization, and interested readers can find relevant information online. This experiment shows that tar is actually a liquid rather than a solid, although it looks like one. What about glass, which looks more like a solid? Some claim that glass is also a liquid, and point to the fact that the glass in medieval church windows is thicker at the bottom than at the top, suggesting that the solid glass has flowed. However, it is not known what state the glass was in when the windows were installed, and most scientists agree that glass should be a solid.

Compressibility of Water

Since water is more resistant to compression, it is usually treated as an incompressible fluid. However, there are at least two cases in which the compressibility of water needs to be considered. One is the presence of extremely high-pressure conditions. For example, the water pressure in a hydraulic mining system that adopts the impact force of a high-pressure waterjet is up to 4,000 atm, and the corresponding water density is 15% higher than that at standard atmospheric pressure. The other case is when the problem being investigated is very sensitive to the change in water density, such as the calculation for the speed of sound in water. Sound is a wave consisting of alternating cycles of compression and expansion. If water were totally incompressible, the theoretical value of the speed of sound in water would be infinite. In reality, the measured speed of sound in water at room temperature and pressure is about 1,500 m/s. The magnitude of compression and expansion caused by the sound wave traveling through the water is very small. Such a weak compressibility is sufficient to allow the sound waves carrying very little energy to travel through the water.

Compressibility of Solids

The compressibility of solids is a problem rarely encountered. Compressibility is not very important for solids. The primary reason is, of course, that solids are just as significantly difficult to compress as liquids. Besides this, since a stationary solid can experience shear stresses, it requires no volume change to resist external pressures acting on it. For example, if a cylindrical object is compressed along its axis, it gets thicker and shorter, but its volume remains essentially the same. Such a stress state is commonly encountered in the study of solid mechanics. The interaction between a solid and the external forces acting on it is usually little affected by the change in its volume, but mainly related to the elastic forces arising from any combination of fundamental load types: tension, compression, shear, bending, and torsion. When the external forces significantly change the volume of the solid, the working stress reaches approximately the limit that causes the material to yield or fail. An exception occurs when a solid is under uniform external pressure. For example, it is necessary to take into account the compressibility of the rocks on the ocean floor, or deep inside Earth's crust, that are subjected to pressure coming from all directions.

Questions

1.1 You put a bottle of water in outer space, and the bottle itself suddenly disappears. Imagine the variations of the water and its final state. What happens if you use air instead of water in this experiment?

1.2 Look up the coefficient of viscosity for air, and use Newton's law of viscosity to discuss why air viscosity is generally not perceived in daily life.

1.3 What happens if you fill a glass bottle with water, tighten the lid, and let it sink to a depth of 5,000 meters under the surface of the sea? For an open glass bottle filled with seawater at a depth of 5,000 meters, what happens if you tighten the lid and bring it to the surface?

2 Forces in a Static Fluid

Ladies and gentlemen, I can crack this barrel with just a little water.

2.1 Analysis of Forces in a Static Fluid

Pressure is the only surface force in a static fluid since there are no shear forces. Therefore, fluid statics focuses largely on the equilibrium relations between pressure and body forces. Since body forces generally include gravitational and inertial forces, fluid statics can be classified into two main categories: fluids at rest in a given gravity field and fluids moving in rigid-body motion.

For any fluid element at rest, the vector sum of its surface forces (pressure forces) exerted by the surrounding fluid elements must cancel out the body forces acting on it. In a coordinate system, the components of the body force and the resultant force from pressure are equal in magnitude but opposite in direction along any direction. We analyze next a fluid element and derive a general form of relations.

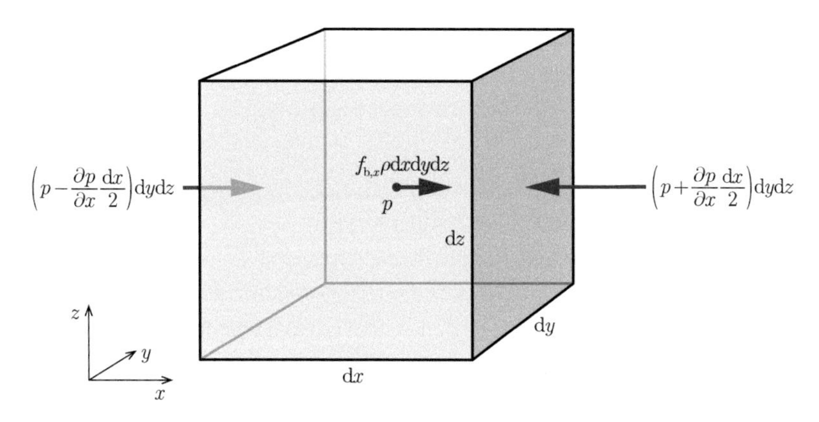

Figure 2.1 Analysis of forces acting on a fluid element at rest.

Consider an infinitesimal cubical element (dx, dy, dz) in a static fluid, as shown in Figure 2.1. The forces acting on it are the surface stresses from its surroundings and gravity. The cubical fluid element is dxdydz in volume, and ρdxdydz in mass. If $\vec{F}_\mathrm{b}$ represents the body force and f_b the body force per unit mass, we obtain:

$$\vec{F}_\mathrm{b} = \vec{f}_\mathrm{b}\rho \mathrm{d}x\mathrm{d}y\mathrm{d}z.$$

In a Cartesian coordinate system this formula can be expressed in component form as:

$$\vec{F}_\mathrm{b} = F_{\mathrm{b},x}\vec{i} + F_{\mathrm{b},y}\vec{j} + F_{\mathrm{b},z}\vec{k} = \left(f_{\mathrm{b},x}\vec{i} + f_{\mathrm{b},y}\vec{j} + f_{\mathrm{b},z}\vec{k}\right)\rho \mathrm{d}x\mathrm{d}y\mathrm{d}z.$$

Pressure is a scalar property, that is to say, the pressure at any point in a fluid remains the same in all directions. As a surface force in equilibrium with the body force, the pressure force is defined as the product of the pressure times the surface area. Thus, the pressure force is a vector quantity, with direction perpendicular to the acting surface. For an object immersed in water, for example, only the pressure acting on the bottom face produces an upward buoyancy force. When a fluid element is in equilibrium state, the pressure of the liquid acting on the immersed body is not the same everywhere, thus producing a pressure gradient force that counteracts the body forces (gravity).

If the fluid element shown in Figure 2.1 experiences a body force in the rightward direction, the surface force acting on its right face must be greater than that acting on its left face, thus providing a pressure gradient force to maintain hydrostatic equilibrium. Assuming that p is the pressure at the center of the fluid element, the pressure acting on the left face is smaller than p, while the pressure acting on the right face is greater than p. So how is the pressure acting on the left and right faces of the fluid element expressed in terms of p? The answer is: by a Taylor series expansion, which is a method commonly used in mechanics. The Tip 2.1 box discusses the Taylor series expansion for easy reference. Using the Taylor series expansion, and neglecting the second- and higher-order terms, the pressure acting on the left and right faces of the fluid element can be written as:

$$p_\mathrm{left} = p - \frac{\partial p}{\partial x}\frac{\mathrm{d}x}{2}; \qquad p_\mathrm{right} = p + \frac{\partial p}{\partial x}\frac{\mathrm{d}x}{2},$$

where $\partial p/\partial x$ represents the rate of pressure change in the x direction, known as pressure gradient in the x direction. As can be seen, the pressure acting on both left and right sides of the fluid element can be expressed by the pressure and the pressure gradient at its center.

Tip 2.1: What Is a Taylor Series Expansion?

In mathematics, a Taylor series is an expansion of a complex function into an infinite sum of polynomials. Based on Taylor series, the value of the function and its derivatives at a single point can be used to estimate the values at its neighboring points. Let us have a look at how this works.

Figure T2.1 shows the variation of a physical quantity y with spatial distance x. If the function is continuously differentiable, its properties at point O can provide an estimate of its value at a nearby point A through the Taylor series expansion. For example, the value of the function at A can be estimated using only the value of the function (y_O) and its slope $\left(\mathrm{d}y/\mathrm{d}x\right)_O$ at point O:

$$y_A \approx y_O + \left(\frac{\mathrm{d}y}{\mathrm{d}x}\right)_O \Delta x.$$

Obviously, this estimation is accurate only if y varies linearly with x. To accurately approximate the values of the function from O to A, an infinite number of the second-, third-, fourth-, etc. higher-order derivatives at O are also required. Therefore, even if the variation of a physical quantity is not linear, the Taylor series expansion will be accurate enough as long as the two points are sufficiently close. As the distance between O and A becomes infinitesimally small, a linear approximation suffices to estimate the values of the function near the point O. In mechanics, since the derivation of differential forms of governing equations is always applied to an infinitesimal element, we can generally ignore the second- and higher-order small quantities, thus obtaining a set of linear relations for the unknowns. Only in some cases, when dealing with elements on a finite scale, are second- or higher-order terms required.

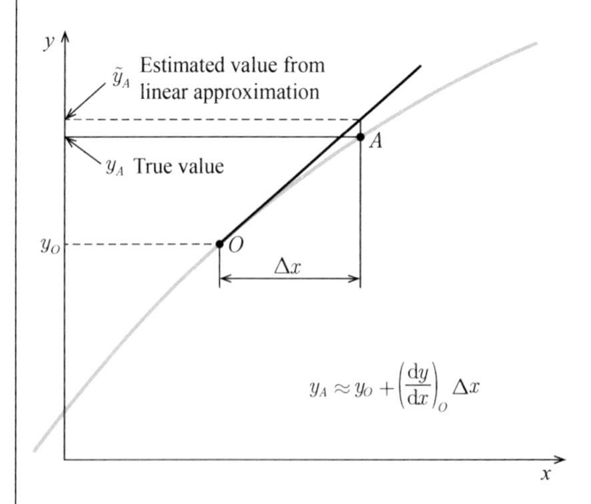

Figure T2.1.

Now that we have obtained the body force in the x direction and the surface forces acting on left and right sides of the fluid element, the force balance equation in the x direction can be written as:

$$\sum F_x = f_{\mathrm{b},x}\rho \mathrm{d}x\mathrm{d}y\mathrm{d}z + \left(p - \frac{\partial p}{\partial x}\frac{\mathrm{d}x}{2}\right)\mathrm{d}y\mathrm{d}z - \left(p + \frac{\partial p}{\partial x}\frac{\mathrm{d}x}{2}\right)\mathrm{d}y\mathrm{d}z = 0.$$

After simplification, we obtain:

$$f_{\mathrm{b},x} = \frac{1}{\rho}\frac{\partial p}{\partial x}.$$

The force balance relations in the y and z directions can be similarly obtained, and the force balance equations in component form can then be rearranged as:

$$f_{\mathrm{b},x} = \frac{1}{\rho}\frac{\partial p}{\partial x}; \quad f_{\mathrm{b},y} = \frac{1}{\rho}\frac{\partial p}{\partial y}; \quad f_{\mathrm{b},z} = \frac{1}{\rho}\frac{\partial p}{\partial z}, \tag{2.1}$$

where $\partial p/\partial x$, $\partial p/\partial y$, $\partial p/\partial z$ represent the pressure gradients in the x, y, and z directions.

According to the definition of gradient, we obtain:

$$\nabla p = \frac{\partial p}{\partial x}\vec{i} + \frac{\partial p}{\partial y}\vec{j} + \frac{\partial p}{\partial z}\vec{k}.$$

Equation (2.1) can be written in vector form as follows:

$$\vec{f}_{\mathrm{b}} = \frac{1}{\rho}\nabla p. \tag{2.1a}$$

The relationship between pressure and body forces in a static fluid, described by Equations (2.1) and (2.1a), is the fundamental equation for fluids at rest. As we have seen, hydrostatic pressure distribution in a static fluid depends only on the body force, and the pressure force increases along the direction of this force. In a gravitational field, the lower layer of fluid has a pressure larger than that of the upper layer; in a centrifugal field, the larger the radius of a position, the higher the fluid pressure. This can also be understood as follows: In a gravitational field, the weight of the upper layer of fluid is completely supported by the lower layer; in a centrifugal field, the centripetal force on the inner layer of fluid is totally exerted by the outer layer.

2.2 Pressure in a Static Fluid under the Action of Gravity

When the fluid in a uniform gravitational field stays at rest or travels in a straight line in uniform motion, hydrostatic pressure distribution depends only on gravity. Figure 2.2 shows the surface forces in a static liquid acting on a liquid element. If the positive z axis points vertically upward, the static equilibrium equation (2.1) can be written as follows:

$$\mathrm{d}p = -\rho g \mathrm{d}z.$$

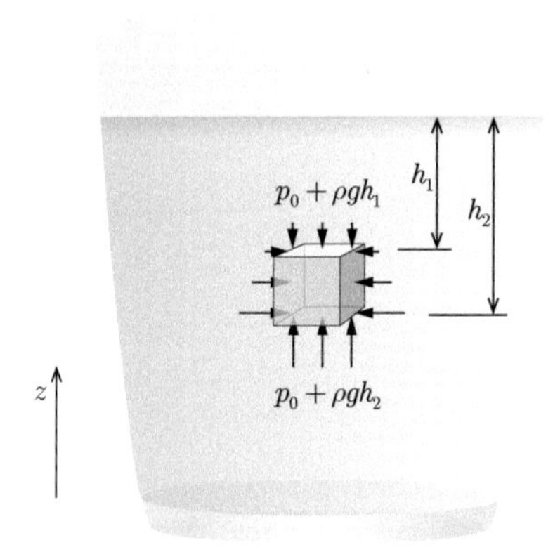

Figure 2.2 Pressure distribution in a static liquid.

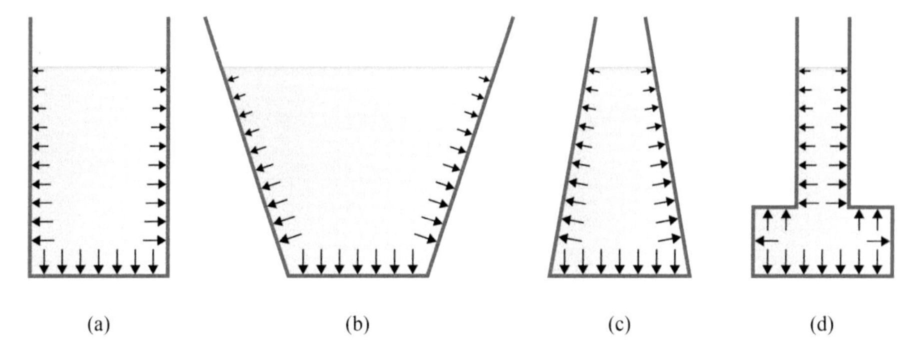

Figure 2.3 The pressure force exerted by the fluid on the same bottom area remains the same.

When acceleration of gravity and density of the liquid are assumed to be invariant with depth, the pressure at a point in the liquid can be obtained by integrating the above formula along the depth:

$$p = p_0 + \rho g h,$$

where p_0 is the atmospheric pressure on the surface of the liquid, and h is the depth below the surface.

As can be seen, the pressure in a static liquid depends only on the atmospheric pressure, the density of the liquid, the depth, and the acceleration of gravity. As shown in Figure 2.3, four vessels of different shapes and having the same bottom area are filled with water to the same height. Clearly, the weight of the water in each vessel varies greatly. However, since the pressure at the bottom is the same, the vertical force exerted by water on the bottom of each vessel is always equal to $(p_0 + \rho g h)A$. This particular behavior of liquids, first discovered by Blaise Pascal (1623–1662) and known as the *hydrostatic paradox*, was very puzzling at the time.

There is surely no contradiction. Because the force exerted by the water on the bottom of each vessel is not directly related to that exerted by the vessel on the table. The water exerts not only a downward force on the vessel, but also an upward force in some cases. The net force exerted by the water on the vessel is just equal to the weight of the water. In Figure 2.3 the small arrows represent the magnitude and direction of the pressure force exerted by the water. Although the weight of the water in vessel (b) is the greatest, this weight beyond the projection of the bottom is supported by the side walls. Although filled with less water, the vessels (c) and (d) include the internal surfaces against which upward hydrostatic pressure is exerted. The pressure forces acting on these surfaces cancel out some of those exerted by the water on the vessel's bottom.

Pascal once carried out an interesting experiment. He inserted the end of a long, narrow pipe through the lid of a closed wooden barrel filled with water. The pipe was also prefilled with water up to a certain height. While he was performing the experiment, he stood on a balcony and poured more water into the pipe. Due to the increase in hydrostatic pressure, the barrel leaked and even burst with the addition of just a glass of water. This is the historically famous Pascal's barrel experiment, which astonished people at that time, and is very eye-catching and impressive even now (as illustrated on the opening page of this chapter). The key in designing a successful experiment is to use a relatively narrow pipe, so that a glass of water will sufficiently increase the depth of the water column to create a large hydrostatic pressure in the barrel.

The atmospheric pressure exerted by the air around us and the high pressure in the depths of the oceans are two examples of enormous hydrostatic pressure due to the force of gravity. Although the air's density is small, the Earth's thick atmosphere creates high surface pressure near the ground. The atmospheric pressure cannot simply be calculated using the formula $p_0 + \rho g h$, because atmospheric density is determined by pressure and absolute temperature, which vary greatly with height. In general, atmospheric temperature is considered to be constant in the stratosphere, and to decrease linearly with height in the troposphere. According to these assumptions, including the equation of state for an ideal gas, the atmospheric pressure at a given altitude can be calculated using the static equilibrium equation (2.1).

At 11,000 meters, which is a typical cruising altitude for most commercial aircraft, the ambient pressure is about one-quarter of the atmospheric pressure at sea level. Since passengers on a plane flying at such an altitude cannot adapt to a low-pressure, low-oxygen environment, the cabin must be pressurized. The cabin pressure at cruising altitude is often set to about 0.8 standard atmospheric pressure, corresponding to a pressure at altitude of about 2 km. Changes in cabin pressure can usually be felt by passengers during takeoff and landing. Cabin pressure is regulated as follows: the pressure inside the aircraft cabin is maintained at 0.8 atm above the altitude of 2 km. When the aircraft descends below that altitude, cabin pressure is gradually adjusted to be equal to external pressure, thereby leading to discomfort or ear pain for many passengers during landing. With the progress in materials science and strength design, new airliners are expected to increase cabin pressure at cruising altitude, and so ease passengers' discomfort.

For human beings, the deep sea is as mysterious as outer space. The main reason is that the great pressure at the seafloor makes it very difficult to explore. The deep sea is even more dangerous than outer space if only the local conditions rather than the

Tip 2.2: Beware of Surface Tension!

In Pascal's barrel experiment, the smaller the diameter of the pipe, the greater the amount of pressure it generates with the same amount of water. However, when the pipe is too narrow (say, less than 1 mm in diameter), the barrel is safe. This is because the surface tension of the water in a very narrow pipe is intense enough to completely counteract the water pressure generated by gravity. As shown in Figure T2.2, there are two vertical pipes filled with water. With both ends of the pipe exposed to atmospheric pressure, the water in the wide pipe flows out immediately, while that in the narrow pipe may not flow out at all. How does the water subjected only to the force of gravity remain at rest in a pipe? In fact, the free water surfaces at both ends of the pipe provide a net upward force caused by surface tension to balance the weight of the water column. It is by a similar force that water can be transported to the top leaves of trees tens of meters in height.

Therefore, it should be clear that the hydrostatics in this chapter applies only to cases in which the surface tension largely does not work. On small scales, ignoring the surface tension may lead to a totally false conclusion.

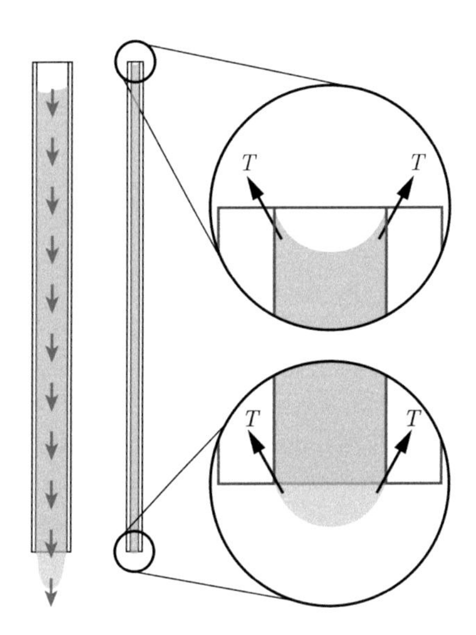

Figure T2.2.

roundtrip journey are taken into account. In outer space, the cabin differential pressure of a spacecraft is only 1 atm, while in the deep sea the cabin differential pressure of a submarine can be hundreds or even thousands of atmospheres, presenting a huge challenge for the sealing and strength of the hull.

Buoyancy is caused by the pressure exerted by the fluid in which an object is completely or partially immersed. It reflects the increase of the pressure with growing depth. Archimedes stated a simple and useful principle for buoyancy: *the magnitude of the buoyancy force is equal to the weight of the fluid displaced by the object.* Actually, the weight of the displaced fluid depends on several variables: the acceleration of gravity g, the density of the fluid ρ, the scale of the object in the direction of depth Δh, and the area of the object projected on a horizontal plane. The product of these variables happens to be the weight of the displaced fluid. According to Archimedes' principle, if the volume of the object is zero, it is not subjected to a buoyancy force. If a zero-thickness plate is immersed horizontally in water, then Δh is equal to zero. Since the pressures acting on the upper and lower surfaces are the same, the magnitude of the buoyancy force is equal to zero. If the plate is immersed vertically in the water, then A is equal to zero. Although the pressures acting on the upper and lower surfaces are different, the magnitude of the buoyancy force is still equal to zero because the action area is zero.

The mechanical analysis of submerged plane and curved surfaces has been widely applied in the design and construction of dams, locks, and ships. In fluid mechanics, hydrostatic forces include the resultant force and resultant moment caused by the pressure loading of a liquid acting on submerged surfaces. However, no matter how complex engineering problems may seem, there is only one condition to be satisfied in hydrostatics: the static equilibrium equation (2.1). If we understand the physical meaning of this equation, we can draw inferences about other cases from one instance and understand all practical problems in hydrostatics.

2.3 Pressure in a Fluid under the Action of Inertial Forces

For a body of fluid moving with constant acceleration, there is no motion between adjacent fluid layers relative to each other. In the reference frame of the accelerating fluid, an additional inertial force is exerted on the fluid. This is still considered as a part of hydrostatics because the whole fluid is moving as a rigid body and the liquid particles are still in equilibrium relative to each other. Two cases of relative equilibrium will be discussed: uniform linear acceleration and uniform rotation about a fixed axis.

Figure 2.4 shows the pressure acting on the liquid elements in the two cases, where both the inertial force and gravity act as the body force. During the calculation, the two force vectors are superposed and then solved by the static equilibrium equation (2.1).

It can be seen from Equation (2.1a) that the change of pressure in any direction depends only on the body force in that direction. Therefore, if the body force and depth in a certain direction are obtained, the pressure can be calculated along that direction without solving the complete equation. As in the case of a static fluid under the action of gravity, the pressure in relative equilibrium under the combined action of gravity and inertial forces is calculated using the equation:

$$p = p_0 + \rho a h,$$

where h is the depth below the liquid surface along the direction of acceleration, a.

The acceleration, a, does not necessarily refer to the total acceleration but to the acceleration in any direction. As shown in Figure 2.5, the pressure at a point in a

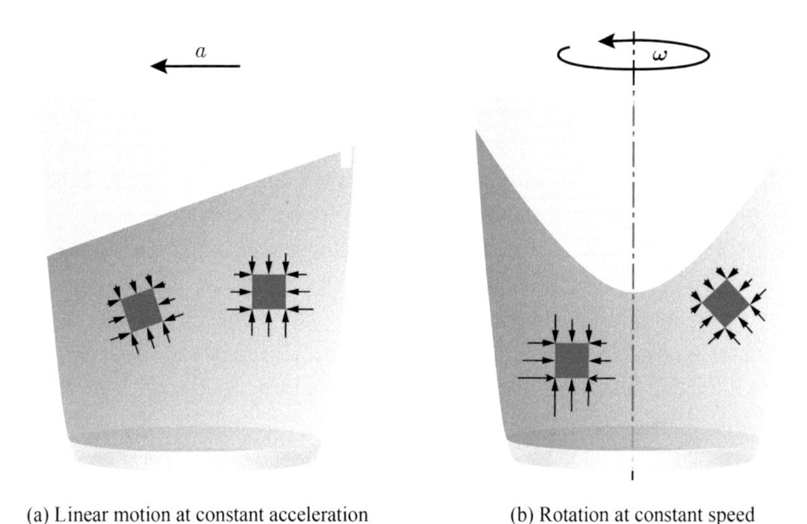

Figure 2.4 Pressure in a moving body of fluid with constant acceleration.

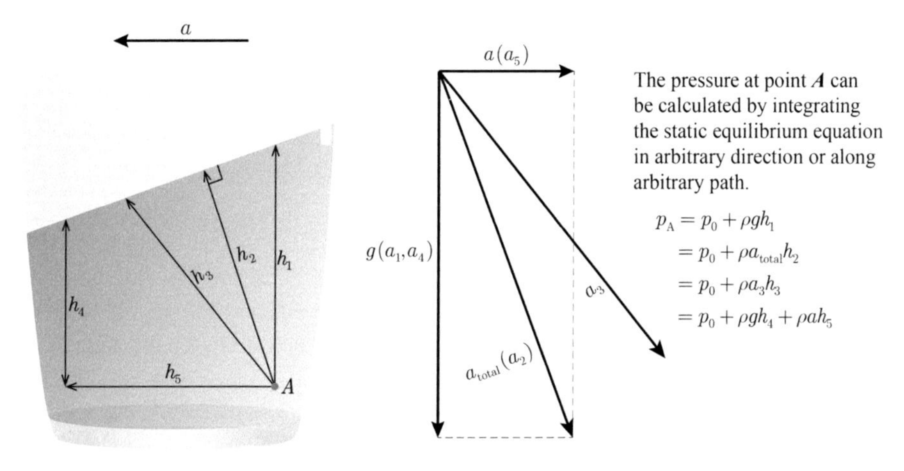

Figure 2.5 The pressure at a point in a liquid under the combined action of gravity and inertial forces.

liquid (under the combined action of gravity and inertial forces) can be computed by integrating the static equilibrium equation in any direction and along any path. The result will be the same, due to the direction-independent characteristics of pressure.

As shown in Figure 2.6, when a tank containing liquid slides down an incline, it is subjected to the combined action of gravity, support, and friction. If the friction is balanced by the component of gravity along the incline, the tank is at rest or sliding down with uniform speed. In this case, the surface of the liquid will remain horizontal, since it is only subjected to gravity. If the tank is sliding with constant acceleration down a frictionless incline, the surface of the liquid will remain parallel to the incline

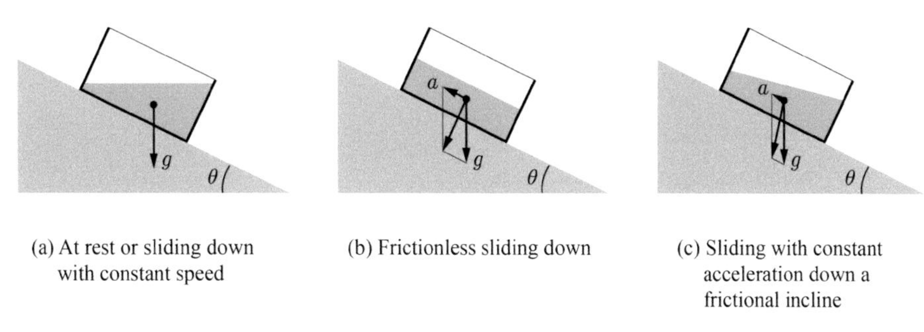

(a) At rest or sliding down with constant speed

(b) Frictionless sliding down

(c) Sliding with constant acceleration down a frictional incline

Figure 2.6 The forces acting on the water in a tank sliding down an incline and the shape of the liquid free surface.

since it is not subjected to any body force along the incline. If the tank is sliding with a smaller constant acceleration down a frictional incline, the shape of the liquid free surface will fall somewhere in between the above two cases.

2.4 Differences and Similarities in the Transfer of Force by Fluids and Solids

The maximum force a person can exert is of the same order as that of his or her own weight. Labor-saving devices such as levers and pulleys allow human beings to move heavier objects. In theory, a labor-saving device can "magically" create as much force as you want, but only if you use the small force on your side over a longer distance, based on the principle that the work or power remains unchanged. The same principle also applies to fluids. Hydraulic actuators are normally used to convert a small force into an extremely large one. For example, a child can easily raise a car using a hydraulic jack with enough input distance, realized by pressing multiple times on the handle.

Work cannot be created out of thin air, and neither is force. The larger force created by labor-saving devices has to come from somewhere. For the lever shown in Figure 2.7(a), the long moment arm is three times as long as the short one. When the downward force applied on the long end is F, the upward force exerted on the weight by the short one is $3F$. Force analysis shows that the upward force exerted on the lever by the fulcrum is $4F$. Therefore, the "magic" additional force is created by the fulcrum.

For fluids, Figure 2.7(b) shows the schematic diagram of a hydraulic labor-saving device, where the cross-sectional area of the large piston is three times as large as that of the small one. If the downward external force exerted on the small piston is F, the upward force exerted on the large piston by the fluid is $3F$, which does not come from nowhere either. The magnitude of the resultant upward force exerted by the entire vessel on the liquid is $4F$.

The essential difference between a solid and a fluid is that a solid at rest can generate shear forces, whereas a fluid at rest cannot, which results in different ways of

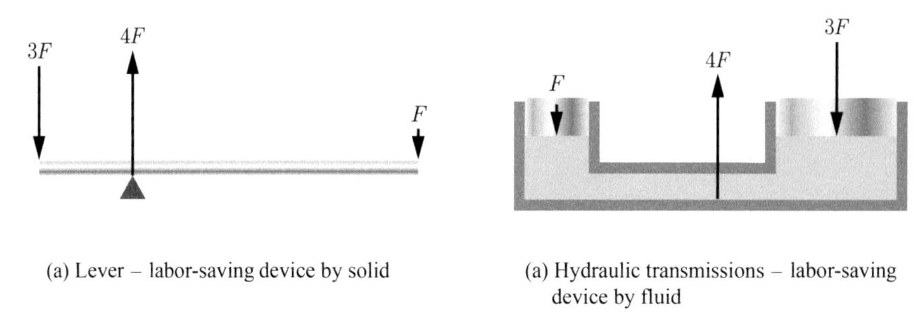

(a) Lever – labor-saving device by solid

(a) Hydraulic transmissions – labor-saving device by fluid

Figure 2.7 Force analysis of levers and hydraulic transmissions.

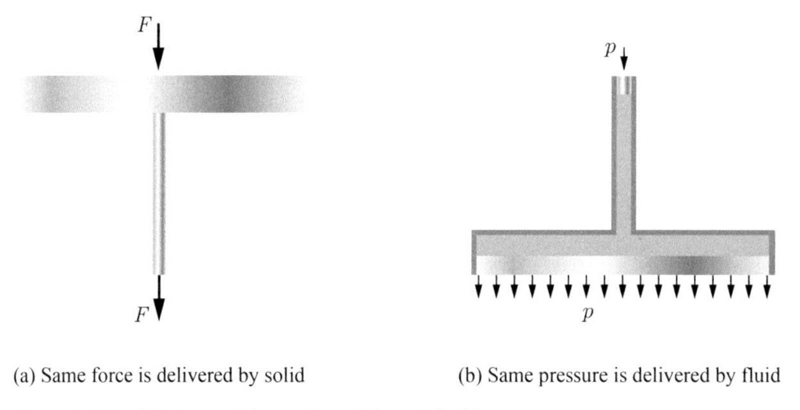

(a) Same force is delivered by solid

(b) Same pressure is delivered by fluid

Figure 2.8 Equilibrium of forces in solids and fluids.

transferring force. This can be summarized as follows: *the amount of force transferred through a solid remains constant, while the amount of pressure (force per unit area) transferred through a fluid remains constant.* Figure 2.8 illustrates how to take advantage of the difference in force transfer between a fluid and a solid. When someone presses a thumbtack into a wooden board, the force exerted by the thumb on the top of the thumbtack equals that of the force exerted by the tip of the thumbtack to the wood. Due to the significant difference in contact area between the tip and the top of the thumbtack, the hand does not get hurt, but the wooden board, which is much harder than the hand, is pierced by the tip of the thumbtack. This is an illustration of the fact that force per unit area is the key cause of material failure. In contrast to the case of a solid, it is pressure rather than force that is transmitted undiminished to all portions of a static fluid. In the hydraulic device shown in Figure 2.8(b), if the hand exerts an input force on the small piston, a higher output force is exerted by the large piston, with the force being pressure times the area.

It is easy to understand why solids transmit force equally, but why is pressure applied to a fluid transmitted equally throughout it? In fact, only in an enclosed fluid is the pressure applied at any point equally transmitted in all directions. Seen from the outside, the vessel, together with the fluid inside it, may be considered as a solid

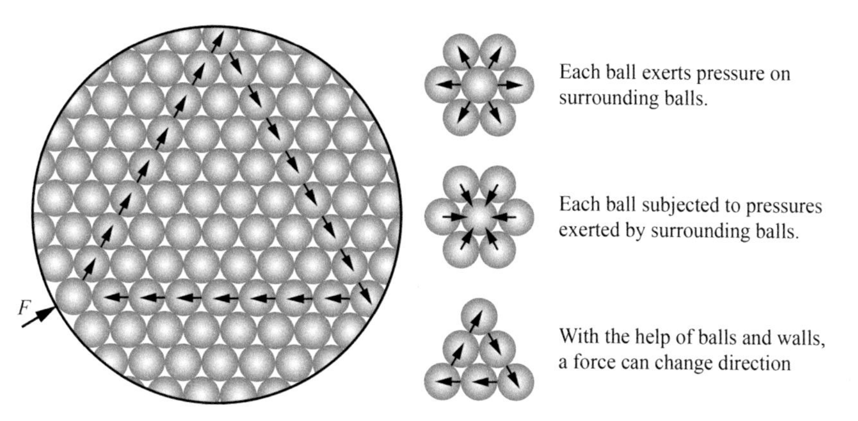

Figure 2.9 Forces transmitted by smooth spheres.

throughout which the force is equally transmitted. The pressure applied at any point of the vessel wall is always perpendicular to the wall. The essential reason why the direction of the force can change in transmission is that fluids at rest cannot generate a shear stress. To facilitate understanding, imagine a vessel is filled with smooth spheres instead of fluid, as shown in Figure 2.9. Ignoring the weight of the balls and the friction between them, the force exerted anywhere is transmitted through these balls without change in magnitude to all portions of the vessel, just as in the case of fluids.

Expanded Knowledge

Atmospheric Pressure

Atmospheric pressure, generated by gravity pulling the air toward the Earth, varies significantly with the altitude, season, weather, and time of day. Taking Beijing, for example: the average atmospheric pressure is about 102 kPa in winter and about 100 kPa in summer. The difference in atmospheric pressure on the same day can be as much as 1 kPa. Atmospheric pressure is generally low at noon and high in the morning and evening. In order to have a unified standard, several atmospheric models were developed between 1920 and 1976, and the standard atmospheric pressure at sea level was defined as 101,325 Pa. In the troposphere, temperature decreases with altitude. In the stratosphere, the temperature remains basically constant. The atmospheric temperature distribution given by the standard atmospheric models is:

$$\begin{cases} T = 288.15 - 0.0065z & 0 \le z < 11{,}000 \\ T = 216.65 & 11{,}000 \le z \le 24{,}000 \end{cases}.$$

This relation roughly represents the mean atmospheric temperature distribution in various climate zones on Earth.

According to this relation and the equation of state for an ideal gas, Equation (2.1a) can be integrated to obtain the distribution of pressure with altitude in the atmosphere.

Interested readers can carry out the appropriate calculations themselves. The results are given below (the gravitational acceleration is treated as a constant):

$$\begin{cases} p = p_0\left(1-\dfrac{0.0065z}{288.15}\right)^{\frac{g}{0.0065R}} & 0 \leq z < 11{,}000 \\ p = 22{,}656\mathrm{e}^{\frac{g(11000-z)}{216.65R}} & 11{,}000 \leq z \leq 24{,}000 \end{cases},$$

where p_0 is the atmospheric pressure exerted against the surface of the Earth, g is the gravitational acceleration, and R is the gas constant of air. All physical quantities are expressed in the International System of Units. The atmospheric pressure at a given altitude can be computed using this relation. Some typical values of atmospheric pressure are given here for reference:

Tibetan Plateau (4,000 m): ~0.6 atm
Summit of Mount Everest (8,848 m): ~0.3 atm
Operational ceiling of a fighter aircraft (20 km): ~0.05 atm

An airplane's maximum operating altitude is limited. The main reason is that atmospheric pressure at high altitude is too low, resulting in insufficient lift and engine thrust.

Pressure Measurement

Pressure measurement is perhaps the most common task in fluid engineering. Pressure is a physical quantity that is easy to measure. All kinds of pressure gauges and sensors work by measuring the force or displacement. In mechanical pressure gauges the fluid directly pushes the diaphragm that drives the pointer. In piezoelectric, capacitive, and quartz oscillator diaphragm pressure sensors, the fluid pushes the diaphragm, which deforms or moves to produce electrical output signals. For historical reasons, there are many units of pressure, such as the commonly used bar, kg/cm^2, PSI, mmH_2O, and so on. At sea level we already experience 1 atm of pressure, which is quite large, but hardly noticeable. When a tire has a serious leak the pressure measured by the gauge is close to 0. This does not mean that there is no air in the tire, but rather that the pressure of the gas inside the tire is equal to the atmospheric pressure outside. The reading shown on the tire pressure gauge is known as gauge pressure, which is equal to absolute pressure minus atmospheric pressure. Almost all kinds of pressure gauges in pipelines show gauge pressure, but in fluid mechanics problems absolute pressure is also needed sometimes.

Unlike other state parameters, such as density, temperature, and velocity, pressure can be measured over great distances through tubes without necessarily putting sensors into a flow field. The explanation of this fact makes use of the principle of hydrostatics. As long as the tubes do not leak, the fluid inside them will stay at rest. Then, for two points at the same height in the pipeline, the pressure will be the same. How much will the pressure being measured be in error if the pressure gauge and the measurement point are not at the same height? For liquids, this kind of error is relatively large; for gases, it depends on the required measurement accuracy. Let us look at some specific examples.

Suppose that the task is to investigate the way in which the static pressure varies over the airfoil surface of a model aircraft in a low-speed wind tunnel. The freestream velocity is 15 m/s, and the typical pressure difference between the upper and lower surfaces of the airfoil is about 30 Pa. If the laboratory is located on the first floor while the pressure sensors are placed on the second floor, the difference in floor height will create a pressure difference greater than 30 Pa. Therefore, in the pressure measurement of low-speed flow fields, it is strongly recommended that the pressure sensor and the measurement point be at the same height. On the other hand, the minimum pressure to be measured in a test bench for aircraft engines has a magnitude of 1.0 atm. In this case, the pressure difference of tens of Pa caused by different floor heights only accounts for 0.05% of the measured pressure. This measurement error is generally smaller than the accuracy of the sensor itself, so it is less important to ensure that the pressure sensor is placed at the same height as the measurement point.

Questions

2.1 Assuming that the total weight of a fully loaded truck is twice that of an empty one, does the tire pressure with a full load need to be twice as high as that without load? Why?

2.2 An aquarium is equipped with a kind of "vacuum suspended fish tank" that allows children to reach in and touch small fish through a small opening on the side. Learn about its structure and explain how it works.

3 Description of Fluid Motion

Air flow over the surface of a racing car, creating a wake behind the car.

3.1 Methods of Describing Fluid Motion

Kinematics concerns change of the position or orientation of an object or system of objects without consideration of the forces involved. In solid mechanics, particle kinematics is the simplest; rigid body kinematics is concerned with the shape of the object, and elastic mechanics and plastic mechanics have to do with the deformation of the object. Fluid motion is more complex than solid motion, and the relative positions of the fluid particles can vary greatly after they move. Therefore, there are two special methods to describe fluid motion.

In solid mechanics, force analysis of an object generally concentrates on the object itself. Everything outside of the object is called the environment. The force between the object and the environment causes a change in the object's state of motion. This is called *the Lagrangian method.* For fluids, the Lagrangian method, although applicable, is not convenient because there are too many fluid particles to be described, and the relative positions of these vary greatly. In addition, most fluid mechanics problems in engineering concentrate on the force exerted by a fluid on a solid, rather than on the state of motion of the fluid itself. Therefore, studies in fluid mechanics usually work with a particular space, focusing on the change of fluid passing through this space and the interaction between them. This is called *the Eulerian method.*

In the Lagrangian method, the independent variables are time, t, and spatial coordinates, $\vec{\xi}$. The spatial coordinates are used to identify individual fluid particles, and different $\vec{\xi}$ represent different particles. As the fluid particle moves, its spatial coordinates vary with time. $\vec{\xi}$ is the position of a fluid particle at time t_0. The fluid particle is "marked" by ξ, and its spatial position, velocity, and acceleration at any time t are described as:

$$\vec{r} = \vec{r}(\vec{\xi}, t); \quad \vec{V} = \left(\frac{\partial \vec{r}}{\partial t}\right)_{\xi}; \quad \vec{a} = \left(\frac{\partial^2 \vec{r}}{\partial t^2}\right)_{\xi}. \tag{3.1}$$

The Lagrangian method focuses on a parcel of fluid particles through space and time. In contrast, the Eulerian method is concerned with a fixed volume of space in a flow field through time. Its independent variables are time t and the fixed coordinates relative to the space, $\vec{r}$. *The Lagrange method describes the fluid motion as follows: "at time t, the velocity of the fluid particle A is ...," while the Eulerian method describes it as: "at time t, the velocity at point A is ..."*

In some flow problems, flow properties at a fixed point in space do not change with time. Consider the flow in a river, for example. If the velocity is the same for any water that passes through the point where the observer stands, the flow is said to be *steady* at that point. If the flow properties remain constant with time at every space point, the whole flow is said to be steady. On the other hand, if the river is constantly rising, the flow properties at the observation point will change with time. In this case one speaks of an *unsteady* flow. If the water flows over a stone, the waves splashing around the stone are also unsteady.

Furthermore, the terms one-, two-, or three-dimensional flow refer to the number of space coordinates required to describe a flow. It may appear that every physical flow in nature is three-dimensional, but for a great number of them the three-dimensional problem may be reduced to a two-dimensional – or even one-dimensional – one by ignoring the changes in other directions. For example, although the flow velocities at the surface and bottom are not equal at any section, the river can be treated as one-dimensional by defining a mean flow speed.

3.2 Pathlines and Streamlines

In solid mechanics we have touched upon the concept of pathlines. The trajectory of a projectile is a pathline, and so is the contrail of an airplane. To put it simply, *A pathline is the trajectory traced by a single particle during its motion, formed by connecting the positions of the particle at each instant in time.* Obviously, a pathline is a Lagrangian method concept, in which we track the path of a single fluid particle as it moves through the flow field. In the Eulerian method we are more concerned with the magnitude and direction of the fluid velocity at a fixed point in space; that is, the velocity of the fluid particle passing through that point at that instant. If the flow is unsteady, a particle that appeared later at a certain point would have a different velocity and move along a different pathline. At any given instant in time there is a velocity vector at each point of the flow field. The vectors at every point and its adjacent points can be connected to form a family of curves, known as *streamlines. A streamline is defined as a curve on which the tangent at any point represents the direction of the fluid velocity at that point.*

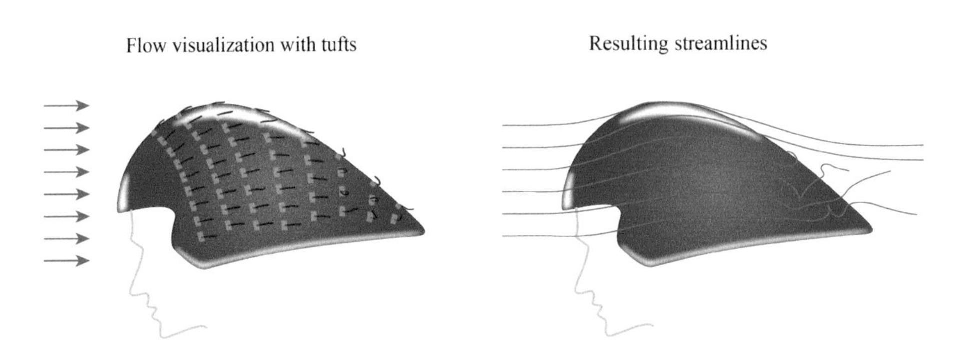

Figure 3.1 Tuft flow visualization of the streamlines on the surface of an object.

If you photograph fireworks, a long exposure time can show the beautiful patterns drawn by many individual sparks. Every curve appearing in these patterns is a pathline. If the exposure time is short, only some very short curves will be captured. Every single short curve is a pathline, and you can draw some long curves by connecting these short curves, which are streamlines.

One way to observe a flow is to stick many soft and short threads of silk or wool on the surface of an object. These threads indicate the local flow direction at a certain instant. If one then draws lines parallel to the threads (i.e., the direction of fluid velocity), the streamlines can be obtained. This method is called *tuft flow visualization*, Figure 3.1 shows the experiment and a family of streamlines obtained in the study of a new type of bicycle helmet.

In a steady flow, the velocity of any fluid particle at a certain point always remains constant; therefore, streamlines are identical to pathlines. In an unsteady flow, two fluid particles starting from a certain point may follow different pathlines, so streamlines and pathlines do not coincide. Figure 3.2 shows the flow caused by a sphere passing through a stationary fluid. At any given instant, the velocity vectors of the perturbed fluid particles are connected to form streamlines that start from the front of the ball and end at the rear of it. If we observe some specific fluid particles we shall find that they do not move along these streamlines, but follow their own pathlines individually. Figure 3.2(c) shows the pathline traced by a fluid particle located above the trajectory of the ball center.

Figure 3.3 shows a self-driven rotating lawn sprinkler. Water flows upwards into the sprinkler through the center pipe, spurts out from two nozzles, and drives the sprinkler to rotate. Viewed directly from above, the two spirals composed of liquid water droplets are neither pathlines nor streamlines. Once ejected, a water droplet experiences a horizontal projectile motion and moves in a straight line when viewed from above. The pathlines of water droplets should be a family of radial straight lines. No streamline appears in Figure 3.3. This is because the water flow is not continuous in space. If the sprinkler nozzle is a continuous circumferential seam, then the streamlines are also to be shown as a family of radial straight lines. The spiral liquid water currents in Figure 3.3 are named as streaklines. *A streakline is the locus of fluid particles that have earlier passed through a prescribed point in the flow field.*

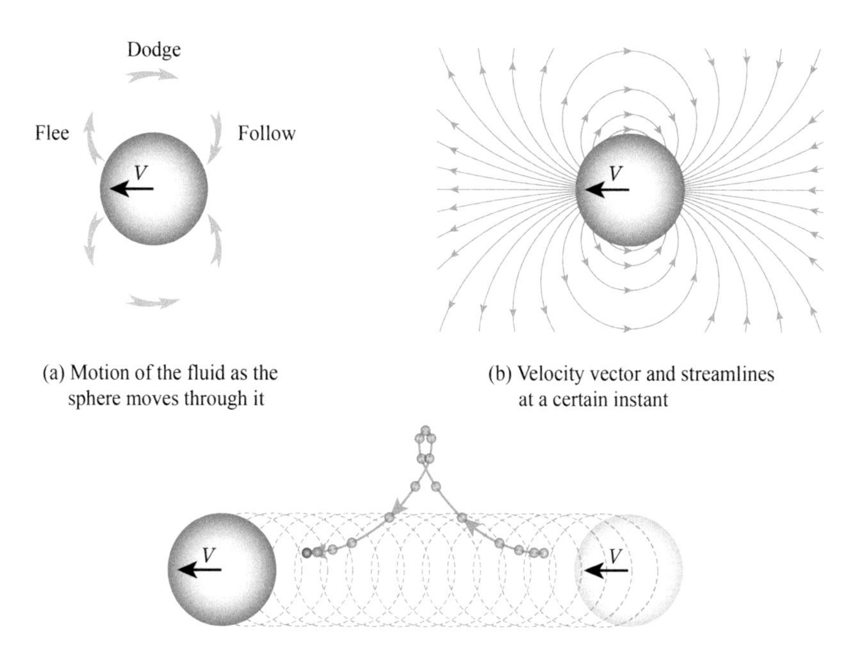

(a) Motion of the fluid as the sphere moves through it

(b) Velocity vector and streamlines at a certain instant

(c) The motion of a fluid particle and its trajectory over a period of time

Figure 3.2 Streamlines and pathlines formed by a sphere passing through a stationary fluid.

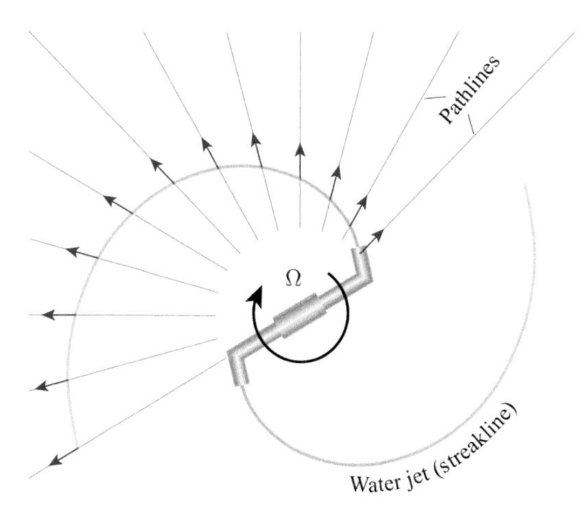

Figure 3.3 Streaklines and pathlines formed by the water flow from a rotating lawn sprinkler.

3.3 Velocity, Acceleration, and Substantial Derivative

When a fluid particle moves in space from one point to another, its instantaneous velocity is the derivative of its spatial coordinates with respect to time, that is:

$$\vec{V} = \mathrm{d}\vec{r}/\mathrm{d}t. \tag{3.2}$$

The instantaneous velocity vector can be expressed in terms of its x, y, and z components in the Cartesian coordinate system:

$$u = \mathrm{d}x/\mathrm{d}t, \quad v = \mathrm{d}y/\mathrm{d}t, \quad w = \mathrm{d}z/\mathrm{d}t. \tag{3.2a}$$

In the Eulerian method, velocity refers to that of a fluid particle at a certain point and at a given instant. Therefore, the velocity depends on both time and space coordinates:

$$\vec{V}(t,x,y,z) = \vec{i}u(t,x,y,z) + \vec{j}v(t,x,y,z) + \vec{k}w(t,x,y,z).$$

The acceleration of a fluid particle is the derivative of its velocity with respect to time:

$$\vec{a} = \frac{\mathrm{d}\vec{V}}{\mathrm{d}t} = \vec{i}\frac{\mathrm{d}u}{\mathrm{d}t} + \vec{j}\frac{\mathrm{d}v}{\mathrm{d}t} + \vec{k}\frac{\mathrm{d}w}{\mathrm{d}t}, \tag{3.3}$$

where the x component of the acceleration is $\mathrm{d}u/\mathrm{d}t$ and the velocity u is a function of spatial coordinates and time. So, we obtain:

$$a_x = \frac{\mathrm{d}u(t,x,y,z)}{\mathrm{d}t} = \frac{\partial u}{\partial t} + \frac{\partial u}{\partial x}\frac{\partial x}{\partial t} + \frac{\partial u}{\partial y}\frac{\partial y}{\partial t} + \frac{\partial u}{\partial z}\frac{\partial z}{\partial t}.$$

Equation (3.2a) states that the derivative of the spatial coordinates with respect to time is the velocity, so the x component of the acceleration can be written as:

$$a_x = \frac{\partial u}{\partial t} + u\frac{\partial u}{\partial x} + v\frac{\partial u}{\partial y} + w\frac{\partial u}{\partial z}. \tag{3.3a}$$

Similarly, the y and z components of the acceleration can be written as:

$$a_y = \frac{\partial v}{\partial t} + u\frac{\partial v}{\partial x} + v\frac{\partial v}{\partial y} + w\frac{\partial v}{\partial z}, \tag{3.3b}$$

$$a_z = \frac{\partial w}{\partial t} + u\frac{\partial w}{\partial x} + v\frac{\partial w}{\partial y} + w\frac{\partial w}{\partial z}. \tag{3.3c}$$

The acceleration can be expressed in vector form as:

$$\vec{a} = \frac{\partial \vec{V}}{\partial t} + \left(\vec{V}\cdot\nabla\right)\vec{V}, \tag{3.4}$$

where $\vec{V}\cdot\nabla$ is a particular mathematical expression, which may be written as:

$$\vec{V}\cdot\nabla = \left(\vec{i}u + \vec{j}v + \vec{k}w\right)\cdot\left(\vec{i}\frac{\partial}{\partial x} + \vec{j}\frac{\partial}{\partial y} + \vec{k}\frac{\partial}{\partial z}\right) = u\frac{\partial}{\partial x} + v\frac{\partial}{\partial y} + w\frac{\partial}{\partial z}.$$

Consequently, in Eulerian coordinates the acceleration of a fluid particle consists of two parts: $\partial\vec{V}/\partial t$ represents the rate of change of velocity only with respect to time at a given point in an unsteady flow field, known as *local acceleration*; $(\vec{V}\cdot\nabla)\vec{V}$ represents the rate of change of velocity due to the fluid particle being convected from one point to another in an inhomogeneous flow field, known as *convective acceleration*.

Equations (3.4) are mathematically equivalent to the total derivative of velocity with respect to time, which can also be applied to other fluid properties, such as pressure:

$$\frac{\mathrm{d}p}{\mathrm{d}t} = \frac{\partial p}{\partial t} + \left(\vec{V}\cdot\nabla\right)p.$$

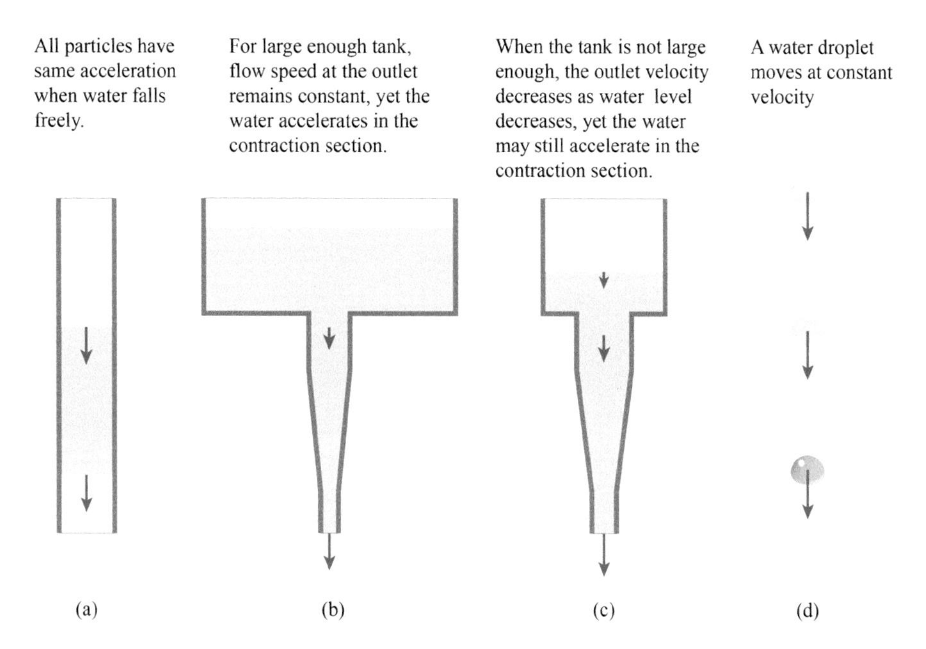

Figure 3.4 Acceleration of fluid in four cases.

In fluid mechanics, the total derivative of the Eulerian variable, known as *substantial derivative* or *material derivative*, is usually denoted by an uppercase differential symbol due to its particularity. Let Φ be a property of some fluid, then the material derivative is generally expressed as:

$$\frac{\mathrm{D}\Phi}{\mathrm{D}t} = \frac{\partial \Phi}{\partial t} + \left(\vec{V} \cdot \nabla\right)\Phi. \tag{3.5}$$

As with the definition of acceleration, the first and second terms on the right-hand side are called the *local term* and *convection term*, respectively.

Velocity and acceleration are concepts for *fluid particles*, whereas there is no notion of velocity or acceleration for spatial points. The total acceleration $\mathrm{d}\vec{V}/\mathrm{d}t$ represents the average acceleration of *a fluid particle* moving from the spatial point (x, y, z) at time t to the point $(x + \mathrm{d}x, y + \mathrm{d}y, z + \mathrm{d}z)$ at time $t + \mathrm{d}t$; and the local term $\partial\vec{V}/\partial t$ represents the change in local velocity of *different fluid particles* passing through the certain spatial point (x, y, z) over that period of time. Therefore, $\mathrm{d}\vec{V}/\mathrm{d}t$ represents the actual acceleration of a fluid particle, while $\partial\vec{V}/\partial t$ *only represents the time rate of change of velocity at a certain spatial point, not the acceleration for a fluid particle.*

To help readers further understand the concept of acceleration and substantial derivative, the physical meaning of both local and convective acceleration is analyzed in the following cases.

Case 1: Unsteady uniform flow, that is $\partial\vec{V}/\partial t \neq 0$, $(\vec{V} \cdot \nabla)\vec{V} = 0$. In this case, all fluid particles have the same acceleration at any given instant.

As shown in Figure 3.4(a), a certain amount of water falls freely down a vertical pipe of constant cross-section, ignoring the air drag and the viscous forces between water and wall. There is no relative motion among different parts of the water. That is to say, they have the same velocity, so we obtain: $(\vec{V}\cdot\nabla)\vec{V}=0$. The acceleration of the water as a whole is gravitational acceleration, g, that is $a=\mathrm{d}V/\mathrm{d}t=\partial V/\partial t=g$. For a given point in space, the local acceleration is supposed to represent the velocity difference of different fluid particles passing through that point over a period of time. However, since all fluid particles have the same velocity, this difference in velocity represents also the velocity change of the specific fluid particle.

Case 2: Steady nonuniform flow, that is $\partial\vec{V}/\partial t=0$, $(\vec{V}\cdot\nabla)\vec{V}\neq 0$. Flows that are called "steady" may still have accelerations due to the convective terms, but the velocity at any spatial point does not change with time.

As shown in Figure 3.4(b), a tank is filled with water and open to the atmosphere at the top. The water drains through a converging pipe at the bottom. The flow is assumed to be inviscid. If the tank volume is very large relative to the drainage discharge, the water surface elevation will remain unchanged. Based on Bernoulli's equation and continuity equations, the velocity at any cross-section of the drain pipe is constant – that is, the flow is steady. The flow accelerates through the converging portion. The acceleration of the fluid at a certain point, $\vec{a}=(\vec{V}\cdot\nabla)\vec{V}$, is entirely composed of convective acceleration. The fluid particle accelerates when moving from the low- to the high-velocity region.

Case 3: Unsteady nonuniform flow, that is $\partial\vec{V}/\partial t\neq 0$, $(\vec{V}\cdot\nabla)\vec{V}\neq 0$. In this case, the total acceleration of the fluid particle is the sum of local and convective acceleration.

Figure 3.4(c) shows an example similar to case 2. Here, the tank volume is not very large, and the water surface elevation will decrease as the water drains. Therefore, the velocity at any cross-section of the pipe decreases with time, that is $\partial\vec{V}/\partial t<0$. Meanwhile, the water should accelerate through the converging portion, that is $(\vec{V}\cdot\nabla)\vec{V}>0$. Whether a fluid particle accelerates or decelerates depends on the relative magnitude of the above two terms.

Case 4: Unsteady nonuniform flow with no acceleration, that is $\partial\vec{V}/\partial t\neq 0$, $(\vec{V}\cdot\nabla)\vec{V}\neq 0$, and $\mathrm{d}\vec{V}/\mathrm{d}t=\partial\vec{V}/\partial t+(\vec{V}\cdot\nabla)\vec{V}=0$.

Figure 3.4(d) shows an example for such a rare case, which is not entirely rigorous. Consider a spatial point on the path of one water drop that moves at a constant speed through a vacuum. At the beginning, the velocity at that point is zero (since no fluid is there). Suddenly, it becomes the velocity of the water drop moving through it, and then goes back to zero again. Obviously, this flow is unsteady and nonuniform. As the water drop passes by, the rate of change of velocity at that spatial point is $\partial\vec{V}/\partial t$, and the rate of change of velocity due to the non-uniformity of velocity in space is $(\vec{V}\cdot\nabla)\vec{V}$. Because the water drop has no acceleration, these two kinds of acceleration must cancel each other out.

The substantial derivative describes the rate of change of certain physical quantities for a fluid particle by tracking it with the Lagrangian method, but in Eulerian coordinates. It can serve as a link between Eulerian and Lagrangian descriptions of fluid flow. By expressing the substantial derivative in terms of Eulerian quantities, various definitions and conservation laws used to describe a fluid particle can be applied in the Eulerian reference frame.

3.4 Reynolds Transport Theorem

The substantial derivative can convert the changes of fluid particles into those at spatial points. However, in most engineering problems it is not necessary to know the state of motion of each fluid particle and the external forces acting on it. Engineers are more concerned with the influence of macroscopic fluid motion, such as the aerodynamic drag of a car or the pressure loss of a fluid traveling through a pipe. Such problems are generally solved through the *integral approach*, in which the formula serving as a link between Eulerian and Lagrangian methods is called the *Reynolds transport theorem*.

In the integral approach, the Lagrangian method concentrates on a chunk of fluid called a *system*. All the mass or region outside the system is referred to as the *surroundings*. In the process of flow, a fluid parcel may be constantly deformed and even split into many parts, but it is always the object of study. The Eulerian method concentrates on a fixed region in space called *control volume*, and observes the mechanical and thermodynamic effects of a fluid flowing through the control volume on its *boundary*. Figure 3.5 schematically shows several systems and control volumes in some typical flow problems. For a one-dimensional flow in a pipe of constant cross-section, the shape of the system does not change after it passes through the control volume. When the pressure relief valve of a pressure tank is opened, the system will still completely occupy the control volume, but a portion of fluid flowing out of the nozzle increases the volume of the system. The air flowing through a jet engine is subjected to complex external forces, and the volume and shape of the system change significantly as

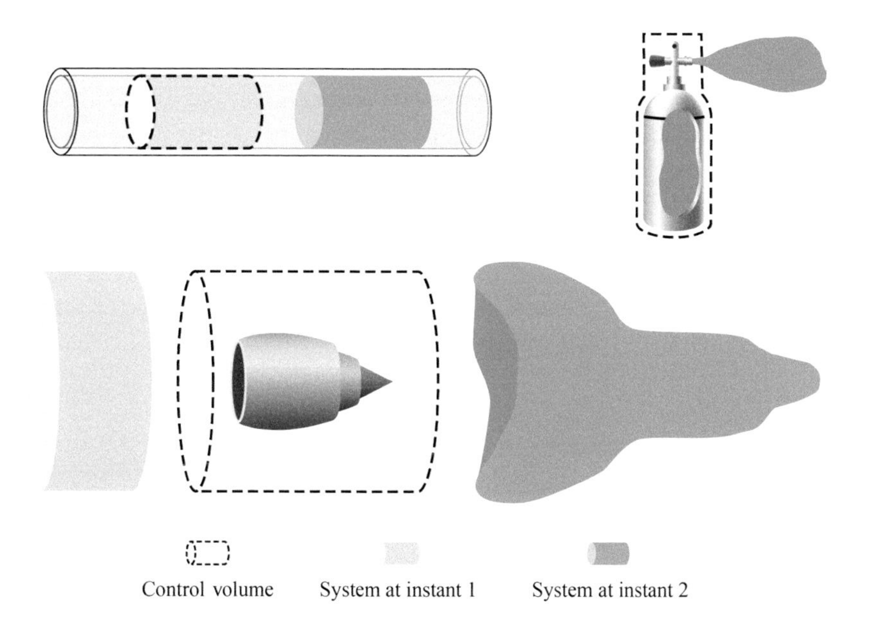

Figure 3.5 Systems and control volumes in several flow problems.

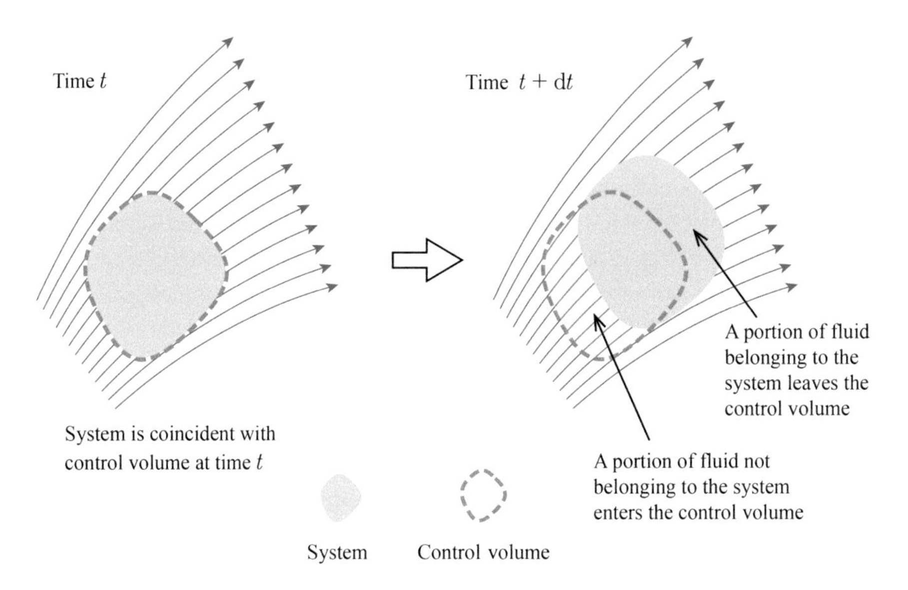

Figure 3.6 System and control volume – the derivation of the Reynolds transport theorem.

it passes through the control volume. For complex flows it is often more convenient to work with the Eulerian control volume method rather than with the Lagrangian system method.

Suppose we want to find the aerodynamic drag on a moving car. If we adopt the Lagrange method in differential form, we will need to sum up the momentum transferred to the car by each fluid element hitting it. However, if the Eulerian integral method is chosen, it will only be necessary to choose a control volume which encloses the car and moves with it at a constant speed. The difference between the air momentum out of the control volume and that into the control volume per unit time will equal the resultant force acting on the control volume. The aerodynamic drag on the car can thus be obtained.

We now derive the Reynolds transport theorem in general form to further understand the relationship between a system and a control volume. As shown in Figure 3.6, a control volume is chosen in the flow field, and the fluid within the control volume at time t is the selected system. Suppose Φ to be some mechanical property of the system (mass, momentum, energy, etc.). Since the system is coincident with the control volume at time t, the property of the system is that of the fluid within the control volume, that is

$$\Phi_{\mathrm{cv}}(t) = \Phi_{\mathrm{sys}}(t),$$

where the subscripts cv and sys represent the control volume and system, respectively.

After a short period of time, $\mathrm{d}t$,[1] a portion of the fluid that was inside the control volume flows out, and the system is no longer coincident with the control volume. The

[1] Strictly speaking, $\mathrm{d}t$ should be written here as δt, and the concept of limit should be used in the following derivation. For the sake of clarity and simplicity, the differential form is directly used, where $\mathrm{d}t$ represents the increment of time and $\mathrm{d}\Phi$ represents that of Φ. This expression will be used many times later in the book.

Φ carried by the fluid flowing out of the control volume is written as $(\mathrm{d}\Phi)_{out}$, while the Φ carried by the fluid flowing into the control volume is written as $(\mathrm{d}\Phi)_{in}$. At time $t + \mathrm{d}t$, the relationship between the amount of property Φ in the control volume and the system, respectively, is:

$$\Phi_{cv}(t+\mathrm{d}t) = \Phi_{sys}(t+\mathrm{d}t) - (\mathrm{d}\Phi)_{out} + (\mathrm{d}\Phi)_{in}.$$

The time rate of change of Φ within the control volume can be derived using the definition of differential as follows:

$$\begin{aligned}
\frac{\mathrm{d}\Phi_{cv}}{\mathrm{d}t} &= \frac{\Phi_{cv}(t+\mathrm{d}t) - \Phi_{cv}(t)}{\mathrm{d}t} \\
&= \frac{\Phi_{sys}(t+\mathrm{d}t) - (\mathrm{d}\Phi)_{out} + (\mathrm{d}\Phi)_{in} - \Phi_{cv}(t)}{\mathrm{d}t} \\
&= \frac{\Phi_{sys}(t+\mathrm{d}t) - \Phi_{cv}(t)}{\mathrm{d}t} - \frac{(\mathrm{d}\Phi)_{out} - (\mathrm{d}\Phi)_{in}}{\mathrm{d}t} \\
&= \frac{\Phi_{sys}(t+\mathrm{d}t) - \Phi_{sys}(t)}{\mathrm{d}t} - \frac{(\mathrm{d}\Phi)_{out} - (\mathrm{d}\Phi)_{in}}{\mathrm{d}t} \\
&= \frac{\mathrm{d}\Phi_{sys}}{\mathrm{d}t} - \frac{(\mathrm{d}\Phi)_{out} - (\mathrm{d}\Phi)_{in}}{\mathrm{d}t}
\end{aligned}.$$

Consequently, we obtain:

$$\frac{\mathrm{d}\Phi_{sys}}{\mathrm{d}t} = \frac{\mathrm{d}\Phi_{cv}}{\mathrm{d}t} + \frac{(\mathrm{d}\Phi)_{out} - (\mathrm{d}\Phi)_{in}}{\mathrm{d}t}. \tag{3.6}$$

Equation (3.6) is the *Reynolds transport theorem*, which describes the relation of time rates of change for some property between a system and a control volume. This formula is used to convert the fundamental laws of physics applied to a system into those applied to a control volume. Therefore, the Reynolds transport theorem is the basis for the Eulerian integral method.

In Equation (3.6), Φ flows into and out of the control volume across its boundary, which is called a *control surface* and denoted by cs. The Reynolds transport theorem in more general form can be written as:

$$\frac{\mathrm{d}\Phi_{sys}}{\mathrm{d}t} = \frac{\mathrm{d}\Phi_{cv}}{\mathrm{d}t} + \iint_{cs} \phi\left(\vec{V}\cdot\vec{n}\right)\mathrm{d}A, \tag{3.6a}$$

where ϕ represents Φ per unit volume; $\mathrm{d}A$ represents the infinitesimal area element on the control surface; and the integral term in the formula represents the net outflow of Φ from the control volume per unit time.

3.5 Relationship between the Reynolds Transport Theorem and Substantial Derivative

The substantial derivative describes the time rate of change of some property by following a fluid particle in Eulerian coordinates. The Reynolds transport theorem describes the time rate of change of some property by following a parcel of fluid (i.e.,

Table 3.1 Comparison between the material derivative and the Reynolds transport theorem

$\frac{\mathrm{D}\Phi}{\mathrm{D}t}=\frac{\partial\Phi}{\partial t}+\left(\vec{V}\cdot\nabla\right)\Phi$	*The material derivative* represents the relationship between the time rates of change of a property for a point in space and for a fluid particle passing through that point.	
$\frac{\mathrm{d}\Phi_{\mathrm{sys}}}{\mathrm{d}t}=\frac{\mathrm{d}\Phi_{\mathrm{cv}}}{\mathrm{d}t}+\iint_{\mathrm{cs}}\phi\left(\vec{V}\cdot\vec{n}\right)\mathrm{d}A$	*The Reynolds transport theorem* represents the relationship between the time rates of change of a property for a finite space (control volume) and for a parcel of fluid (system) passing through that space.	
The time rate of change of property Φ	$\frac{\mathrm{D}\Phi}{\mathrm{D}t}$	The time rate of change of property Φ for a fluid particle
	$\frac{\mathrm{d}\Phi_{\mathrm{sys}}}{\mathrm{d}t}$	The time rate of change of property Φ for a system
Local term (unsteady term)	$\frac{\partial\Phi}{\partial t}$	The time rate of change of property Φ for a point in space
	$\frac{\mathrm{d}\Phi_{\mathrm{cv}}}{\mathrm{d}t}$	The time rate of change of property Φ for a control volume
Convection term (nonuniform term)	$(\vec{V}\cdot\nabla)\Phi$	The net outflow of property Φ from a point in space per unit time
	$\iint_{\mathrm{cs}}\phi\left(\vec{V}\cdot\vec{n}\right)\mathrm{d}A$	The net outflow of property Φ from a control volume per unit time

the system) in Eulerian coordinates. The two concepts express essentially the same thing. The only difference is that one is used for a particle and the other for a group of particles. Table 3.1 shows the meaning of the terms in the substantial derivative formulas (3.5) and the Reynolds transport theorem (3.6a).

Mathematically, the substantial derivative is given in differential form, while the Reynolds transport theorem is expressed in integral form. Next, we'll prove that the two formulas are equivalent.

Equations (3.6a) are written as a general form:

$$\frac{\mathrm{d}}{\mathrm{d}t}\iiint_{\mathrm{sys}}\phi\,\mathrm{d}B=\frac{\mathrm{d}}{\mathrm{d}t}\iiint_{\mathrm{cv}}\phi\,\mathrm{d}B+\iint_{\mathrm{cs}}\phi\left(\vec{V}\cdot\vec{n}\right)\mathrm{d}A, \tag{3.6b}$$

where $\mathrm{d}B$ represents infinitesimal volume. For the first term on the right-hand side of Equation (3.6b), the volume of the control volume B is an invariant, so the derivative with respect to the product of ϕ multiplied by B only leaves the partial derivative with respect to ϕ, expressed as:

$$\frac{\mathrm{d}}{\mathrm{d}t}\iiint_{\mathrm{cv}}\phi\,\mathrm{d}B=\iiint_{\mathrm{cv}}\frac{\partial\phi}{\partial t}\mathrm{d}B.$$

Applying Gauss's theorem to the second term on the right-hand side of Equation (3.6b), we obtain:

$$\iint_{\mathrm{cs}}\phi\left(\vec{V}\cdot\vec{n}\right)\mathrm{d}A=\iiint_{\mathrm{cv}}\nabla\cdot(\phi\vec{V})\mathrm{d}B,$$

where the expression in the integral sign on the right-hand side can be expanded as:

$$\nabla\cdot\left(\phi\vec{V}\right)=\left(\vec{V}\cdot\nabla\right)\phi+\phi\left(\nabla\cdot\vec{V}\right).$$

We can rearrange the right-hand side items of Equation (3.6b) as:

$$\frac{\mathrm{d}}{\mathrm{d}t}\iiint_{\mathrm{cv}}\phi\,\mathrm{d}B+\iint_{\mathrm{cs}}\phi\left(\vec{V}\cdot\vec{n}\right)\mathrm{d}A=\iiint_{\mathrm{cv}}\left[\frac{\partial\phi}{\partial t}+\left(\vec{V}\cdot\nabla\right)\phi+\phi\left(\nabla\cdot\vec{V}\right)\right]\mathrm{d}B. \tag{3.7}$$

For the term on the left-hand side of Equation (3.6b), the differential sign cannot be directly applied to the term within the integral sign since the volume of the system is not constant. The derivation considering the change in the volume of the system is:

$$\begin{aligned}\frac{\mathrm{d}}{\mathrm{d}t}\iiint_{\mathrm{sys}}\phi\,\mathrm{d}B &= \iiint_{\mathrm{cv}}\frac{\mathrm{d}\phi}{\mathrm{d}t}\mathrm{d}B+\iiint_{\mathrm{sys}}\phi\,\frac{\mathrm{d}(\delta B)}{\mathrm{d}t}\\ &= \iiint_{\mathrm{cv}}\frac{\mathrm{d}\phi}{\mathrm{d}t}\mathrm{d}B+\iiint_{\mathrm{sys}}\phi\,\frac{\mathrm{d}(\delta B)}{\mathrm{d}t}\frac{1}{\delta B}\delta B\\ &= \iiint_{\mathrm{cv}}\frac{\mathrm{d}\phi}{\mathrm{d}t}\mathrm{d}B+\iiint_{\mathrm{cv}}\phi\left(\nabla\cdot\vec{V}\right)\mathrm{d}B\\ &= \iiint_{\mathrm{cv}}\left[\frac{\mathrm{d}\phi}{\mathrm{d}t}+\phi\left(\nabla\cdot\vec{V}\right)\right]\mathrm{d}B\end{aligned}\quad . \tag{3.8}$$

The above derivation is based on the concept that the divergence of the velocity represents the change in the volume of the system:

$$\nabla\cdot\vec{V}=\frac{1}{\delta B}\frac{\mathrm{d}(\delta B)}{\mathrm{d}t}.$$

Substituting Equations (3.7) and (3.8) into Equation (3.6b) in that order:

$$\iiint_{\mathrm{cv}}\left[\frac{\mathrm{d}\phi}{\mathrm{d}t}+\phi\left(\nabla\cdot\vec{V}\right)\right]\mathrm{d}B=\iiint_{\mathrm{cv}}\left[\frac{\partial\phi}{\partial t}+\left(\vec{V}\cdot\nabla\right)\phi+\phi\left(\nabla\cdot\vec{V}\right)\right]\mathrm{d}B.$$

Since the control volume is arbitrarily chosen, the integral signs on both sides of the above equation can be removed. So, we obtain:

$$\frac{\mathrm{D}\phi}{\mathrm{D}t}=\frac{\partial\phi}{\partial t}+\left(\vec{V}\cdot\nabla\right)\phi.$$

This is the formula of the substantial derivative – that is, Equation (3.5).

The above derivation is not rigorous. Readers can look for a more rigorous one in relevant books. The main purpose of this rather informal derivation is to make the physical concepts easier to understand.

3.6 The Incompressibility Hypothesis

In this section we will further discuss the incompressibility hypothesis. This hypothesis states that a fluid element neither shrinks nor swells in the process of flow, so that the only criterion to judge the compressibility of the flow is whether the density of the fluid element changes. The incompressibility hypothesis is defined by the Lagrangian method. According to the expression of the substantial derivative, the time rate of change of density of the fluid element can be expressed as:

$$\frac{\mathrm{D}\rho}{\mathrm{D}t}=\frac{\partial\rho}{\partial t}+\left(\vec{V}\cdot\nabla\right)\rho. \tag{3.9}$$

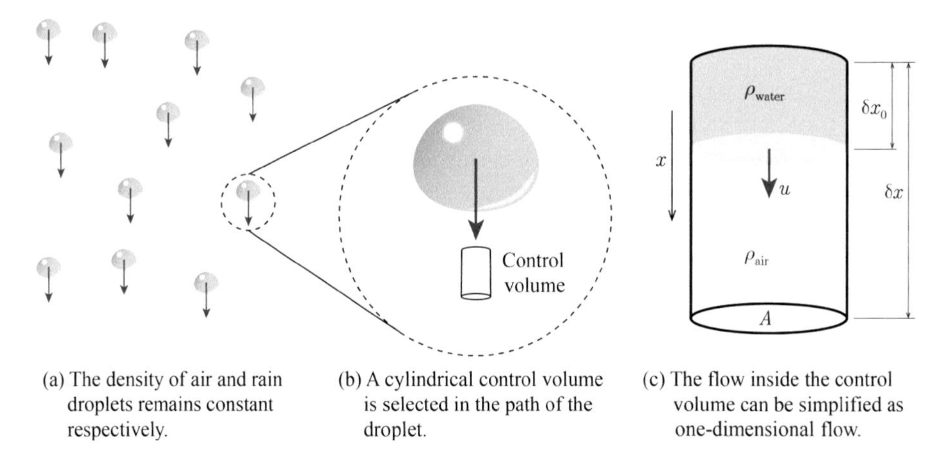

(a) The density of air and rain droplets remains constant respectively.

(b) A cylindrical control volume is selected in the path of the droplet.

(c) The flow inside the control volume can be simplified as one-dimensional flow.

Figure 3.7 Incompressible characteristic of a falling raindrop.

In other words, the time rate of change of density of the fluid element in Eulerian coordinates is composed of a local term $\partial\rho/\partial t$ and a convective term $(\vec{V}\cdot\nabla)\rho$. The local term represents the time rate of change of density at a certain point in space, while the convection term represents the difference in density between adjacent points in space. For an incompressible flow, these two terms do not have to be both zero, but just cancel each other out.

Let us look at the example of rain. The densities of air and raindrops are both invariant values, and the density at a certain point in space depends on whether air or water is passing through that point at that instant. As shown in Figure 3.7, the front edge of a raindrop just enters an infinitesimal control volume. During a short period of time, $\mathrm{d}t$, the density of the fluid within the control volume increases with time. The local term in Equation (3.9) is positive: $\partial\rho/\partial t > 0$. Since the density of the air in the front is smaller than that of the raindrop in the back, the density gradient along the direction of velocity is negative: $\partial\rho/\partial x < 0$.

Now, the motion of a raindrop into a control volume is simplified as one-dimensional flow to quantitatively compare the magnitudes of the local and convective terms. As shown in Figure 3.7(c), the water enters an infinitesimal cylindrical control volume at a velocity of u. The control volume is taken with a base area of A and a length of δx. The density of the water is ρ_{water}, and the density of the air is ρ_{air}. Then, the convection term in the formula of the substantial derivative can be expressed as:

$$\left(\vec{V}\cdot\nabla\right)\rho = u\frac{\partial u}{\partial x} = u\frac{\rho_{\text{air}} - \rho_{\text{water}}}{\delta x}. \tag{3.10}$$

For the local term, assuming that the length of the water into the control volume at time t is ρx_0, the average density of the fluid mixture within the control volume is:

$$\rho_t = u\frac{\delta x_0 \rho_{\text{water}} + (\delta x - \delta x_0)\rho_{\text{air}}}{\delta x}.$$

During a short period of dt, the volume of the water added to the control volume is Audt. Therefore, at time t + dt, the average density of the fluid mixture within the control volume becomes

$$\rho_{t+\mathrm{d}t} = u\frac{(\delta x_0 + u\mathrm{d}t)\rho_{\mathrm{water}} + (\delta x - \delta x_0 - u\mathrm{d}t)\rho_{\mathrm{air}}}{\delta x}.$$

From the above two formulas, we obtain the local term:

$$\frac{\partial \rho}{\partial t} = \frac{\rho_{t+\mathrm{d}t} - \rho_t}{\mathrm{d}t} = u\frac{\rho_{\mathrm{water}} - \rho_{\mathrm{air}}}{\delta x}. \tag{3.11}$$

As can be seen from Equations (3.10) and (3.11), the convection term and local term exactly cancel each other out, which is in accordance with the fact that the densities of both water and air remain unchanged. In other words, the density of any fluid element does not change with time. The whole flow is an incompressible flow.

3.7 Motion and Deformation of a Fluid Element

Section 3.3 discussed the law of motion for fluid particles. In this section, let us consider the motion and deformation of a fluid element. Figure 3.8 shows the motion of a fluid passing through a two-dimensional converging channel. During a period of time, a rectangular fluid element not only moves a certain distance, but also deforms considerably. It is obvious that rigid body kinematics cannot describe this seemingly complex kind of motion, so how do we describe it?

No matter how complex the fluid motion may be, the deformation on the scale of a single element can be considered to be linear (that is to say, small quantities of the second and higher orders are negligible). As shown in Figure 3.9, a complex fluid motion can therefore be decomposed into four types of simple motion and deformation, namely *translation*, *rotation*, *linear deformation* (shrinking or swelling), and *angular deformation* (shear deformation).

The deformation of the fluid element is equivalent to the change of the spatial position of each particle inside it. As long as the relations describing the spatial positions of each particle are established, the kinematics of a particle introduced in Section 3.3 can be used to describe the deformation. As shown in Figure 3.9, choosing two points inside the fluid element for observation, point P acts as the reference point, and point A represents a point adjacent to point P. The distance between the points A and P along the three Cartesian directions is δx, δy, and δz, respectively. The velocity of point P at any instant can be expressed as:

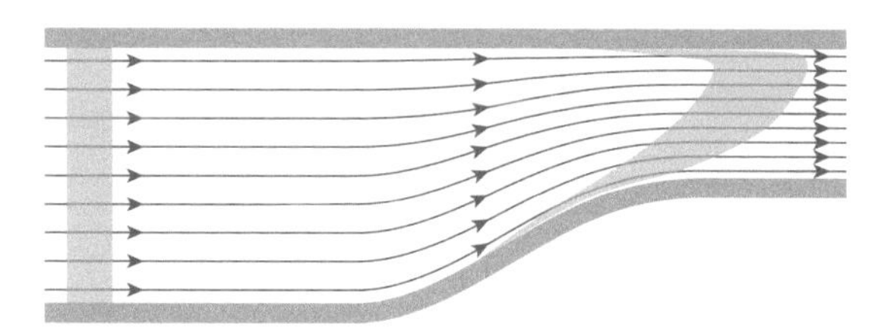

Figure 3.8 Transforms of the fluid flow through a two-dimensional passage.

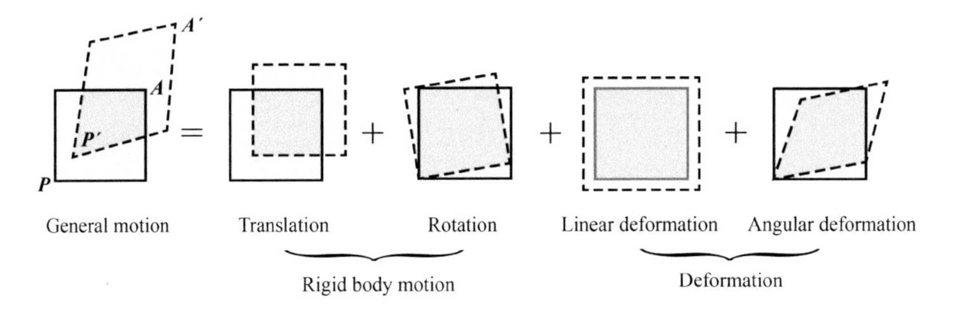

Figure 3.9 Decomposition of a general fluid motion.

$$\vec{V}_{\mathrm{P}}(t,x,y,z),$$

whereas the velocity of point A can be expressed as:

$$\vec{V}_{\mathrm{A}}(t,x+\delta x,y+\delta y,z+\delta z).$$

Through the Taylor expansion, the relationship between the velocities of the two points can be established as:

$$\vec{V}_{\mathrm{A}}(t,x+\delta x,y+\delta y,z+\delta z)=\vec{V}_{\mathrm{P}}(t,x,y,z)+\left(\frac{\partial\vec{V}}{\partial x}\right)_{\mathrm{P}}\delta x+\left(\frac{\partial\vec{V}}{\partial y}\right)_{\mathrm{P}}\delta y+\left(\frac{\partial\vec{V}}{\partial z}\right)_{\mathrm{P}}\delta z\,.$$

Neglecting the small quantities of the second and higher orders, the above formula is accurate only if the distance between the two points is infinitesimal. The velocity difference between the two points can be written in component form as:

$$\begin{aligned}\vec{V}_{\mathrm{A}}-\vec{V}_{\mathrm{P}} &= \left[\left(\frac{\partial u}{\partial x}\right)_{\mathrm{P}}\delta x+\left(\frac{\partial u}{\partial y}\right)_{\mathrm{P}}\delta y+\left(\frac{\partial u}{\partial z}\right)_{\mathrm{P}}\delta z\right]\vec{i}\\ &+\left[\left(\frac{\partial v}{\partial x}\right)_{\mathrm{P}}\delta x+\left(\frac{\partial v}{\partial y}\right)_{\mathrm{P}}\delta y+\left(\frac{\partial v}{\partial z}\right)_{\mathrm{P}}\delta z\right]\vec{j}\\ &+\left[\left(\frac{\partial w}{\partial x}\right)_{\mathrm{P}}\delta x+\left(\frac{\partial w}{\partial y}\right)_{\mathrm{P}}\delta y+\left(\frac{\partial w}{\partial z}\right)_{\mathrm{P}}\delta z\right]\vec{k}\end{aligned}$$

As can be seen from the above equation, the velocity variation of point A relative to point P can be expressed by the nine partial derivatives of the velocity components, that is the rate of change of velocity components of point P in the x, y, and z directions. These nine partial derivatives represent all the deformation modes of the fluid element and constitute a second-order *tensor* as follows (see Tip 3.1 for the tensor notation):

$$\frac{\partial u_j}{\partial x_i}=\begin{bmatrix}\dfrac{\partial u}{\partial x} & \dfrac{\partial u}{\partial y} & \dfrac{\partial u}{\partial z}\\ \dfrac{\partial v}{\partial x} & \dfrac{\partial v}{\partial y} & \dfrac{\partial v}{\partial z}\\ \dfrac{\partial w}{\partial x} & \dfrac{\partial w}{\partial y} & \dfrac{\partial w}{\partial z}\end{bmatrix} \tag{3.12}$$

Tip 3.1: Understanding Tensor Notation

A tensor is a generalized mathematical form of scalar and vector. In simple theory, the number of tensor components in Cartesian coordinates is denoted by 3^N, where N is the order of the tensor. A scalar is a tensor of order zero, and a vector is a tensor of order one. What we're talking about here is a tensor of order two with nine components. Some of the most basic representations are as follows:

(1) Let 1, 2, and 3 represent the three coordinates, then x_1, x_2, and x_3 represent x, y, and z; u_1, u_2, and u_3 represent u, v, and w.

(2) Both a_i and a_k represent the same vectors, where i and k are called free indices, and given 1, 2, and 3, respectively. The vectors in component form are expressed as:

$$u_i = u_1\vec{i} + u_2\vec{j} + u_3\vec{k}\ ; \quad \frac{\partial p}{\partial x_i} = \frac{\partial p}{\partial x_1}\vec{i} + \frac{\partial p}{\partial x_2}\vec{j} + \frac{\partial p}{\partial x_3}\vec{k}\ .$$

(3) The two free indices in the same term are given the values 1, 2, and 3, respectively. Then, this term becomes a tensor of order two with nine components, represented by the following matrix:

$$\tau_{ij} = \begin{bmatrix} \tau_{11} & \tau_{12} & \tau_{13} \\ \tau_{21} & \tau_{22} & \tau_{23} \\ \tau_{31} & \tau_{32} & \tau_{33} \end{bmatrix}\ ; \quad \frac{\partial u_i}{\partial x_k} = \begin{bmatrix} \partial u_1/\partial x_1 & \partial u_1/\partial x_2 & \partial u_1/\partial x_3 \\ \partial u_2/\partial x_1 & \partial u_2/\partial x_2 & \partial u_2/\partial x_3 \\ \partial u_3/\partial x_1 & \partial u_3/\partial x_2 & \partial u_3/\partial x_3 \end{bmatrix}.$$

(4) If there are two identical indices in the same term, one has to sum from 1 to 3, which sometimes is equivalent to the dot product of two vectors – for example:

$$u_i x_i = u_1 x_1 + u_2 x_2 + u_3 x_3\ ; \quad \frac{\partial u_k}{\partial x_k} = \frac{\partial u_1}{\partial x_1} + \frac{\partial u_2}{\partial x_2} + \frac{\partial u_3}{\partial x_3}\ .$$

(5) If there are identical and different indices in the same term, both (3) and (4) should be simultaneously satisfied. For example, the following formula actually represents a vector:

$$\begin{aligned} u_j \frac{\partial u_i}{\partial x_j} &= \left(u_1 \frac{\partial u_1}{\partial x_1} + u_2 \frac{\partial u_1}{\partial x_2} + u_3 \frac{\partial u_1}{\partial x_3}\right)\vec{i} \\ &+ \left(u_1 \frac{\partial u_2}{\partial x_1} + u_2 \frac{\partial u_2}{\partial x_2} + u_3 \frac{\partial u_2}{\partial x_3}\right)\vec{j} \\ &+ \left(u_1 \frac{\partial u_3}{\partial x_1} + u_2 \frac{\partial u_3}{\partial x_2} + u_3 \frac{\partial u_3}{\partial x_3}\right)\vec{k} \end{aligned}.$$

Equation (3.12) describes all types of relative motion between any two adjacent points inside the fluid element, including rotation, linear deformation, and angular deformation. A general motion is the superposition of the three types of motion. Therefore, it should be possible to rewrite Equation (3.12) into the superposition of three terms that represent the three types of motion, respectively. The formula is rearranged as:

$$\frac{\partial u_j}{\partial x_i}=\begin{bmatrix}\frac{\partial u}{\partial x} & 0 & 0\\ 0 & \frac{\partial v}{\partial y} & 0\\ 0 & 0 & \frac{\partial w}{\partial z}\end{bmatrix}+\begin{bmatrix}0 & \frac{1}{2}\left(\frac{\partial u}{\partial y}-\frac{\partial v}{\partial x}\right) & \frac{1}{2}\left(\frac{\partial u}{\partial z}-\frac{\partial w}{\partial x}\right)\\ \frac{1}{2}\left(\frac{\partial v}{\partial x}-\frac{\partial u}{\partial y}\right) & 0 & \frac{1}{2}\left(\frac{\partial v}{\partial z}-\frac{\partial w}{\partial y}\right)\\ \frac{1}{2}\left(\frac{\partial w}{\partial x}-\frac{\partial u}{\partial z}\right) & \frac{1}{2}\left(\frac{\partial w}{\partial y}-\frac{\partial v}{\partial z}\right) & 0\end{bmatrix}$$
$$+\begin{bmatrix}0 & \frac{1}{2}\left(\frac{\partial u}{\partial y}+\frac{\partial v}{\partial x}\right) & \frac{1}{2}\left(\frac{\partial u}{\partial z}+\frac{\partial w}{\partial x}\right)\\ \frac{1}{2}\left(\frac{\partial u}{\partial y}+\frac{\partial v}{\partial x}\right) & 0 & \frac{1}{2}\left(\frac{\partial v}{\partial z}+\frac{\partial w}{\partial y}\right)\\ \frac{1}{2}\left(\frac{\partial u}{\partial z}+\frac{\partial w}{\partial x}\right) & \frac{1}{2}\left(\frac{\partial v}{\partial z}+\frac{\partial w}{\partial y}\right) & 0\end{bmatrix}. \quad (3.12a)$$

In Equation (3.12a), the first matrix is a diagonal matrix with three independent variables. The second matrix is an anti-symmetric matrix also with three independent variables. The third matrix is a symmetric matrix with three independent variables as well. Therefore, after Equation (3.12) is transformed into Equation (3.12a), the deformation of the fluid element is still composed of nine independent components. Next, we will prove, in that order, that the first matrix in Equation (3.12a) represents the linear deformation of the fluid element, the second represents its rotation, and the third its angular deformation.

3.7.1 Linear Deformation of a Fluid Element

Figure 3.10 shows the first type of deformation. A rectangular fluid element extends only in the x direction. Assuming that the left side of the fluid element is moving at a velocity of u, the velocity of the right side can be expressed as:

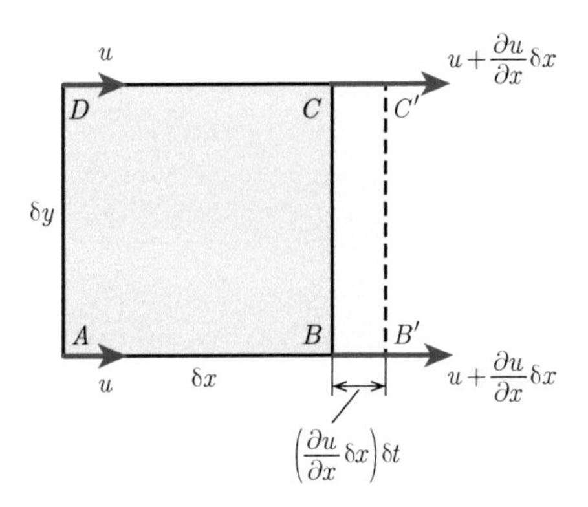

Figure 3.10 Linear deformation of a fluid element.

$$u+\frac{\partial u}{\partial x}\delta x.$$

During a short period of time dt, the right side moves more than the left side a distance:

$$\left(\frac{\partial u}{\partial x}\delta x\right)\delta t.$$

Therefore, the relative elongation of the fluid element in the x direction is:

$$\left(\frac{\partial u}{\partial x}\delta x\right)\delta t\Big/\delta x=\frac{\partial u}{\partial x}\delta t.$$

The relative elongation per unit time is:

$$\left(\frac{\partial u}{\partial x}\delta t\right)\Big/\delta t=\frac{\partial u}{\partial x}.$$

This represents the rate of linear deformation of the fluid element in the x direction.

Similarly, the rate of linear deformation in the y and z directions can be obtained, so the rate of linear deformation of the fluid element can be expressed as:

$$\frac{\partial u}{\partial x},\ \frac{\partial v}{\partial y},\ \frac{\partial w}{\partial z},$$

which are the three quantities in the first matrix of Equation (3.12a). That is to say, this matrix represents the linear deformation of the fluid element.

For the case shown in Figure 3.10, the fluid element extends only in the x direction, but does not change in the other two directions. Thus, the volume of the fluid element increases, corresponding to a compressible flow. If the flow is incompressible, the fluid element elongates in one direction and shrinks in at least one of the other two. This deformation may be expressed simply as follows: stretching makes the object longer and narrower, while compression makes the object shorter and wider. Next, we shall discuss what linear deformation looks like in an incompressible flow.

For the deformation shown in Figure 3.10, the volume change caused by the linear deformation in the x direction is:

$$d(\delta B)_x=\delta y\delta z\left(\frac{\partial u}{\partial x}\delta x\right)\delta t.$$

Similarly, the volume changes caused by the linear deformation in the y and z directions are:

$$d(\delta B)_y=\delta z\delta x\left(\frac{\partial v}{\partial y}\delta y\right)\delta t,$$

$$d(\delta B)_z=\delta x\delta y\left(\frac{\partial w}{\partial z}\delta z\right)\delta t,$$

The total volume change is the sum of these three terms:

$$\begin{aligned}d(\delta B)&=d(\delta B)_x+d(\delta B)_y+d(\delta B)_z\\&=\left(\frac{\partial u}{\partial x}\delta x\delta y\delta z+\frac{\partial v}{\partial y}\delta x\delta y\delta z+\frac{\partial w}{\partial z}\delta x\delta y\delta z\right)\delta t\end{aligned}.$$

The relative volume change per unit time is called the rate of volume change, expressed as:

$$\begin{aligned}\frac{1}{\delta B}\frac{d(\delta B)}{dt} &= \frac{1}{\delta x\delta y\delta z}\left(\frac{\partial u}{\partial x}\delta x\delta y\delta z+\frac{\partial v}{\partial y}\delta x\delta y\delta z+\frac{\partial w}{\partial z}\delta x\delta y\delta z\right) \\ &= \frac{\partial u}{\partial x}+\frac{\partial v}{\partial y}+\frac{\partial w}{\partial z} \\ &= \nabla\cdot\vec{V}\end{aligned}.$$

Notice that the rate of volume change of a fluid element is the divergence of its velocity, $\nabla\cdot\vec{V}$. Clearly, for an incompressible flow, the divergence of the velocity field should be zero everywhere.

3.7.2 Rotation of a Fluid Element

Figure 3.11 shows the second type of deformation. A rectangular fluid element is rotating about a fixed axis in the counter-clockwise (defined as positive) direction. Assuming that the two velocity components of point A are u and v, respectively, then the velocity component of point B in the y direction can be expressed as:

$$v_{\mathrm{B}} = v+\frac{\partial v}{\partial x}\delta x\,.$$

The angular velocity of the side AB around point A is:

$$\Omega_{\mathrm{AB}} = \frac{v_{\mathrm{B}}-v}{\delta x} = \frac{\frac{\partial v}{\partial x}\delta x}{\delta x} = \frac{\partial v}{\partial x}.$$

The velocity component of point D in the x direction can be expressed as:

$$v_{\mathrm{D}} = u+\frac{\partial u}{\partial y}\delta y\,.$$

The angular velocity of the side AD around point A is:

$$\Omega_{\mathrm{AD}} = \frac{u-u_{\mathrm{D}}}{\delta y} = \frac{-\frac{\partial u}{\partial y}\delta y}{\delta y} = -\frac{\partial u}{\partial y}\,.$$

If the fluid element is rotating as a rigid body, the two angular velocities should be equal: $\Omega_{\mathrm{AB}} = \Omega_{\mathrm{AD}}$. For general cases, the fluid element experiences an angular deformation as well, and its angular velocity should be expressed as the average value of

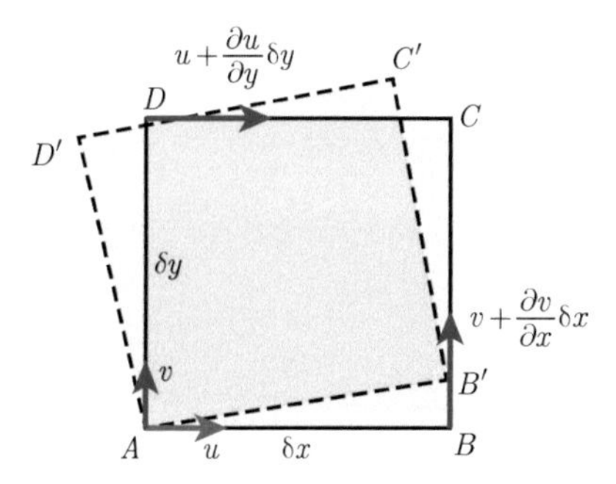

Figure 3.11 Rotation of a fluid element.

the angular velocities of all the particles inside the fluid element. The average value of the angular velocities of any two perpendicular lines inside the fluid element can represent all the particles. Therefore, the average value of the angular velocity of the fluid element about the z axis is expressed as:

$$\Omega_z = \frac{1}{2}(\Omega_{\mathrm{AB}} + \Omega_{\mathrm{AD}}) = \frac{1}{2}\left(\frac{\partial v}{\partial x} - \frac{\partial u}{\partial y}\right).$$

Similarly, the angular velocities of the fluid element about the x and y axes can be expressed as:

$$\Omega_x = \frac{1}{2}\left(\frac{\partial w}{\partial y} - \frac{\partial v}{\partial z}\right),$$

$$\Omega_y = \frac{1}{2}\left(\frac{\partial u}{\partial z} - \frac{\partial w}{\partial x}\right).$$

These three angular velocities are the three independent components in the second matrix of Equation (3.12a), which represent the rotation of the fluid element.

3.7.3 Angular Deformation of a Fluid Element

Figure 3.12 shows the third type of deformation. Both sides AB and AD rotate by a certain angle, producing a change in the angle $\angle BAD$. As a result, the fluid element experiences an angular, or shear, deformation. As mentioned above, the velocities of points B and D are:

$$v_{\mathrm{B}} = v + \frac{\partial v}{\partial x}\delta x,$$

$$u_{\mathrm{D}} = u + \frac{\partial u}{\partial y}\delta y.$$

The time rates of change of the angle $\angle BAD$ due to the rotation of sides AB and AD are:

$$\delta\alpha_{\mathrm{B}} = \frac{(v_{\mathrm{B}} - v)\cdot\delta t}{\delta x}\bigg/\delta t = \frac{\partial v}{\partial x},$$

$$\delta\alpha_{\mathrm{D}} = \frac{(u_{\mathrm{D}} - u)\cdot\delta t}{\delta y}\bigg/\delta t = \frac{\partial u}{\partial y},$$

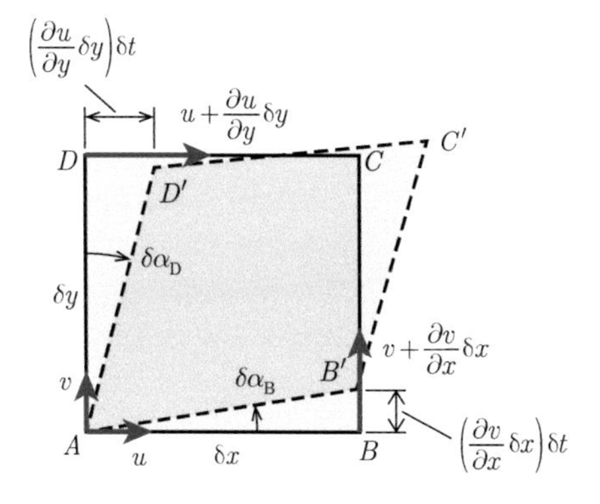

Figure 3.12 Angular deformation of a fluid element.

The total change in the angle $\angle BAD$ per unit time is the sum of these two terms:

$$\delta\alpha = \delta\alpha_{\mathrm{B}} + \delta\alpha_{\mathrm{D}} = \frac{\partial u}{\partial y} + \frac{\partial v}{\partial x},$$

which represents the angular deformation of the fluid element in the x–y plane. The angular deformation per unit time is called the rate of angular deformation, generally denoted by ε. The above derivation shows that the rates of angular deformation of the fluid element in the three coordinate planes are:

$$\varepsilon_{xy} = \frac{\partial u}{\partial y} + \frac{\partial v}{\partial x}; \quad \varepsilon_{yz} = \frac{\partial v}{\partial z} + \frac{\partial w}{\partial y}; \quad \varepsilon_{zx} = \frac{\partial w}{\partial x} + \frac{\partial u}{\partial z}.$$

The three rates of angular deformation are twice as large as the three independent components in the third matrix of Equation (3.12a), from which it can be inferred that this matrix represents the angular deformation of the fluid element.

Expanded Knowledge

Streaklines and Their Applications

A streakline is the locus of fluid particles that have earlier passed through a prescribed point in the flow field. In a steady flow, streamlines, pathlines, and streaklines are identical; none of them coincide in an unsteady flow. Although the physical definition of streaklines is not as clear as those of pathlines and streamlines, they are the easiest to perceive and are often used as a means of observation in experiments. As illustrated on the title page of this chapter, the curves passing through the surface of a car represent some sort of streaklines. These curves correspond to the continuous release of marked smoke particles from a pipe in front of the car model. If the experiment is conducted in water, coloring agents are often used. The spiral flows shown in Figure 3.3 can also be regarded as a type of streakline. However, the strict definition is not satisfied, as water flows out of a rotating nozzle rather than a fixed point.

Streamline Coordinates

It is sometimes easier to analyze the flow problem by using the coordinates relative to the local flow direction. For the flow shown in Figure 3.13, for example, either the Cartesian or the streamline coordinates may be adopted. Streamline coordinates consist of two components, namely, along the streamline, s, and perpendicular to it, n. The unit vectors in these two directions are denoted by $\vec{s}$ and $\vec{n}$, respectively. The biggest challenge in determining streamline coordinates is that the streamline must be known beforehand. This is a contradiction, because the streamline can only be obtained after solving the flow problem. Usually, a streamline is assumed first and the final solution is obtained by iteration. Another application of streamline coordinates is

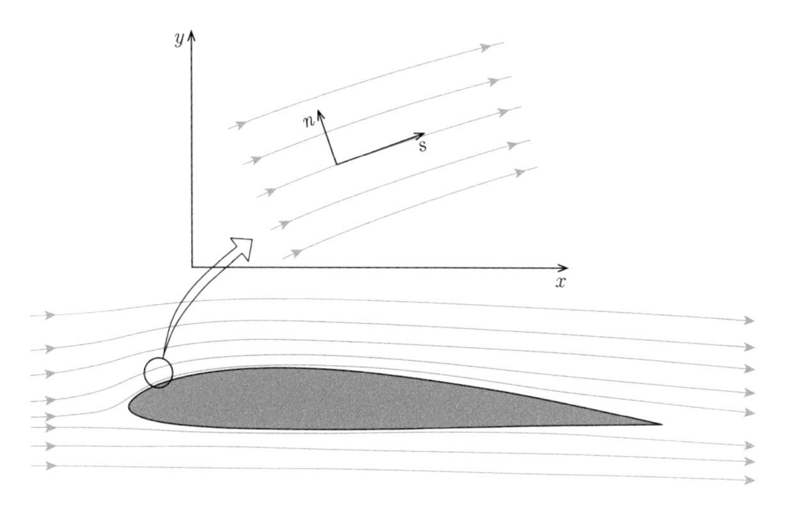

Figure 3.13 Flow around an airfoil and streamline coordinates.

judging the flow qualitatively. In this case, it is not necessary to know the streamlines accurately.

With the unit tangential and normal vectors defined, the velocity vector can be expressed as:

$$\vec{V} = V\vec{s}.$$

Similarly, the acceleration can be expressed as:

$$\vec{a} = \frac{D\vec{V}}{Dt} = a_s\vec{s} + a_n\vec{n}.$$

Depending on the effect it produces on the velocity, the acceleration can be classified into tangential acceleration (if it changes the magnitude of the velocity vector) and normal or centripetal acceleration (if it changes its direction). For a steady flow, the tangential and normal components of the acceleration can be obtained through some transformation:

$$a_s = V\frac{\partial V}{\partial s}; \quad a_n = \frac{V^2}{R},$$

where R represents the local curvature radius of the streamline.

The tangential component of the acceleration represents the change in magnitude of velocity, and its normal component that in direction of velocity – that is, centripetal acceleration. In Figure 3.13 the streamline bends toward the airfoil, so the normal component of acceleration is directed toward the surface of the airfoil. The centripetal force is due to the difference between the higher pressure in regions of upper streamlines and the lower pressure in regions of lower streamlines. This would explain why the pressure on the upper surface of the airfoil is lower than the external undisturbed atmospheric pressure.

Questions

3.1 Reflect on the changes in the magnitude and direction of the velocity of the fluid particle in Figure 3.2(c), as well as the changes of pressure difference force exerted on the particle with time.

3.2 Fluid flows through a converging conical tube. Given the inlet and outlet diameters and the length of the converging portion, derive the changes in magnitude of the acceleration along the streamline (assume that the flow is incompressible – that is, that the flow velocity is inversely proportional to the pipe cross-sectional area).

3.3 In a vessel filled with water, a drop of oil rises at constant speed from the bottom to the surface. If the vessel is chosen as a control volume, is this flow steady or unsteady? Compressible or incompressible? Is there an acceleration throughout the entire flow field?

3.4 The linear deformation shown in Figure 3.10 actually contains an angular deformation. What is the amount of this deformation? What does a pure linear deformation without angular deformation look like? Does a pure linear deformation exist in an incompressible flow?

4 Basic Equations of Fluid Dynamics

Wind can generate lift and drag on an umbrella, thereby resulting in disaster.

4.1 Integral and Differential Approach

Just like solids, fluids obey three fundamental laws of physics: the law of conservation of mass, the momentum theorem (Newton's second law of motion), and the first law of thermodynamics. All these laws are inherently Lagrangian, since they apply to objects (particles or systems). In fluid mechanics it is necessary to perform the transformation of the fundamental laws from a Lagrangian particle or system to a Eulerian spatial point or control volume by the substantial derivative or the Reynolds transport theorem. *The approach aiming at calculating the features of a particle or at a spatial point is called the differential approach, and that aiming at computing the overall features of a system or through a control volume is called the integral approach.*

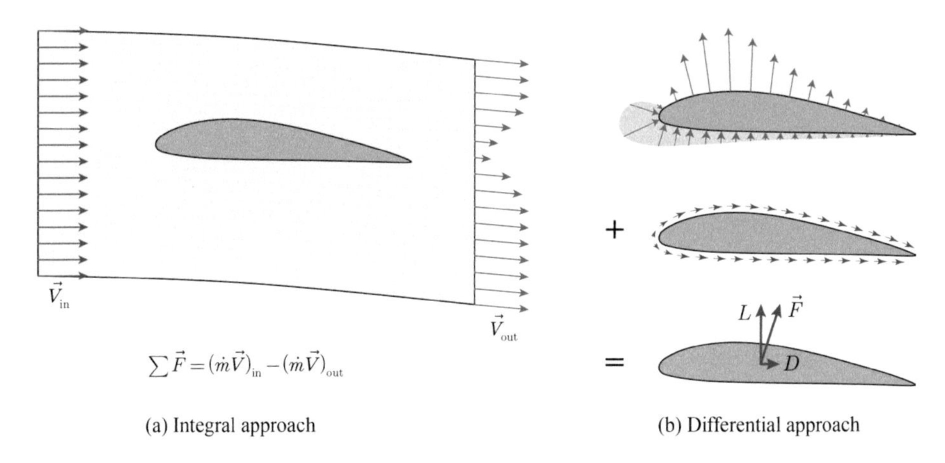

Figure 4.1 Aerodynamic forces acting on an airfoil-integral approach and differential approach.

The Eulerian integral approach is also called control volume analysis. In a given fluid problem, once a control volume is appropriately chosen, only the forces acting on the control volume and the properties of fluids in and out of the control volume are analyzed, while the changes in those properties inside the control volume are neglected. The integral approach is the most widely used method for one-dimensional flows. However, for more complex two- and three-dimensional flows, it is difficult to obtain useful results by using only the integral approach. It is necessary to use the differential approach to study the detailed flow field. Although the differential approach can better reflect the nature of the flow, the more complex governing differential equations are difficult to solve.

Figure 4.1 shows the two approaches used to solve the lift and drag of an airfoil. The integral approach, as shown in Figure 4.1(a), computes the force exerted by the airfoil on the airflow by measuring the change of momentum in the airflow passing over the airfoil. The differential approach, as shown in Figure 4.1(b), aims at measuring the pressure and viscous forces at every point on the airfoil surface, and integrates the product of these forces and the infinitesimal area to deduce the resultant force. For this problem, the integral approach is relatively simple and feasible, while the differential approach is difficult to implement. However, if we wish to find the highest and lowest pressure points on the airfoil surface, the integral approach is ineffective, and these tasks can only be fulfilled by the differential approach.

In mathematics, the integral and differential approaches can be transformed into each other. A differential equation can be converted into an integral equation by integrating it within a finite control volume, and an integral equation can be converted into a differential one by applying it to an infinitesimal control volume or a fluid particle. In this chapter we will deduce the integral and differential forms of the three fundamental laws of physics in fluid mechanics and analyze their physical meaning in depth.

4.2 Continuity Equation

4.2.1 Continuity Equation: Integral Form

For a system of constant mass, the continuity equation is expressed as:

$$\frac{\mathrm{d}m_{\mathrm{sys}}}{\mathrm{d}t} = \frac{\mathrm{d}}{\mathrm{d}t}\iiint_{\mathrm{sys}} \rho \mathrm{d}B = 0.$$

By applying the Reynolds transport theorem, this equation can be converted into an integral relation suitable for a control volume. Let ϕ in Equation (3.6b) of the Reynolds transport theorem represent the mass per unit volume (i.e., the density ρ); then:

$$\frac{\mathrm{d}}{\mathrm{d}t}\iiint_{\mathrm{sys}} \rho \mathrm{d}B = \frac{\partial}{\partial t}\iiint_{\mathrm{cv}} \rho \mathrm{d}B + \iint_{\mathrm{cs}} \rho\left(\vec{V}\cdot\vec{n}\right)\mathrm{d}A.$$

From the above two equations we can obtain the equation of conservation of mass for a control volume:

$$\frac{\partial}{\partial t}\iiint_{\mathrm{cv}} \rho \mathrm{d}B + \iint_{\mathrm{cs}} \rho\left(\vec{V}\cdot\vec{n}\right)\mathrm{d}A = 0. \tag{4.1}$$

This integral relation is called *the integral continuity equation for a control volume.*

In Equation (4.1) the first term represents the time rate of change of mass within the control volume, and the second one the rate at which mass leaves the control volume through the control surface. The physical meaning of this equation is perfectly clear: *the mass gained within the control volume can only come from the net flow across the volume's boundary.*

For a steady flow, the amount of mass in the control volume remains constant; thus the first term in Equation (4.1) is zero. We have:

$$\iint_{\mathrm{cs}} \rho\left(\vec{V}\cdot\vec{n}\right)\mathrm{d}A = 0.$$

That is to say, the amount of mass flowing into and out of the control volume is dynamically balanced. At any given time, the total amount of mass into the control volume from any direction must be equal to that out of this volume from other directions.

For a one-dimensional steady flow, the continuity equation can be rewritten in a more practical form:

$$\dot{m}_{\mathrm{in}} = \dot{m}_{\mathrm{out}},$$

where $\dot{m} = \mathrm{d}m/\mathrm{d}t$ represents the amount of mass passing through a cross-section per unit time, called *mass flow rate.*

Figure 4.2 shows a one-dimensional flow through a variable cross-section circular pipe. The mass flow rate at any section can be expressed as:

$$\dot{m} = \frac{\rho \mathrm{d}B}{\mathrm{d}t} = \frac{\rho A \mathrm{d}x}{\mathrm{d}t} = \rho A V.$$

This relation can also be obtained directly from the second term in Equation (4.1). For one-dimensional flows, all flow parameters vary only in the streamwise direction. The

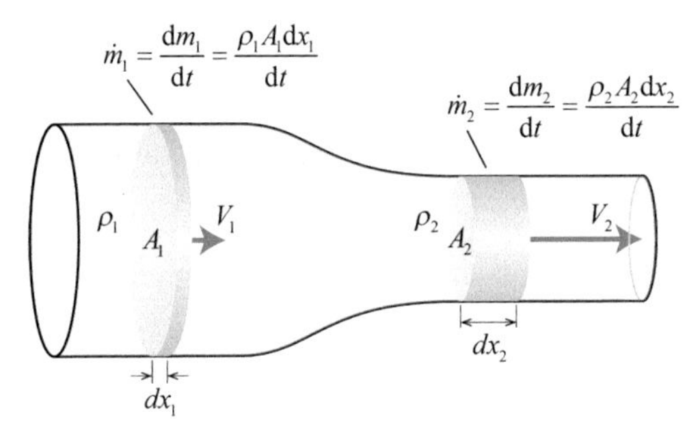

Figure 4.2 Flow in a variable cross-section pipe.

density and velocity at any cross-section are uniform, and can be moved outside the integral sign in the following way:

$$\dot{m} = \iint_{cs} \rho\left(\vec{V}\cdot\vec{n}\right)\mathrm{d}A = \rho AV.$$

So, the continuity relation for one-dimensional steady flows can be written in the form commonly used in engineering:

$$\rho_1 A_1 V_1 = \rho_2 A_2 V_2, \tag{4.2}$$

where subscripts 1 and 2 represent two different cross-sectional planes normal to the flow direction, and usually represent the inlet and outlet of a control volume. This equation indicates that *for one-dimensional steady flows, the product of cross-sectional area, density, and velocity remains constant at any cross-section.*

If the selected cross-section is located in the converging or diverging portion of a pipe, Equation (4.2) will contain a certain error. This error may come from two sources: one is that the velocity is not uniform on the cross-section; the other is that the velocity is not everywhere perpendicular to the cross-section.

As long as the flow at the inlet and outlet of the control volume is one-dimensional, the use of Equation (4.2) is assumed to be accurate enough for practical applications, regardless of what the flow inside the control volume is.

When the flow is incompressible, a simpler form of Equation (4.2) is

$$A_1 V_1 = A_2 V_2\,. \tag{4.3}$$

Equation (4.3) shows that *when the fluid density remains unchanged, the velocity is inversely proportional to the cross-sectional area.* The fluid accelerates in the converging portion of the pipe and decelerates in the diverging one. As the pipe contracts, a fluid element of constant volume is laterally compressed, thereby resulting in the increase of streamwise scale and velocity, as shown in Figure 4.2.

The integral form of the continuity equation is a useful tool only for one-dimensional flows, while the differential continuity equation is often used for solving three-dimensional problems. In mathematics, the differential continuity equation can be obtained by directly transforming the integral continuity equation or, alternatively, by applying the Reynolds transport theorem to an elemental control volume.

4.2.2 Conversion from Integral to Differential Equation

The integral continuity equation is rewritten as:

$$\frac{\partial}{\partial t}\iiint_{\text{cv}} \rho \mathrm{d}B + \iint_{\text{cs}} \rho\left(\vec{V}\cdot\vec{n}\right)\mathrm{d}A = 0.$$

After applying Gauss's divergence theorem, the second term of the above equation can be transformed into a volume integral:

$$\iint_{\text{cs}} \rho\left(\vec{V}\cdot\vec{n}\right)\mathrm{d}A = \iiint_{\text{cv}} \nabla\cdot\left(\rho\vec{V}\right)\mathrm{d}B.$$

Consequently, the continuity equation can be transformed as follows:

$$\frac{\partial}{\partial t}\iiint_{\text{cv}} \rho \mathrm{d}B + \iiint_{\text{cv}} \nabla\cdot\left(\rho\vec{V}\right)\mathrm{d}B = 0.$$

For a fixed control volume, the differential sign can be moved inside the integral sign, and the above equation then written as:

$$\iiint_{\text{cv}} \frac{\partial\rho}{\partial t}\mathrm{d}B + \iiint_{\text{cv}} \nabla\cdot\left(\rho\vec{V}\right)\mathrm{d}B = 0.$$

It follows that

$$\iiint_{\text{cv}} \left[\frac{\partial\rho}{\partial t} + \nabla\cdot\left(\rho\vec{V}\right)\right]\mathrm{d}B = 0.$$

In order for the above equation to be true for any control volume, the term inside the integral sign must be zero:

$$\frac{\partial\rho}{\partial t} + \nabla\cdot\left(\rho\vec{V}\right) = 0. \tag{4.4}$$

Equation (4.4) is the *differential form of the continuity equation*, which has the same meaning as the integral continuity equation, but for a point in space. The term $\partial\rho/\partial t$ represents the time rate of change of mass at a certain point in space, while $\nabla\cdot(\rho\vec{V})$ represents the rate at which mass leaves that point.

Tip 4.1: Gauss's Theorem in Fluids

Gauss's theorem, also called the divergence theorem, is stated as follows:

$$\iiint_{\text{B}} \left(\nabla\cdot\vec{F}\right)\mathrm{d}B = \oiint_{\text{A}} \left(\vec{F}\cdot\vec{n}\right)\mathrm{d}A.$$

The left-hand side is the volume integral of the divergence of any vector $\vec{F}$ held in a bounded region in space, which represents the net quantity of the vector gained within that region. The right-hand side represents the net flux of the vector through the closed surface enclosing that region in space. Gauss's theorem describes the conservation of certain physical properties; it can be interpreted as stating that the net quantity gained within a region in space is equal to the net outflow through its boundary.

When applied to the flow field, if we let the vector in Gauss's theorem represent velocity, we obtain:

$$\iiint_{\text{cv}} \left(\nabla \cdot \vec{V}\right) \mathrm{d}B = \iint_{\text{cs}} \left(\vec{V} \cdot \vec{n}\right) \mathrm{d}A .$$

The left-hand side represents the volume integral of the divergence of the velocity field, or, in physical terms, the incremental volume per unit time. The right-hand side represents the surface integral of the normal component of the vector point function over the control surface, that is, the volume flow rate $Q = VA$. In this case, Gauss's theorem demonstrates that the increased volume of the fluid within the control volume per unit time is equal to the rate at which volume leaves the control volume through the control surface.

4.2.3 Differential Equation for an Elemental Control Volume

We next deduce the differential continuity equation by considering a cubical elemental control volume with its six control surfaces perpendicular to the three coordinate axes, as shown in Figure 4.3. If the fluid velocity at the left control surface is u, then the mass flow rate into the control volume across this surface is:

$$\dot{m}_{\text{left}} = \rho u \mathrm{d}A = \rho u \mathrm{d}y \mathrm{d}z .$$

Applying the Taylor expansion, the mass flow rate out of the control volume across the right control surface can be expressed as:

$$\dot{m}_{\text{right}} = \left[\rho u + \frac{\partial(\rho u)}{\partial x} \mathrm{d}x\right] \mathrm{d}y \mathrm{d}z .$$

The net mass flow rate out of the control volume across these two control surfaces is:

$$\Delta \dot{m}_{\text{out},x} = \dot{m}_{\text{right}} - \dot{m}_{\text{left}} = \left[\rho u + \frac{\partial(\rho u)}{\partial x} \mathrm{d}x\right] \mathrm{d}y \mathrm{d}z - \rho u \mathrm{d}y \mathrm{d}z ,$$

which can be arranged as:

$$\Delta \dot{m}_{\text{out},x} = \frac{\partial(\rho u)}{\partial x} \mathrm{d}x \mathrm{d}y \mathrm{d}z .$$

Similarly, the net mass flow rate out of the control volume across the y (front and back) and the z (bottom and top) control surfaces are:

$$\Delta \dot{m}_{\text{out},y} = \frac{\partial(\rho v)}{\partial y} \mathrm{d}x \mathrm{d}y \mathrm{d}z ,$$

$$\Delta \dot{m}_{\text{out},z} = \frac{\partial(\rho w)}{\partial z} \mathrm{d}x \mathrm{d}y \mathrm{d}z .$$

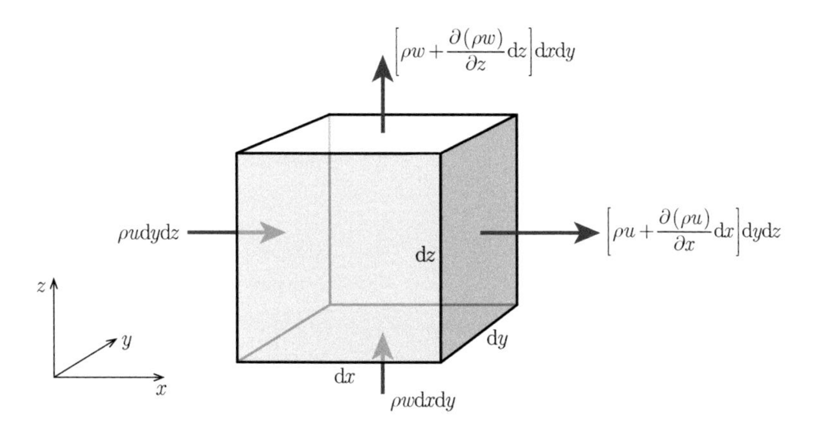

Figure 4.3 The mass flow rate in and out of the elemental control volume.

Therefore, the total mass flow rate out of the control volume is expressed as:

$$\begin{aligned}\Delta\dot{m}_{\text{out}} &= \Delta\dot{m}_{\text{out},x} + \Delta\dot{m}_{\text{out},y} + \Delta\dot{m}_{\text{out},z} \\ &= \frac{\partial(\rho u)}{\partial x}\mathrm{d}x\mathrm{d}y\mathrm{d}z + \frac{\partial(\rho v)}{\partial y}\mathrm{d}x\mathrm{d}y\mathrm{d}z + \frac{\partial(\rho w)}{\partial z}\mathrm{d}x\mathrm{d}y\mathrm{d}z \\ &= \left[\frac{\partial(\rho u)}{\partial x} + \frac{\partial(\rho v)}{\partial y} + \frac{\partial(\rho w)}{\partial z}\right]\mathrm{d}x\mathrm{d}y\mathrm{d}z\end{aligned}.$$

On the other hand, the incremental mass within the control volume per unit time can be expressed as:

$$\Delta\dot{m} = \frac{\partial m}{\partial t} = \frac{\partial \rho}{\partial t}\mathrm{d}x\mathrm{d}y\mathrm{d}z .$$

According to the law of conservation of mass, the decreased mass within the control volume per unit time must be equal to the total mass flow rate out of the control volume. Therefore, we have:

$$-\frac{\partial \rho}{\partial t}\mathrm{d}x\mathrm{d}y\mathrm{d}z = \left[\frac{\partial(\rho u)}{\partial x} + \frac{\partial(\rho v)}{\partial y} + \frac{\partial(\rho w)}{\partial z}\right]\mathrm{d}x\mathrm{d}y\mathrm{d}z ,$$

rearranged as:

$$\frac{\partial \rho}{\partial t} + \frac{\partial(\rho u)}{\partial x} + \frac{\partial(\rho v)}{\partial y} + \frac{\partial(\rho w)}{\partial z} = 0 ,$$

or rewritten in vector form as:

$$\frac{\partial \rho}{\partial t} + \nabla\cdot\left(\rho\vec{V}\right) = 0 .$$

This is exactly the same as Equation (4.4), which was obtained by integral transformation.

In Equation (4.4), the first term $\partial\rho/\partial t$ is the partial derivative of the density with respect to time, representing the unsteady term of the flow. For a steady flow, this term should be zero. Therefore, the second term $\nabla\cdot(\rho\vec{V})$, namely the total mass flow rate out of the control volume, should also be zero:

$$\nabla \cdot \left(\rho \vec{V}\right) = 0. \tag{4.5}$$

This is *the continuity equation for steady flow*.

The following transformation can also be applied to the two terms in the continuity equation (4.4):

$$\frac{\partial \rho}{\partial t} + \nabla \cdot \left(\rho \vec{V}\right) = \frac{\partial \rho}{\partial t} + \left(\vec{V} \cdot \nabla\right)\rho + \rho\left(\nabla \cdot \vec{V}\right) = \frac{\mathrm{D}\rho}{\mathrm{D}t} + \rho\left(\nabla \cdot \vec{V}\right),$$

and the continuity equation can thus be rearranged as:

$$\frac{\mathrm{D}\rho}{\mathrm{D}t} + \rho\left(\nabla \cdot \vec{V}\right) = 0. \tag{4.4a}$$

Equations (4.4) and (4.4a) are mathematically equivalent, but their physical meaning is different. Equation (4.4) refers to the elemental control volume, and can be interpreted as follows: *the decreased mass within the control volume is equal to the mass leaving the control volume*. Equation (4.4a), on the other hand, refers to the elemental system, where the first term $\mathrm{D}\rho/\mathrm{D}t$ represents the incremental density per unit time, and the velocity divergent $\nabla \cdot \vec{V}$ in the second term $\rho(\nabla \cdot \vec{V})$ is the incremental volume per unit time. Therefore, Equation (4.4a) can be interpreted as follows: *for a system, the increase in density is accompanied by the decrease in volume*.

For incompressible flows we have $\mathrm{D}\rho/\mathrm{D}t = 0$, so that Equation (4.4a) can be rearranged as

$$\nabla \cdot \vec{V} = 0, \tag{4.6}$$

or rewritten in component form as

$$\frac{\partial u}{\partial x} + \frac{\partial v}{\partial y} + \frac{\partial w}{\partial z} = 0. \tag{4.6a}$$

This is *the continuity equation for an incompressible flow*.

Since the divergence of the velocity field represents the rate of change of volume of a fluid element, Equation (4.6) implies that for an incompressible flow, the volume of the fluid element remains unchanged.

Next, we will apply Equation (4.6) to the analysis of velocity variations in a simple flow. As shown in Figure 4.4, a constant-density fluid flows in the converging two-dimensional channel. The continuity equation for an incompressible two-dimensional flow is:

$$\frac{\partial u}{\partial x} + \frac{\partial v}{\partial y} = 0.$$

If the flow is considered to be quasi-one-dimensional, the integral continuity equation for incompressible, one-dimensional flow can expressed as:

$$A_1 V_1 = A_2 V_2.$$

Obviously, the velocity in the streamwise direction increases as the channel contracts (i.e., $\partial u/\partial x > 0$). By substituting it into the continuity equation for two-dimensional

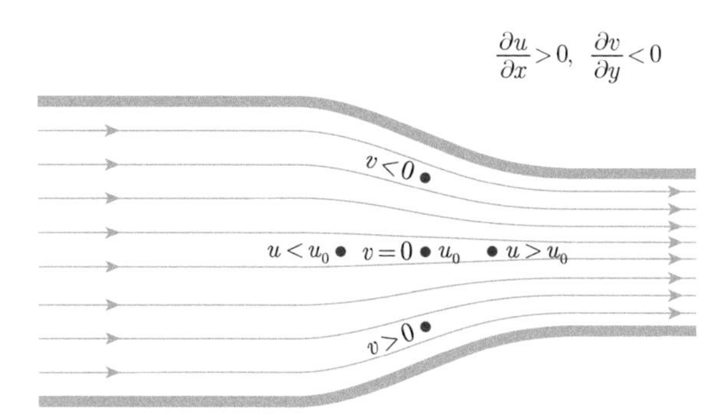

Figure 4.4 Velocity variations in a converging, two-dimensional channel.

flows, we obtain $\partial v/\partial y < 0$. In Figure 4.4 the fluid below the center-line has upward velocity (i.e., $v > 0$). The fluid on the center-line has no vertical velocity (i.e., $v = 0$), while the fluid above this line has downward velocity (i.e., $v < 0$.) From the lower wall to the upper wall, the velocity in the y direction goes from positive to zero, then to negative. That is to say, we must have $\partial v/\partial y < 0$ in the converging section. In conclusion, we can arrive at the same conclusion by applying the continuity equation for one- and two-dimensional flows.

In practical engineering problems no flow is perfectly one-dimensional. The above analysis states that if the flow velocity direction does not change, the flow magnitude remains the same. This kind of flow is a uniform flow and does not need to be investigated. Many so-called one-dimensional flows are actually three-dimensional, since the velocity of the fluid passing through a cross-section with variable area is nonuniform. The one-dimensional calculation uses the mean velocity in a cross-section to approximate the real flow.

4.3 Momentum Equation

4.3.1 Integral Form of the Momentum Equation

The momentum equation, known as Newton's second law of motion, is mathematically expressed as:

$$\sum \vec{F} = \frac{\mathrm{d}\left(m\vec{V}\right)}{\mathrm{d}t}.$$

For a system composed of fluid particles, its more general expression is:

$$\sum \vec{F} = \frac{\mathrm{d}}{\mathrm{d}t} \iiint_{\text{sys}} \vec{V} \rho \mathrm{d}B. \tag{4.7}$$

The left-hand side represents the resultant force acting on the system. If the physical space occupied by the system at a certain instant is chosen as the control volume, the external forces acting on the system are equal to those acting on the control volume:

$$\sum \vec{F}_{cv} = \sum \vec{F}_{sys}. \tag{4.8}$$

The change in momentum of the system can be converted into that of the fluid within the control volume by means of the Reynolds transport theorem. If we let the ϕ in Equation (3.6b) represent the momentum per unit volume, $\rho\vec{V}$, we obtain:

$$\frac{d}{dt}\iiint_{sys} \vec{V}\rho dB = \frac{\partial}{\partial t}\iiint_{cv} \vec{V}\rho dB + \iint_{cs} \vec{V}\rho\left(\vec{V}\cdot\vec{n}\right)dA. \tag{4.9}$$

where:

$\frac{d}{dt}\iiint_{sys} \vec{V}\rho dB$_ is the rate of change of the momentum for the system;

$\frac{\partial}{\partial t}\iiint_{cv} \vec{V}\rho dB$ is the rate of change of the momentum for the control volume; and

$\iint_{cs} \vec{V}\rho\left(\vec{V}\cdot\vec{n}\right)dA$ is the momentum flow rate out of the control volume.

Substituting Equations (4.8) and (4.9) into Equation (4.7), we obtain:

$$\sum \vec{F}_{cv} = \frac{\partial}{\partial t}\iiint_{cv} \vec{V}\rho dB + \iint_{cs} \vec{V}\rho\left(\vec{V}\cdot\vec{n}\right)dA. \tag{4.10}$$

Equation (4.10) is *the integral momentum equation for a control volume*. This equation is particularly useful in theoretical derivations. In engineering, it is most commonly used for quasi-one-dimensional flow. Then, the density and velocity terms on the right-hand side can be represented by their mean values. Therefore, the momentum equation for one-dimensional flow can be simplified, resulting in:

$$\sum \vec{F} = \frac{\partial\left(m\vec{V}\right)}{\partial t} + \left(\dot{m}\vec{V}\right)_{out} - \left(\dot{m}\vec{V}\right)_{in}.$$

The physical meaning of the above equation can be expressed as follows: *the resultant force acting on the control volume may produce two effects. On the one hand, the momentum of the fluid within the control volume increases; on the other hand, a portion of momentum is 'pushed' out of the control volume.* If the last two terms on the right-hand side are zero, it is equivalent to enclosing the control volume with no momentum in or out. Then, the forces acting on the control volume only cause the change of momentum of the fluid inside, and such a control volume is equivalent to a system. If the first term on the right-hand side is zero, it is equivalent to a steady flow. The momentum of the fluid within the control volume remains unchanged, and the newly added momentum due to the forces acting on the control volume is completely pushed out of it.

Many common flows in engineering are steady flows, and so the momentum equation may be simplified, becoming:

$$\sum \vec{F} = \left(\dot{m}\vec{V}\right)_{out} - \left(\dot{m}\vec{V}\right)_{in}. \tag{4.11}$$

Equation (4.11) is a useful tool to solve many flows of practical interest, as long as the flow at the inlet and outlet can be regarded as one-dimensional. However, some flows have complex inlet and outlet boundary conditions, and the mean inlet and outlet velocities cannot be obtained. In some cases, the flow does not even have clear inlet and outlet. A more general formula is required. Especially, in order to further understand the relationship between the properties of a specific position in the flow field and the external forces, we should use the momentum equations in differential form rather than those in integral form.

4.3.2 Differential Momentum Equation

As in the derivation of the differential continuity equation, the differential momentum equation can be obtained directly from the integral momentum equation by integral transformation, or more explicitly, in physical terms, by analyzing an elemental control volume.

As shown in Figure 4.5, Newton's second law of motion is applied to a cubical fluid element moving with other fluids to obtain:

$$\vec{F} = m\vec{a} = \rho \mathrm{d}x\mathrm{d}y\mathrm{d}z\frac{\mathrm{D}\vec{V}}{\mathrm{D}t}. \tag{4.12}$$

The forces acting on the fluid element can be classified into body and surface forces:

$$\vec{F} = \vec{F}_{\mathrm{body}} + \vec{F}_{\mathrm{surface}}. \tag{4.13}$$

where the body force is expressed as the product of the body force per unit mass and the mass of fluid element:

$$\vec{F}_{\mathrm{body}} = \vec{f}_{\mathrm{b}}\rho \mathrm{d}x\mathrm{d}y\mathrm{d}z. \tag{4.14}$$

The surface forces acting on the six sides of the fluid element are more complex. Figure 4.5 shows only the surface forces acting on the two x (left and right) faces and the two z (bottom and top) faces. To avoid cluttering the drawing, the surface forces acting on the two y (back and front) faces and the body forces have been omitted. By convention, the normal stress is assumed to be positive for tension and negative for compression. The surface force acting on the left face is the product of the surface stress and the area. Given that the surface stress is $\vec{\Gamma}_x$, we have

$$\vec{F}_{\mathrm{s,left}} = \vec{\Gamma}_x \mathrm{d}y\mathrm{d}z.$$

The surface force acting on the right face is expressed as:

$$\vec{F}_{\mathrm{s,right}} = \left(\vec{\Gamma}_x + \frac{\partial \vec{\Gamma}_x}{\partial x}\mathrm{d}x\right)\mathrm{d}y\mathrm{d}z.$$

The resultant force acting on these two surfaces is expressed as:

$$\vec{F}_{\mathrm{s,right}} - \vec{F}_{\mathrm{s,left}} = \left(\vec{\Gamma}_x + \frac{\partial \vec{\Gamma}_x}{\partial x}\mathrm{d}x\right)\mathrm{d}y\mathrm{d}z - \vec{\Gamma}_x\mathrm{d}y\mathrm{d}z = \frac{\partial \vec{\Gamma}_x}{\partial x}\mathrm{d}x\mathrm{d}y\mathrm{d}z.$$

Similarly, the resultant forces acting on the two y (front and back) faces and the two z (bottom and top) faces are expressed as:

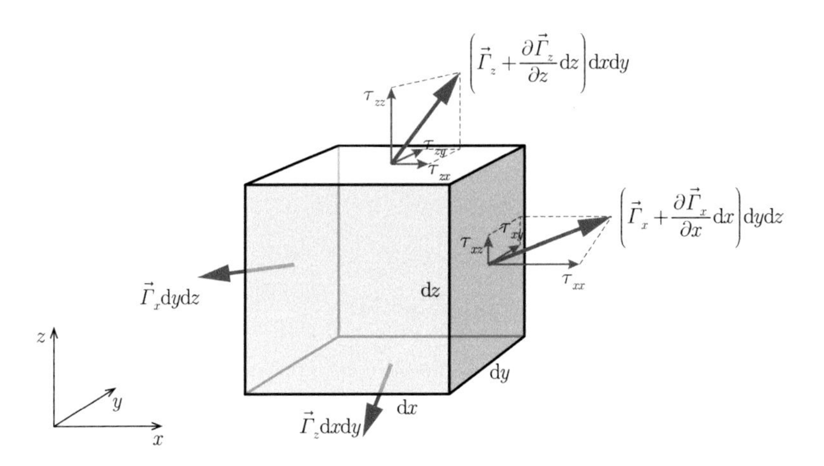

Figure 4.5 Surface forces acting on a fluid element.

$$\frac{\partial \vec{\Gamma}_y}{\partial y}\mathrm{d}x\mathrm{d}y\mathrm{d}z, \quad \frac{\partial \vec{\Gamma}_z}{\partial z}\mathrm{d}x\mathrm{d}y\mathrm{d}z.$$

The total resultant forces acting on the six sides of the fluid element are expressed as:

$$\vec{F}_{\text{surface}} = \left(\frac{\partial \vec{\Gamma}_x}{\partial x} + \frac{\partial \vec{\Gamma}_y}{\partial y} + \frac{\partial \vec{\Gamma}_z}{\partial z}\right)\mathrm{d}x\mathrm{d}y\mathrm{d}z. \tag{4.15}$$

Substituting Equations (4.13), (4.14), and (4.15) into Newton's second law of motion (4.12), we obtain *the momentum equation in stress form* for the fluid element:

$$\frac{\mathrm{D}\vec{V}}{\mathrm{D}t} = \vec{f}_\mathrm{b} + \frac{1}{\rho}\left(\frac{\partial \vec{\Gamma}_x}{\partial x} + \frac{\partial \vec{\Gamma}_y}{\partial y} + \frac{\partial \vec{\Gamma}_z}{\partial z}\right). \tag{4.16}$$

The physical interpretation of this equation should be clear. The term on the left-hand side is the change of momentum per unit mass of fluid (i.e., acceleration). The first term on the right-hand side is the body force per unit mass of fluid, and the second is the surface force per unit mass of fluid.

When applying the momentum equation to flow problems, the surface force should be expressed as flow-related form. As shown in Figure 4.5, there are three stress components into which a surface can be decomposed – one normal stress and two shear stresses:

$$\vec{\Gamma}_x = \tau_{xx}\vec{i} + \tau_{xy}\vec{j} + \tau_{xz}\vec{k}. \tag{4.17}$$

$$\vec{\Gamma}_y = \tau_{yx}\vec{i} + \tau_{yy}\vec{j} + \tau_{yz}\vec{k}. \tag{4.18}$$

$$\vec{\Gamma}_z = \tau_{zx}\vec{i} + \tau_{zy}\vec{j} + \tau_{zz}\vec{k}. \tag{4.19}$$

where τ is the component of the surface stress tensor. The first subscript represents the plane on which the stress acts, and the second the direction in which it acts.

For example, τ_{xz} is the stress component on the two x (left and right) surfaces along the z axis (parallel to the surface).

We substitute Equations (4.17)–(4.19) into Equation (4.16) to rearrange the component form of the momentum equation:

$$\begin{aligned}\frac{\mathrm{D}\vec{V}}{\mathrm{D}t} = \vec{f}_{\mathrm{b}} &+ \frac{1}{\rho}\left(\frac{\partial \tau_{xx}}{\partial x} + \frac{\partial \tau_{yx}}{\partial y} + \frac{\partial \tau_{zx}}{\partial z}\right)\vec{i} \\ &+ \frac{1}{\rho}\left(\frac{\partial \tau_{xy}}{\partial x} + \frac{\partial \tau_{yy}}{\partial y} + \frac{\partial \tau_{zy}}{\partial z}\right)\vec{j} \\ &+ \frac{1}{\rho}\left(\frac{\partial \tau_{xz}}{\partial x} + \frac{\partial \tau_{yz}}{\partial y} + \frac{\partial \tau_{zz}}{\partial z}\right)\vec{k}.\end{aligned} \tag{4.20}$$

Equation (4.20) is expressed more concisely in tensor notation:

$$\frac{\mathrm{D}u_i}{\mathrm{D}t} = f_{\mathrm{b},i} + \frac{1}{\rho}\left(\frac{\partial \tau_{ji}}{\partial x_j}\right). \tag{4.20a}$$

It can be proved that the nine stress components satisfy the following relationship (see the angular momentum equation for a similar proof):

$$\tau_{xy} = \tau_{yx}, \quad \tau_{yz} = \tau_{zy}, \quad \tau_{zx} = \tau_{xz}.$$

Therefore, there are six independent stress components in total.

Equation (4.20), originally due to Claude-Louis Navier (1785–1836) and Simeon-Denis Poisson (1781–1840), is not very useful for actual flow problems, because the six stress components are usually unknown. Until these stresses could be computed, real-life fluid problems were solved by experience and experimentation. Leonhard Euler (1707–1783) had already given the equation of motion of a fluid in 1755. He stated that for fluids, viscous forces are negligible compared to pressure forces. Therefore, it should be possible to ignore the shear stress and make the normal stress equal to the pressure, expressed as:

$$\begin{aligned}&\tau_{xy} = \tau_{yx} = 0, \quad \tau_{yz} = \tau_{zy} = 0, \quad \tau_{zx} = \tau_{xz} = 0. \\ &\tau_{xx} = \tau_{yy} = \tau_{zz} = -p\end{aligned}$$

Substituting the above relations into Equation (4.20), we obtain:

$$\frac{\mathrm{D}\vec{V}}{\mathrm{D}t} = \vec{f}_{\mathrm{b}} - \frac{1}{\rho}\nabla p. \tag{4.21}$$

This is the *momentum equation for inviscid flow* (generally called the *Eulerian equation*, named after Euler, who is credited with its derivation).

As seen from Equation (4.21), the term on the left-hand side is the change of momentum per unit mass of fluid, the first term on the right-hand side is the body force, and the second the differential pressure. Its physical meaning is that for inviscid flow, a change of momentum is caused only by body force and differential pressure force. *For an inviscid flow under the action of gravity, an accelerating fluid particle is either falling or flowing from a high- to a low-pressure zone.*

For a fluid at rest, there is no viscous force and no change of momentum, and therefore Equation (4.21) can be simplified as:

$$\vec{f}_{\mathrm{b}} = \frac{1}{\rho}\nabla p.$$

This is the static equilibrium equation (2.1a) already presented in Chapter 2.

For viscous flow, the stresses in Equation (4.20) still need to be computed. Since the difference between fluids and solids is not considered in the derivation of Equation (4.20), this equation is equally applicable to the motion of solids. In solid mechanics, a great number of materials exhibit a fixed relationship between stress and strain. Thus, Equation (4.20) can be transformed into force–strain relations. Viscous forces in fluids are not determined by strain, as in solids, but induced by fluid flow, and they are a consequence of velocity differences within the fluid. Newtonian fluids exhibit a proportional relationship between viscous force and strain rate. While discussing the viscosity of a fluid in Chapter 1, we had stated the relationship between viscous stress and strain rate for parallel flow –that is, Newton's law of internal friction:

$$\tau = \mu\frac{\partial u}{\partial y}.$$

This formula is based on the flow shown in Figure 1.7. Strictly speaking, the τ should be written as τ_{yx}, that is, the component in the x direction acting on a plane perpendicular to the y axis. If the fluid does not flow in the x direction, the resulting shear stress is no longer due only to the change of velocity component u, but also to the change of velocity component v. For the shearing flow shown in Figure 4.6, the shear stresses in Newtonian fluids can be expressed as:

$$\begin{aligned} \tau_{yx} &= \tau_{xy} = \mu\left(\frac{\partial u}{\partial y} + \frac{\partial v}{\partial x}\right) \\ \tau_{zy} &= \tau_{yz} = \mu\left(\frac{\partial v}{\partial z} + \frac{\partial w}{\partial y}\right) \\ \tau_{xz} &= \tau_{zx} = \mu\left(\frac{\partial w}{\partial x} + \frac{\partial u}{\partial z}\right). \end{aligned} \tag{4.22}$$

Normal stresses are not as easy to compute as shear stresses. It can be proved that normal stresses are not solely a consequence of pressure within the fluid, but are also partly generated by viscosity. Otherwise, the equilibrium of forces on a fluid element cannot be satisfied. The formulas of three normal stresses were first proposed by George G. Stokes (1819–1903). These are:

$$\begin{aligned} \tau_{xx} &= 2\mu\frac{\partial u}{\partial x} - \frac{2}{3}\mu\left(\nabla\cdot\vec{V}\right) - p \\ \tau_{yy} &= 2\mu\frac{\partial v}{\partial y} - \frac{2}{3}\mu\left(\nabla\cdot\vec{V}\right) - p \\ \tau_{zz} &= 2\mu\frac{\partial w}{\partial z} - \frac{2}{3}\mu\left(\nabla\cdot\vec{V}\right) - p. \end{aligned} \tag{4.22a}$$

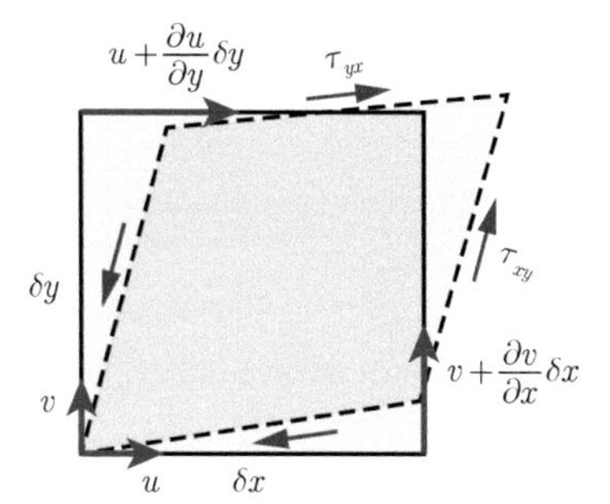

Figure 4.6 Deformation of a fluid element and the shear stresses acting on it.

Equation (4.22a) shows that the viscous forces contribute to the normal stresses. For example, consider the viscous normal stress from τ_{xx}:

$$\tau_{\text{viscous},xx} = 2\mu\frac{\partial u}{\partial x} - \frac{2}{3}\mu\left(\nabla\cdot\vec{V}\right) = \frac{4}{3}\mu\frac{\partial u}{\partial x} - \frac{2}{3}\mu\frac{\partial v}{\partial y} - \frac{2}{3}\mu\frac{\partial w}{\partial z}.$$

In incompressible flows we have $\nabla\cdot\vec{V} = 0$, and the viscous normal stress is proportional to the strain rate in the x direction, $\partial u/\partial x$. In compressible flows, the viscous normal stress is also related to the change of volume. However, the viscous force due to the change of volume is generally much smaller than that induced by elongation, so, in some books, this term is simply ignored and the viscous normal stress rewritten as:

$$\tau_{\text{viscous},xx} = 2\mu\frac{\partial u}{\partial x}.$$

In most cases, the viscous normal stress, expressed as tension, is positive. *In Newtonian fluids, the viscous normal stress is almost always far smaller than the pressure and thus its effect can be ignored.*

As the *constitutive equations for Newtonian fluids*, Equations (4.22) and (4.22a) represent the relationship between stress and strain rate in such fluids under any state of flow. This is also known as *the generalized Newton's law of internal friction*. It should be noted that, for normal stress, Equation (4.22a) is not completely accurate, and some assumptions have been introduced by Stokes. However, for common flows, Equation (4.22a) yields rather exact predictions.

In the constitutive equations, the nine stress components constitute a second-order tensor, expressed as:

$$\Gamma_{ij} = \left[\tau_{ij}\right] = \begin{bmatrix} \tau_{xx} & \tau_{xy} & \tau_{xz} \\ \tau_{yx} & \tau_{yy} & \tau_{yz} \\ \tau_{zx} & \tau_{zy} & \tau_{zz} \end{bmatrix}. \tag{4.23}$$

The nine components of rotation and strain rate also constitute a second-order tensor, expressed as:

$$D_{ij} = \left[d_{ij}\right] = \begin{bmatrix} \partial u/\partial x & \partial u/\partial y & \partial u/\partial z \\ \partial v/\partial x & \partial v/\partial y & \partial v/\partial z \\ \partial w/\partial x & \partial w/\partial y & \partial w/\partial z \end{bmatrix}. \tag{4.24}$$

The generalized Newton's law of internal friction establishes the relationship between stress and strain rate and is therefore called the constitutive equation. The constitutive equations for solids describe the relationship between stress and strain, while for some non-Newtonian fluids the constitutive equations may be related to both strain and strain rate, or even to the length of action time.

Substitute the constitutive equations (4.22) and (4.22a) into the momentum equation in stress form (4.20) to obtain the momentum equation in the final form:

$$\frac{\mathrm{D}\vec{V}}{\mathrm{D}t}=\vec{f}_{\mathrm{b}}-\frac{1}{\rho}\nabla p+\frac{\mu}{\rho}\nabla^2\vec{V}+\frac{1}{3}\frac{\mu}{\rho}\nabla\left(\nabla\cdot\vec{V}\right). \tag{4.25}$$

This is *the Navier–Stokes equation (N-S equation)*. The physical meaning of each term is:

$\frac{\mathrm{D}\vec{V}}{\mathrm{D}t}$: the rate of change of momentum of a fluid, also called the inertia force term;

$\vec{f}_{\mathrm{b}}$: body force term;

$-\frac{1}{\rho}\nabla p$: differential pressure term; and

$\frac{\mu}{\rho}\nabla^2\vec{V}+\frac{1}{3}\frac{\mu}{\rho}\nabla\left(\nabla\cdot\vec{V}\right)$: viscous force term.

The N-S equation can be expanded as:

$$\begin{cases}
\rho\frac{\partial u}{\partial t}+\rho u\frac{\partial u}{\partial x}+\rho v\frac{\partial u}{\partial y}+\rho w\frac{\partial u}{\partial z}=\rho f_{\mathrm{b},x}-\frac{\partial p}{\partial x}+2\frac{\partial}{\partial x}\left(\mu\frac{\partial u}{\partial x}\right) \\
\quad -\frac{2}{3}\frac{\partial}{\partial x}\left[\mu\left(\frac{\partial u}{\partial x}+\frac{\partial v}{\partial y}+\frac{\partial w}{\partial z}\right)\right]+\frac{\partial}{\partial y}\left[\mu\left(\frac{\partial u}{\partial y}+\frac{\partial v}{\partial x}\right)\right]+\frac{\partial}{\partial z}\left[\mu\left(\frac{\partial w}{\partial x}+\frac{\partial u}{\partial z}\right)\right] \\
\rho\frac{\partial v}{\partial t}+\rho u\frac{\partial v}{\partial x}+\rho v\frac{\partial v}{\partial y}+\rho w\frac{\partial v}{\partial z}=\rho f_{\mathrm{b},y}-\frac{\partial p}{\partial y}+2\frac{\partial}{\partial y}\left(\mu\frac{\partial v}{\partial y}\right) \\
\quad -\frac{2}{3}\frac{\partial}{\partial y}\left[\mu\left(\frac{\partial u}{\partial x}+\frac{\partial v}{\partial y}+\frac{\partial w}{\partial z}\right)\right]+\frac{\partial}{\partial z}\left[\mu\left(\frac{\partial w}{\partial y}+\frac{\partial v}{\partial z}\right)\right]+\frac{\partial}{\partial x}\left[\mu\left(\frac{\partial u}{\partial y}+\frac{\partial v}{\partial x}\right)\right] \\
\rho\frac{\partial w}{\partial t}+\rho u\frac{\partial w}{\partial x}+\rho v\frac{\partial w}{\partial y}+\rho w\frac{\partial w}{\partial z}=\rho f_{\mathrm{b},z}-\frac{\partial p}{\partial z}+2\frac{\partial}{\partial z}\left(\mu\frac{\partial w}{\partial z}\right) \\
\quad -\frac{2}{3}\frac{\partial}{\partial z}\left[\mu\left(\frac{\partial u}{\partial x}+\frac{\partial v}{\partial y}+\frac{\partial w}{\partial z}\right)\right]+\frac{\partial}{\partial x}\left[\mu\left(\frac{\partial w}{\partial x}+\frac{\partial u}{\partial z}\right)\right]+\frac{\partial}{\partial y}\left[\mu\left(\frac{\partial w}{\partial y}+\frac{\partial v}{\partial z}\right)\right].
\end{cases}$$

Tip 4.2: Non-Newtonian Fluids

Viscous forces are always present when there is relative motion between adjacent fluid layers. A fluid is said to be Newtonian if its viscous stress is proportional to the strain rate. There is actually a large number of fluids that do not satisfy this relationship; these are called non-Newtonian fluids.

Non-Newtonian fluids are generally classified into three categories:

(1) Non-Newtonian fluids that are independent of time: These fluids display a time-independent viscosity and can be further subdivided into three types: the first one is characterized by a viscosity that is independent of stress, such as serum and custard; for the second type, viscosity increases with an increase in stress, such as water suspended with starch or silt; and for the third type, viscosity decreases with an increase in stress, such as nail polish, ketchup, syrup, latex, and blood.

(2) Non-Newtonian fluids that are dependent of time: These non-Newtonian fluids display a time-dependent increase in viscosity and can be further subdivided into two types: the first one is characterized by a viscosity that increases with time, such as printing ink and gypsum paste; the second is characterized by a viscosity that decreases with time, such as yogurt, gelatin gel, joint fluid, hydrogenated sesame oil, drilling mud, and certain kinds of paints.

(3) Viscoelastic fluids: These non-Newtonian fluids have both viscous and elastic properties, such as silly putty and some lubricants.

Figure T4.2 illustrates the difference between Newtonian and non-Newtonian fluids. The diagram on the left shows the relationship between shear stress and flow, and that on the right shows the change of shear stress with time.

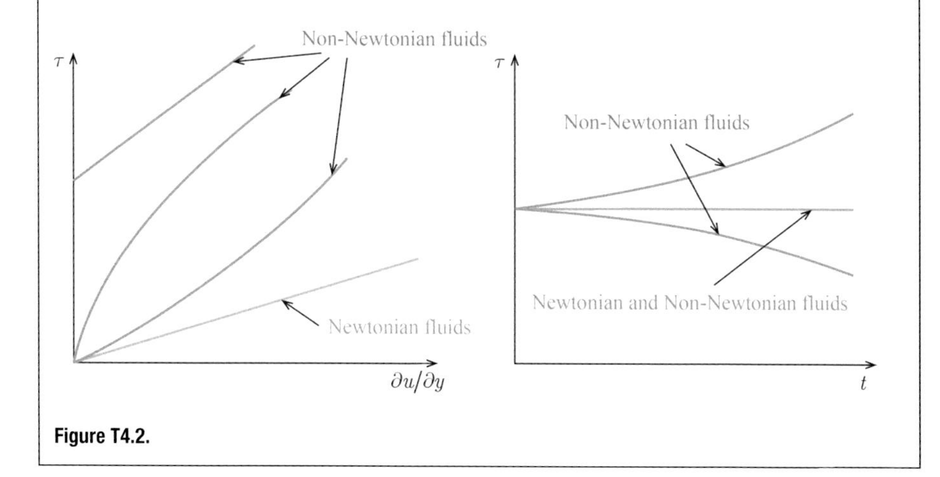

Figure T4.2.

These differential equations may appear complicated, but this is not an obstacle to solving them. The principal mathematical difficulty in flow analysis is that the convective acceleration terms are nonlinear in Eulerian coordinates. Furthermore, the viscous force terms in Equation (4.25) are also nonlinear. Ignoring the change of the coefficient of viscosity with temperature, the viscous force terms can be considered to be linear.

In practical applications, provided the fluid is not moving in a strongly compressed state (for example, the flow inside the strong shock wave region), the last term of Equation (4.25) can be ignored. Therefore, in many books the N-S equation is written as:

$$\frac{D\vec{V}}{Dt} = \vec{f}_b - \frac{1}{\rho}\nabla p + \frac{\mu}{\rho}\nabla^2\vec{V}.$$

4.4 Bernoulli's Equation

As one of the most popular equations in fluid mechanics, Bernoulli's equation can be regarded as a statement of the conservation of mechanical energy principle for a flowing fluid. Because it is derived before the formula for the conservation of energy, and it allows intuitive understanding of flow phenomena, Bernoulli's equation is very popular among engineers and scientists. However, there are restrictions to its validity, and it is particularly important to know them before applying the equation to fluid flows.

Bernoulli's equation for incompressible flow can be derived from the Eulerian equation of motion (4.21) under rather severe restrictions. The Eulerian equation for one-dimensional steady flow in the z direction is:

$$w\frac{dw}{dz} = \vec{f}_{b,z} - \frac{1}{\rho}\frac{dp}{dz}.$$

Assuming that the fluid experiences only the body force of gravity, the positive z axis describes the "upward" direction, and after substituting w by V, the above equation can be rewritten as:

$$\frac{dp}{\rho} + g dz + V dV.$$

For an incompressible flow, this new equation can be easily converted into Bernoulli's equation by integration:

$$\frac{p}{\rho} + gz + \frac{V^2}{2} = \text{const}. \tag{4.26}$$

where

$\frac{p}{\rho}$: pressure potential energy per unit mass (or push work per unit mass);

gz : gravitational potential energy per unit mass; and

$\frac{V^2}{2}$: kinetic energy per unit mass.

Bernoulli's equation states that the sum of these three energies remains unchanged along a streamline. Since these three types of kinetic and potential energy constitute the total mechanical energy of a fluid, it may not come as much of a surprise that *Bernoulli's equation is the result of applying conservation of mechanical energy to a flowing fluid.*

As mentioned above, *there are some restrictions for the applicability of Bernoulli's equation. These are: the flow should be (1) along a streamline, (2) steady, (3) inviscid, and (4) incompressible.* The first three restrictions are imposed by the steady Eulerian equation, and the last one by the process of integration. They aim at guaranteeing the conservation of mechanical energy applicable, and are summarized in the following points:

(1) Bernoulli's equation can only be applied to two points along a single streamline. In a steady flow, a fluid element moves along a single streamline, with its mechanical energy conserved. However, the mechanical energy of the fluid elements in two different streamlines could be different.
(2) Bernoulli's equation can only be applied to a steady flow. In an unsteady flow, the fluid elements at two points along a single streamline may be from different starting points and the conservation of mechanical energy is not applicable between them. In addition, the pressure pulsation at a certain point can do work on the fluid element passing through it, changing its total mechanical energy. Therefore, in an unsteady flow, the mechanical energy of a fluid element is not conserved along a streamline.
(3) Bernoulli's equation can only be applied to an inviscid flow. This restriction is easy to understand. Since the viscous shear force within a fluid is equivalent to friction force between two solid surfaces in motion, the mechanical energy is not conserved due to its irreversible frictional dissipation into internal energy.
(4) Bernoulli's equation can only be applied to an incompressible flow. As we know, compressing a gas increases not only pressure and density, but also the temperature of the gas. Even if it is inviscid and adiabatic compression, a certain amount of mechanical energy will be converted into internal energy, so that the mechanical energy of the gas is not conserved. Different from the effect of viscosity, the conversion of mechanical energy into internal energy caused by an inviscid adiabatic compression is reversible, and the reverse process can be achieved through expansion.

Figure 4.7 shows four different types of fluid flow that violate one of the above four restrictions respectively. In the case of the rotating fluid in a cup, the fluid elements at the same height but in different circles of radius are moving along different streamlines; in the case of the spiral pump, unsteady work is done on the fluid; in the case of long pipes of small diameter, the viscous effect is not negligible; in the case of the supersonic flow over a sphere, the fluid across a shock wave is strongly compressed.

Many actual flows satisfy the conditions of steadiness and incompressibility, but are more or less viscous. Strictly speaking, no flow completely conforms to the restrictions on Bernoulli's equation. However, for a flow with small shear deformation, the mechanical energy loss caused by the viscosity of the fluid is small enough to be neglected, so that Bernoulli's equation can make accurate predictions when used in engineering problems.

The gravity of a gas is usually much smaller than the inertial and differential pressure forces acting on it, and can be ignored. Then, Bernoulli's equation has the simpler form

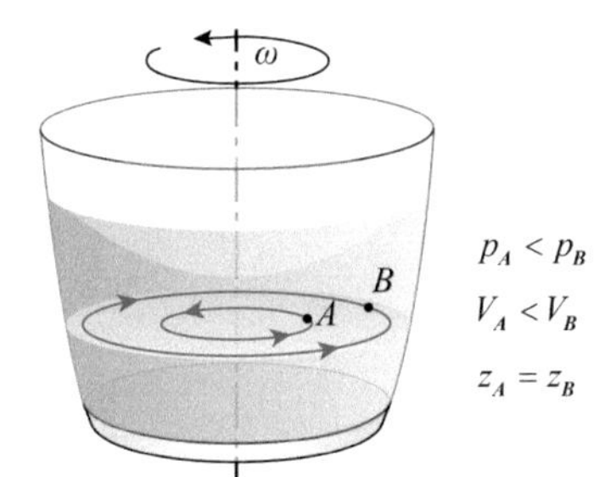

(a) In rigid-rotating water, both pressure and velocity are lower at small radius than those in a large radius.

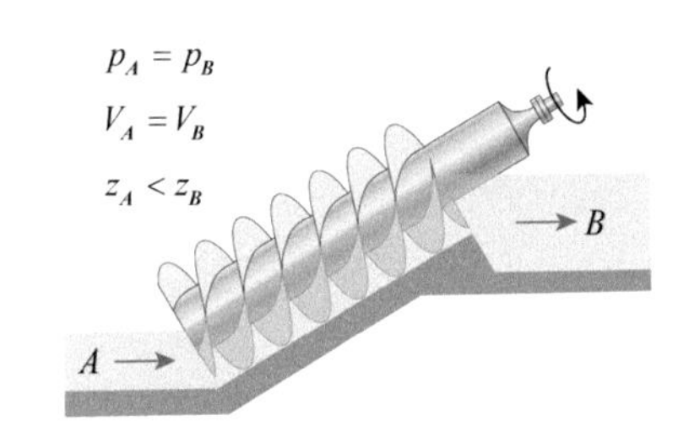

(b) A spiral pump lifts water at a constant speed and pressure.

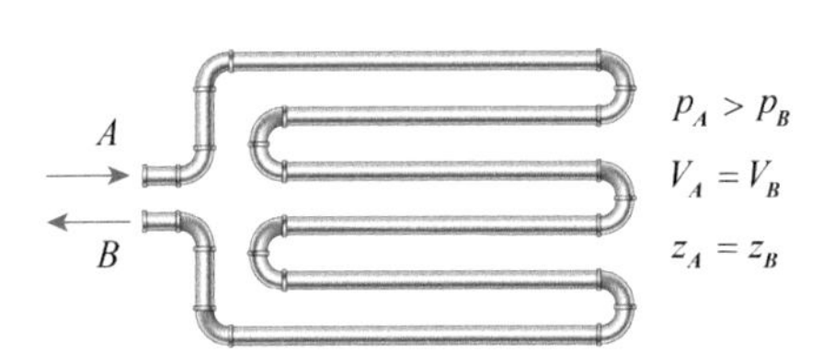

(c) Water flowing through constant cross-section pipe produces pressure losses while flow speed remains unchanged.

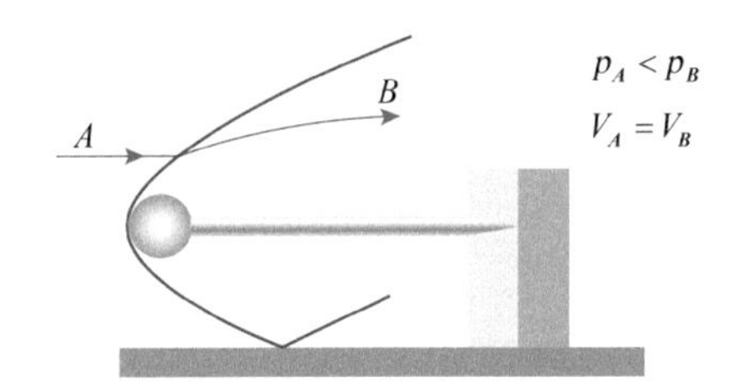

(d) Across a shockwave and a series of expansion waves, the pressure of a gas increases, yet velocity magnitude may remain unchanged.

Figure 4.7 Four different types of fluid flow to which Bernoulli's equation cannot be applied.

$$\frac{p}{\rho} + \frac{V^2}{2} = \text{const}. \tag{4.26a}$$

This formula states that, as the airflow slows down, some of the kinetic energy is converted into pressure potential energy, causing the static pressure to increase. When the velocity of the gas is reduced to zero, pressure is at its maximum value at a stagnation point. This static pressure is called *stagnation pressure*. In gas dynamics, while ignoring gravity, the stagnation pressure, also called *total pressure*, is the maximum pressure that can be achieved for a flow with no power input, defined as:

$$p_t = p + \frac{1}{2}\rho V^2. \tag{4.27}$$

The term on the left-hand side of Equation (4.27) is the total pressure. The first term on the right-hand side is the pressure, called *static pressure* specifically. The second term on the right-hand side represents the additional pressure due to deceleration, called *dynamic pressure*.

Total pressure remains constant along a streamline for a steady, inviscid, and incompressible gas flow. So, *total pressure represents the magnitude of total mechanical energy of a gas flow*. The so-called stagnation pressure or total pressure is not really a property of a fluid. Static pressure is the pressure exerted by air on a body, while the dynamic and total pressure are imaginary and used mainly for calculation.

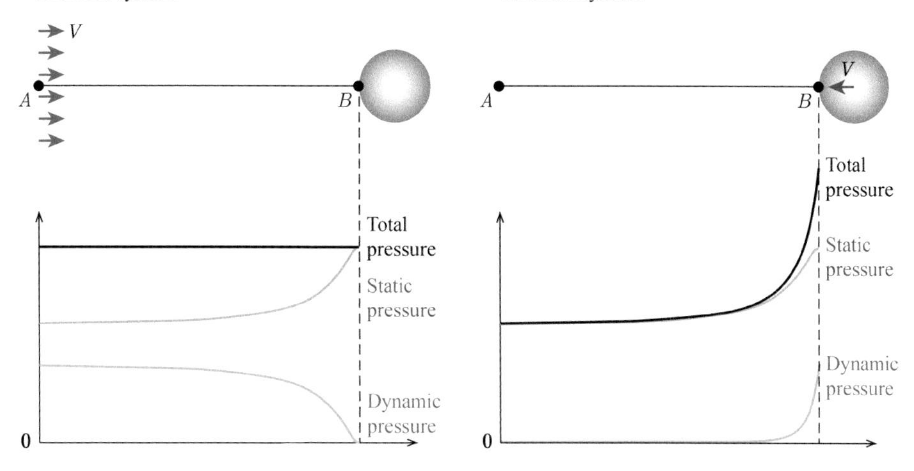

Figure 4.8 The relationship among static, dynamic, and total pressure in different types of reference coordinate systems.

The static pressure is invariant with coordinates, while the dynamic and total pressure depend on the relative velocity of the fluid. To vividly illustrate this point, the change of air pressure against the leading side of an object is shown in Figure 4.8. As can be seen, the total pressure along a streamline is conserved only when the selected coordinates appear stationary with respect to the moving object. When the selected coordinates appear stationary with respect to the airflow (equivalent to an object passing through still air), the static, dynamic, and total pressure all increase from left to right. This is because the moving object does work on the airflow through unsteady pressure, causing an increase in the mechanical energy of the airflow.

Bernoulli's equation for a compressible flow can be derived by keeping the other three restrictions (steady, inviscid, along a streamline) and adding an adiabatic condition. When there is no internal heat generation due to friction and no heat transfer to or from the fluid, the flow is isentropic. According to fundamental thermodynamic relationships, the pressure, density, and temperature of a gas in an isentropic compression/expansion process satisfy the following relations:

$$\frac{p}{\rho^k} = \text{const}, \quad \frac{T}{\rho^{k-1}} = \text{const}, \quad \frac{T}{p^{\frac{k-1}{k}}} = \text{const} .$$

As can be seen, temperature increases with pressure and density in an isentropic compression. In other words, as the gas is compressed isentropically, a portion of mechanical work is converted into internal energy.

By substituting the isentropic relation, $p/\rho^k = \text{const}$, into the one-dimensional Eulerian equation and ignoring gravity, we obtain the following relation for any two points along a streamline:

$$\frac{k}{k-1}\frac{p_1}{\rho_1}\left[\left(p_2/p_1\right)^{\frac{k-1}{k}}-1\right]+\frac{V_2^2-V_1^2}{2}=0.$$

According to the equation of state for an ideal gas, $p=\rho RT$, the above equation can be rewritten as:

$$\frac{k}{k-1}RT_1\left[\left(p_2/p_1\right)^{\frac{k-1}{k}}-1\right]+\frac{V_2^2-V_1^2}{2}=0. \tag{4.28}$$

This extended Bernoulli's equation is often called *Bernoulli's equation for incompressible flow*. Note that there is a temperature term in this formula, that is to say, the effect of internal energy is taken into account. Therefore, Bernoulli's equation for compressible flow no longer describes the conservation of mechanical energy. What kind of energy equation is it?

By applying the isentropic relation,

$$\left(\frac{p_2}{p_1}\right)^{\frac{k-1}{k}}=\frac{T_2}{T_1},$$

and the isobaric heat capacity relation,

$$c_p=\frac{k}{k-1}R,$$

Equation (4.28) can be further rewritten as:

$$c_p\left(T_2-T_1\right)+\frac{V_2^2-V_1^2}{2}=0.$$

The product of the isobaric specific heat capacity times the temperature is enthalpy, that is $c_pT=h$, so the compact form of the above equation is:

$$h+\frac{V^2}{2}=\text{const}. \tag{4.29}$$

This equation clearly states that the sum of the enthalpy and the kinetic energy in a gas flow remains constant.

Tip 4.3: Why Does the Temperature of a Gas Increase as It Is Compressed?

When a gas is compressed, its density will increase. The increase in gas pressure is easy to understand because the pressure of a gas is a measure of the collective impact of the collisions of the gas molecules with the vessel walls. As the density of the gas increases, the number of molecules hitting a unit area of the wall per unit time also increases, so that the pressure of the gas increases as well, even if the temperature remains constant (i.e., the exchange of momentum between a single molecule and the wall remains unchanged). However, it's not so easy to understand why the temperature of the gas increases. The temperature increase by friction can be due to the irreversible conversion of mechanical energy into internal energy, so why does the temperature of a gas increase in a frictionless, isentropic compression? Here's a simple model to explain this phenomenon.

Figure T4.3 shows gas enclosed in an adiabatic cylinder with a movable piston. Assuming that there is no friction between the piston and the cylinder wall, we can analyze the situation as follows:

(1) When the gas molecules collide elastically with the resting piston, the direction of the velocity will change, without changing the magnitude. So, the gas temperature remains unchanged.
(2) When the piston moves inward to compress the gas, the gas molecules will gain extra momentum from the piston, and bounce back with greater speed than before, since the piston is moving toward them. So, the average speed of the gas molecules will increase, and the gas will get hotter.
(3) When the piston moves outward and the gas begins to expand, the gas molecules will lose a certain amount of momentum, and bounce back with less speed than before, since the piston is moving away from them. So, the average speed of the gas molecules will decrease, and the gas will get colder.

As long as the adiabatic expansion and adiabatic compression processes maintain the thermodynamic equilibrium, the speed of the moving piston does not affect the final temperature of the gas. Although the change of kinetic energy of a molecule from a single collision decreases when the piston moves slowly, the whole action time becomes longer, and thus the kinetic energy of a larger number of molecules is changed, or the kinetic energy of a single molecule is changed multiple times. The final effect is that the temperature rise has nothing to do with the piston's speed, but only with the distance it traveled.

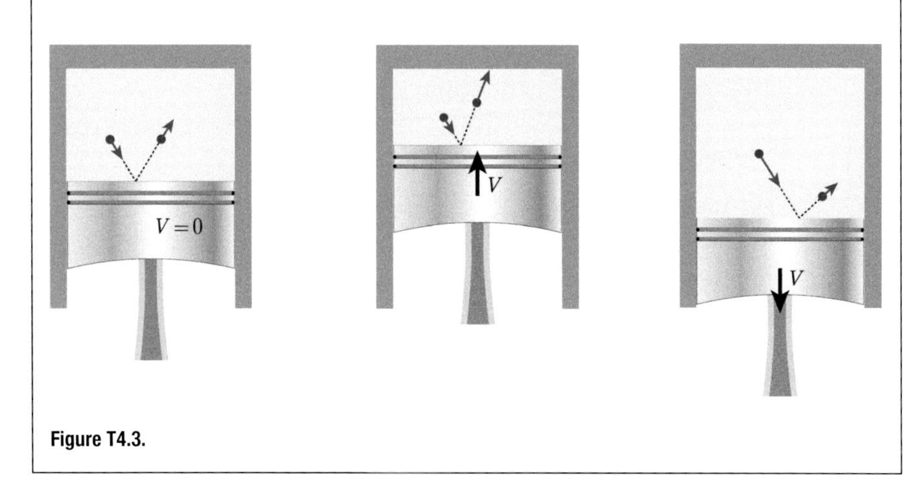

Figure T4.3.

By applying the relationship between internal energy and enthalpy,

$$h = \hat{u} + \frac{p}{\rho},$$

Equation (4.29) becomes

$$\hat{u} + \frac{p}{\rho} + \frac{V^2}{2} = \text{const.} \tag{4.29a}$$

In contrast to Bernoulli's equation for an incompressible flow, Equation (4.26a), the internal energy term $\hat{u}$ is added in Equation (4.29a), which means that *Bernoulli's equation for a compressible flow represents the conservation of total energy (internal and mechanical energy) of the fluid.*

The restrictions of Bernoulli's equation for a compressible flow, Equation (4.28), are: along a streamline, steady, and inviscid. Under these restrictions, the sum of the mechanical and internal energies remains constant, and the conversion between them is reversible. Equation (4.29) is not equivalent to Equation (4.28) in that it is also applicable to irreversible processes. That is to say, the flow can be viscous. How is the inviscid restriction removed from Bernoulli's equation?

In fact, Equation (4.29) can be derived in a more general way without the need for isentropic relations. Although added in the previous derivation of Bernoulli's equation (4.28) for a compressible flow, the isentropic condition had actually been removed during the transformation into Equation (4.29). So Equations (4.29) and (4.29a) represent more general forms of the energy equation, which are valid as long as there is *no exchange of heat and work between the fluid and its surroundings.*

The total energy of a fluid element is not conserved along a streamline during unsteady flow because there may be some exchange of work between any two streamlines. The total energy of a system is conserved, provided the boundary that encloses it does not move. Detailed energy relations in fluids can be seen in Section 4.6, and compressible fluid flow analysis will be further discussed in Chapter 7.

4.5 Angular Momentum Equation

4.5.1 Integral Angular Momentum Equation

The angular momentum equation, also called the moment of momentum equation, is actually the extended Newton's second law of motion. It states that the rate of change of total angular momentum of a system of particles is equal to the sum of all external torques acting on the system:

$$\sum \vec{T} = \frac{\mathrm{d}}{\mathrm{d}t} \iiint_{\text{sys}} \left(\vec{r} \times \vec{V}\right) \rho \,\mathrm{d}B,$$

where T is the torque and r is the distance from the centroid of the system to the origin.

The above relation can be converted into a form suitable for the control volume by means of the Reynolds transport theorem:

$$\sum \vec{T} = \frac{\partial}{\partial t} \iiint_{\text{cv}} \left(\vec{r} \times \vec{V}\right) \rho \,\mathrm{d}B + \iint_{\text{cs}} \left(\vec{r} \times \vec{V}\right) \rho \left(\vec{V} \cdot \vec{n}\right) \mathrm{d}A. \tag{4.30}$$

For the planar rotational flow as shown in Figure 4.9, the angular momentum equation in cylindrical coordinates can be simplified as:

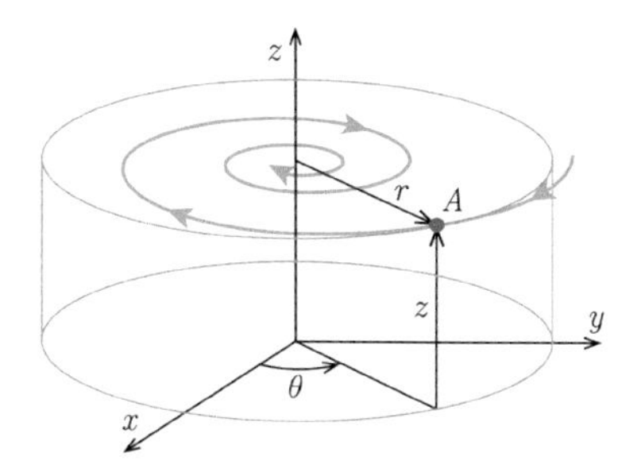

(a) A rotational flow can be simplified to quasi-one-dimensional flow by cylindrical coordinates and angular momentum equations.

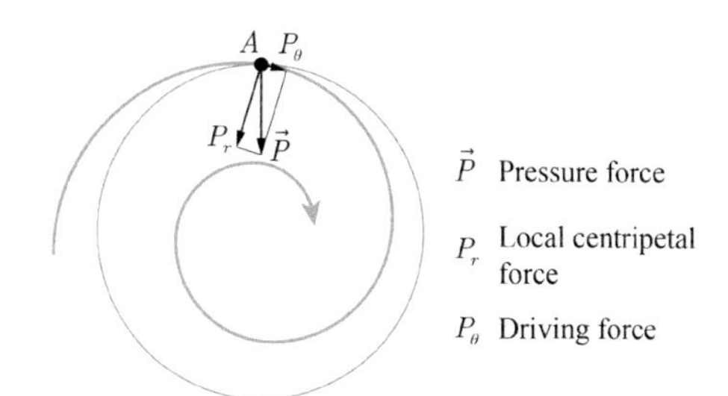

(b) As the rotation radius of the fluid element gradually decreases, the differential pressure provides not only centripetal force, but also provides driving force to accelerate the fluid element.

Figure 4.9 Representation and analysis of a planar rotational flow.

$$\sum \vec{T}_z = \dot{m}\left(r_2 V_{2\theta} - r_1 V_{1\theta}\right). \tag{4.31}$$

where the subscript z represents the longitudinal direction and the subscript θ the circumferential direction. Equation (4.31) states that the change in angular momentum of the fluid passing through the control volume is equal to the sum of all external torques acting on it.

If there is no resultant external torque acting on the control volume, the magnitudes of the angular momentum through the inlet and outlet surface of the control volume are equal:

$$r_2 V_{2\theta} = r_1 V_{1\theta}.$$

This relation reveals the following flow phenomenon: A fluid element speeds up when it moves toward the center along a spiral without the action of torque.

In general, only viscous forces can generate torques. Since the viscous forces are much smaller than the inertial and pressure forces, most flows can be regarded as flows without torques. Observe the flow in a pool being drained. When the water reaches the drain port, it usually has a large rotational speed. Similarly, the wind speeds of most tornadoes and hurricanes are high because the fluid converges from the region of larger radius to that of smaller radius. Quantitatively speaking, if the radius is halved, the tangential velocity is doubled, and the angular velocity is four times greater. This acceleration effect is very strong.

For any fluid element that is spirally accelerated, the streamwise acceleration is caused by the forces in the direction of the streamline. What provides that accelerating force? Since the flow is inviscid, it must be the differential pressure between the inner and outer fluid layers, as shown and explained in Figure 4.9(b).

4.5.2 Differential Angular Momentum Equation

Figure 4.10 shows a rectangular fluid element and the shear stresses acting on it. Let θ be the angle of rotation about the centroid of the fluid element. The only stresses that

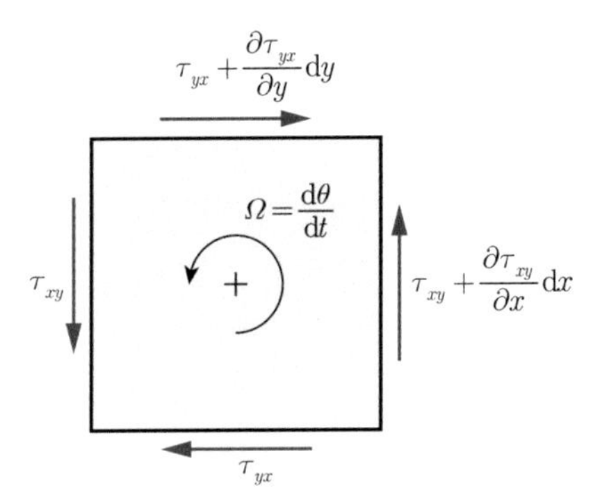

Figure 4.10 Rotation of a fluid element.

generate moments about the centroid are the shear stresses τ_{xy} and τ_{yx}. For a detailed derivation of the angular momentum differential equation, the reader is referred to fluid mechanics or elastic mechanics textbooks. Only the simplified angular momentum equation is given here, namely:

$$\left[\left(\tau_{xy} + \frac{1}{2}\frac{\partial}{\partial x}\tau_{xy}\mathrm{d}x\right) - \left(\tau_{yx} + \frac{1}{2}\frac{\partial}{\partial y}\tau_{yx}\mathrm{d}y\right)\right]\mathrm{d}x\mathrm{d}y\mathrm{d}z = \frac{1}{12}\rho\mathrm{d}x\mathrm{d}y\mathrm{d}z\left[(\mathrm{d}x)^2 + (\mathrm{d}y)^2\right]\frac{\mathrm{d}^2\theta}{\mathrm{d}t^2}.$$

The left-hand side terms contain finite-value and small first-order quantities. The angular acceleration $\mathrm{d}^2\theta/\mathrm{d}t^2$ on the right-hand side is a finite value, and the differential term of $(\mathrm{d}x)^2 + (\mathrm{d}y)^2$ is a small quantity of the second order. So, the right-hand terms can be ignored, and we obtain:

$$\tau_{xy} + \frac{1}{2}\frac{\partial}{\partial x}\tau_{xy}\mathrm{d}x = \tau_{yx} + \frac{1}{2}\frac{\partial}{\partial y}\tau_{yx}\mathrm{d}y.$$

Further neglecting the small quantities of the first order, we obtain:

$$\tau_{xy} = \tau_{yx}.$$

Had we added moments about axes parallel to y or x, we would have obtained analogous results:

$$\tau_{xy} = \tau_{yx}, \quad \tau_{yz} = \tau_{zy}, \quad \tau_{zx} = \tau_{xz}. \tag{4.32}$$

This is the *differential angular momentum equation*. This result has already been used in the previous derivation of the momentum equation, where the stress was simply said to be a symmetric tensor.

The angular momentum equation for a fluid element has been contained in the momentum equation. Therefore, the differential angular momentum equation is scarcely mentioned. Its physical meaning is also very clear: *the torque produced by the shear stresses is negligible for a fluid element*. The reason is that for an infinitesimal fluid element, any finite amount of torque will produce an infinite amount of angular acceleration.

4.6 Energy Equation

4.6.1 Integral Energy Equation

The so-called energy equation is the application of the first law of thermodynamics to fluids. *For a system, the first law of thermodynamics states that heat transfer from the surroundings and the work being done to the system are the only two ways of adding energy into the system*, expressed as:

$$\frac{\mathrm{d}E}{\mathrm{d}t} = \dot{Q}_{\text{in}} - \dot{W}_{\text{out}},$$

where E is the total energy of the system; $\dot{Q}_{\text{in}}$ is the rate of heat transferred into the system from its surroundings; and $\dot{W}_{\text{out}}$ is the rate of work done by the system to its surroundings. A more general form of the energy equation is expressed as:

$$\frac{\mathrm{D}}{\mathrm{D}t}\iiint_{\text{sys}} e\rho \mathrm{d}B = \dot{Q}_{\text{in}} - \dot{W}_{\text{out}},$$

where e is the total energy per unit mass.

We first derive the one-dimensional energy equation for the flow model in Figure 4.11. The control volume has a single inlet and a single outlet, and the fluid within the control volume interacts with its surroundings through heat transfer and shaft work. The one-dimensional energy equation can be derived from the first law of thermodynamics by applying the Reynolds transport theorem:

$$\dot{m}(e_2 - e_1) = \dot{Q}_{\text{in}} - \dot{W}_{\text{out}}. \tag{4.33}$$

where the total energy of a fluid is the sum of the internal, kinetic, and potential energies, written as:

$$e = \hat{u} + \frac{V^2}{2} + gz. \tag{4.34}$$

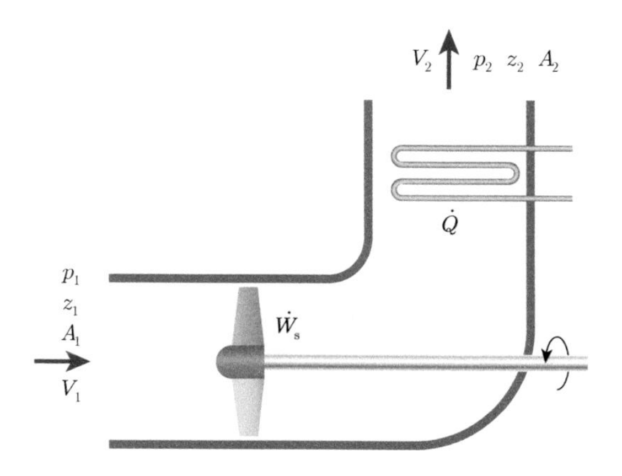

Figure 4.11 A model for one-dimensional flow energy exchange.

The rate of work done by the fluid within the control volume on its surroundings, W_{out}, has two essential elements: a force acting on the boundary, and the boundary must move. Both body and surface forces can do work on the fluid. Now let's consider the case where gravity is the only body force. Since the work done by gravity has already been written as the change of gravitational potential energy, body force work need not be added again.

The work done by surface forces is slightly more complicated. In Figure 4.11 there are three types of interfaces between the fluid within the control volume and its surroundings: inlet and outlet, fixed walls, and impeller surfaces. Assuming that p_1 is the pressure at the inlet and A_1 is the inlet area, then the force exerted by the fluid within the control volume on its surroundings at the inlet is:

$$F_1 = p_1 A_1.$$

The rate of work done by the fluid within the control volume on its surroundings at the inlet is expressed as:

$$\dot{W}_1 = -F_1 V_1 = -p_1 A_1 V_1.$$

where the negative sign indicates that the forces are always in the direction opposite to its velocity. That is to say, the rate of work done by the fluid within the control volume on its surroundings at the inlet is negative. Or, the surrounding fluid works to push fluid into the control volume.

Similarly, the rate of work done at the outlet is:

$$\dot{W}_2 = p_2 A_2 V_2.$$

Thus, the rate of total work done by the fluid within the control volume on its surroundings at the inlet and outlet is expressed as:

$$\dot{W}_p = \dot{W}_2 + \dot{W}_1 = p_2 A_2 V_2 - p_1 A_1 V_1.$$

Applying the flow rate formula $\dot{m} = \rho A V$, and the continuity equation $\dot{m}_{in} = \dot{m}_{out}$, the above equation can be rewritten as:

$$\dot{W}_p = \dot{m}\left(\frac{p_2}{\rho_2} - \frac{p_1}{\rho_1}\right).$$

This work, being the sum of the work done to push fluid in and out of the control volume, is called *flow work*.

At the fixed walls, no work is done by pressure forces because the fluid moves parallel to the walls and pressure force is normal to them. Although the viscous stress is parallel to the flow direction, the fixed walls cannot do work on the fluid since there is no relative motion between the walls and the fluid immediately in contact with the walls:

$$\dot{W}_v = F_{viscous} V_{wall} = 0.$$

Of course, there is work done by viscous forces between adjacent fluid layers, which converts mechanical energy into internal energy.

The rotating impeller can do work on the fluid in contact with it through pressure and viscous forces. Since this type of work is generally transferred across the boundary by a rotating shaft, it is called *shaft work* and denoted by W_s. Not only rotational motion but also reciprocating motion can do shaft work, like stirring a cup of water using a spoon. All of the shaft work done by pressure forces is realized through unsteady motion. A steady pressure force can only do flow work, not shaft work. If one of the walls is continuously moving, such as a conveyor belt, it will do work on the fluid by steady viscous forces. This type of work can also be classified as shaft work.

The rate of total work done by the fluid within the control volume on its surroundings is the sum of the above terms:

$$\dot{W}_{out} = \dot{W}_p + \dot{W}_v + \dot{W}_s = \dot{m}\left(\frac{p_2}{\rho_2} - \frac{p_1}{\rho_1}\right) + \dot{W}_s. \tag{4.35}$$

Substituting Equations (4.34) and (4.35) into Equation (4.33) gives:

$$\dot{Q} - \left[\dot{W}_s + \dot{m}\left(\frac{p_2}{\rho_2} - \frac{p_1}{\rho_1}\right)\right] = \dot{m}\left[\left(\hat{u}_2 + \frac{V_2^2}{2} + gz_2\right) - \left(\hat{u}_1 + \frac{V_1^2}{2} + gz_1\right)\right].$$

Using $\dot{q}$ and $\dot{w}_s$ to express the rate of heat transfer per unit mass and the rate of shaft work per unit mass, the energy equation for one-dimensional steady flow can be rearranged as:

$$\dot{q} - \dot{w}_s = \left(\hat{u}_2 + \frac{p_2}{\rho_2} + \frac{V_2^2}{2} + gz_2\right) - \left(\hat{u}_1 + \frac{p_1}{\rho_1} + \frac{V_1^2}{2} + gz_1\right). \tag{4.36}$$

This is the *one-dimensional integral energy equation.*

The push work done by pressure forces at the inlet and outlet can be understood as a type of energy –that is, the pressure potential energy that we have discussed in Bernoulli's equation. It is the exchange of energy inside the fluid. In general engineering applications, only shaft work is regarded as the exchange of work between the fluid and its surroundings. The term on the left-hand side of Equation (4.36) may be interpreted as the heat transferred into the fluid from, and the work done by the fluid on, its surroundings; the term on the right-hand side represents the change of energy of the fluid. The total energy consists of four kinds of energy:

$\hat{u}$: internal energy per unit mass;

p/ρ: pressure potential energy per unit mass;

$V^2/2$: kinetic energy per unit mass; and

gz: gravitational potential energy per unit mass.

Tip 4.4: Push Work, Flow Work, and Volume Work

These concepts of work generally appear in books on engineering thermodynamics. For the one-dimensional flow shown in Figure T4.4(a), the physical meaning of each of these three types of work are the following:

Push work: As a local concept, this refers to the work done by pressure forces at a certain point on the local fluid to push it to move. Push work occurs on the control surfaces. The push work done by the fluid within the control volume on its surroundings at the inlet and outlet is:

$$W_{\text{push},1} = -p_1 A_1 \cdot (\delta x)_1, \quad W_{\text{push},2} = p_2 A_2 \cdot (\delta x)_2.$$

Flow work: As a holistic concept, this refers to the sum of the work required to push fluid into the inlet and out of the outlet of a control volume. It reflects the energy transfer between the fluid within the control volume and its surroundings in the form of push work:

$$W_{\text{flow}} = p_2 A_2 \cdot (\delta x)_2 - p_1 A_1 \cdot (\delta x)_1.$$

For a steady incompressible flow in a pipe of constant diameter, the push work at the inlet and outlet of the control volume cancel each other out, resulting in zero flow work.

Volume work: As a holistic concept, this refers to the difference in push work due to the volume change of a fluid. If the inlet and outlet pressures and areas are equal, the volume work done by the fluid within the control volume on its surroundings is expressed as:

$$W_{\text{volume}} = pA \cdot \left[(\delta x)_2 - (\delta x)_1 \right].$$

If the pressure is different at the inlet and outlet, the sum of the push work is decomposed into volume work and *displacement work*. The latter part is done by pressure forces on the whole fluid to push it over a displacement of δx, expressed as:

$$W_{\text{move}} = (p_2 \quad p_1) A \cdot \delta x.$$

When applied to infinitesimal fluid elements, the above formulas are summarized in a common expression in thermodynamics, shown in Figure T4.4(b). (Here, $v = 1/\rho$ is the specific volume of the fluid within the control volume.)

Figure T4.4(a).

$$d(pv) = pdv + vdp$$

Flow work — Volume work — Displacement work

Figure T4.4(b).

The general form of the energy equation, Equation (4.36), becomes Bernoulli's equation if there is no exchange of heat or work between the fluid and its surroundings and the internal energy of the fluid remains unchanged. The validity of Bernoulli's equation is restricted by the fulfillment of the following conditions: flow along a streamline, steady flow, inviscid flow, and incompressible flow. These four restrictions do not allow for work to be done on or by the fluid, and there is no conversion between mechanical and internal energy.

However, there is no restriction on the exchange of heat between the fluid and its surroundings. It may seem that even if there is heat transfer into or out of the fluid, its mechanical energy can remain unchanged. Is that correct?

Yes, it is: As long as the restrictions on Bernoulli's equation are maintained, heat transfer affects internal energy but not mechanical energy. This will be further analyzed in Section 4.6.2 on differential energy equation.

For gases, the energy equation (4.36) is usually written in a more general form:

$$\dot{q} - \dot{w}_s = (h_2 - h_1) + \frac{1}{2}\left(V_2^2 - V_1^2\right) + g(z_2 - z_1). \tag{4.37}$$

where $h = \hat{u} + p/\rho$ is the enthalpy of the flowing fluid.

When there is no exchange of heat or work between the gas and its surroundings, and gravity is neglected, we have:

$$(h_2 - h_1) + \frac{1}{2}\left(V_2^2 - V_1^2\right) = 0,$$

or rewritten as:

$$h + \frac{V^2}{2} = \text{const}.$$

This is the energy equation (4.29) derived from Bernoulli's equation for a compressible flow.

The conversion of kinetic energy into internal energy by compression is reversible. However, the conversion of kinetic energy into internal energy by viscous force is irreversible. This is known as viscous dissipation or friction loss. If we wish to learn more about the energy conversion inside the fluid, the integral energy equation is clearly not appropriate; derivation of the differential energy equation is needed.

4.6.2 Differential Energy Equation

Let us consider the heat conduction through the six faces of a cubical fluid element, as shown in Figure 4.12. Heat flux, $\dot{q}$, is defined as the heat transferred per unit time and per unit surface area. By Fourier's law of heat conduction:

$$\dot{q} = -\lambda \frac{\partial T}{\partial n}.$$

Assuming that the heat flux into the left face is $\dot{q}_{\mathrm{left}} = \dot{q}_x$, then the heat flux out of the right face can be expressed as:

$$\dot{q}_{\mathrm{right}} = \dot{q}_x + \frac{\partial \dot{q}_x}{\partial x}\mathrm{d}x.$$

The net heat flux into the fluid element across the left and right faces is:

$$\dot{Q}_x = \left(\dot{q}_{\mathrm{left}} - \dot{q}_{\mathrm{right}}\right)\mathrm{d}y\mathrm{d}z = -\frac{\partial \dot{q}_x}{\partial x}\mathrm{d}x\mathrm{d}y\mathrm{d}z.$$

Similarly, the net heat fluxes into the fluid element across the two y (front and back) faces and two z (bottom and top) faces are:

$$\dot{Q}_y = -\frac{\partial \dot{q}_y}{\partial y}\mathrm{d}x\mathrm{d}y\mathrm{d}z, \quad \dot{Q}_z = -\frac{\partial \dot{q}_z}{\partial z}\mathrm{d}x\mathrm{d}y\mathrm{d}z.$$

The rate of total heat transfer into the fluid element through heat conduction is:

$$\dot{Q}_{\mathrm{conduction}} = \dot{Q}_x + \dot{Q}_y + \dot{Q}_z = -\left(\frac{\partial \dot{q}_x}{\partial x} + \frac{\partial \dot{q}_y}{\partial y} + \frac{\partial \dot{q}_z}{\partial z}\right)\mathrm{d}x\mathrm{d}y\mathrm{d}z.$$

We can apply Fourier's law of conduction to rearrange the above formula as:

$$\dot{Q}_{\mathrm{conduction}} = \left[\frac{\partial}{\partial x}\left(\lambda \frac{\partial T}{\partial x}\right) + \frac{\partial}{\partial y}\left(\lambda \frac{\partial T}{\partial y}\right) + \frac{\partial}{\partial z}\left(\lambda \frac{\partial T}{\partial z}\right)\right]\mathrm{d}x\mathrm{d}y\mathrm{d}z.$$

Besides the heat conduction passing through the six faces, the heat transfer into the fluid element also includes radiation heat transfer. Assuming that the rate of radiation heat transfer of the fluid per unit mass is $\dot{q}$, then the rate of total radiation heat transfer into the fluid element is:

$$\dot{Q}_{\mathrm{radiation}} = \dot{q}\rho\mathrm{d}x\mathrm{d}y\mathrm{d}z.$$

Thus, the rate of total heat transfer into the fluid element is

$$\begin{aligned}\dot{Q} &= \dot{Q}_{\mathrm{conduction}} + \dot{Q}_{\mathrm{radiation}} \\ &= \left[\frac{\partial}{\partial x}\left(\lambda \frac{\partial T}{\partial x}\right) + \frac{\partial}{\partial y}\left(\lambda \frac{\partial T}{\partial y}\right) + \frac{\partial}{\partial z}\left(\lambda \frac{\partial T}{\partial z}\right) + \rho\dot{q}\right]\mathrm{d}x\mathrm{d}y\mathrm{d}z\cdot\end{aligned} \tag{4.38}$$

Let us now look at the rate of work done by the fluid element on its surroundings. The rate of work done by body forces is relatively simple. It is written as:

$$\begin{aligned}\dot{W}_{\mathrm{body}} &= -\left(\vec{f}_{\mathrm{b}}\rho\mathrm{d}x\mathrm{d}y\mathrm{d}z\right)\cdot\vec{V} \\ &= -\left(f_{\mathrm{b},x}\vec{i} + f_{\mathrm{b},y}\vec{j} + f_{\mathrm{b},z}\vec{k}\right)\cdot\left(u\vec{i} + v\vec{j} + w\vec{k}\right)\rho\mathrm{d}x\mathrm{d}y\mathrm{d}z. \\ &= -\left(f_{\mathrm{b},x}u + f_{\mathrm{b},y}v + f_{\mathrm{b},z}w\right)\rho\mathrm{d}x\mathrm{d}y\mathrm{d}z\end{aligned} \tag{4.39}$$

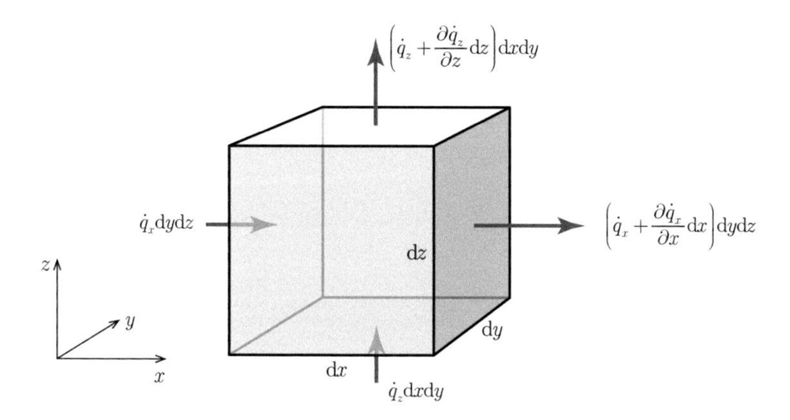

Figure 4.12 Rate of heat conduction through the six sides of a cubical fluid element.

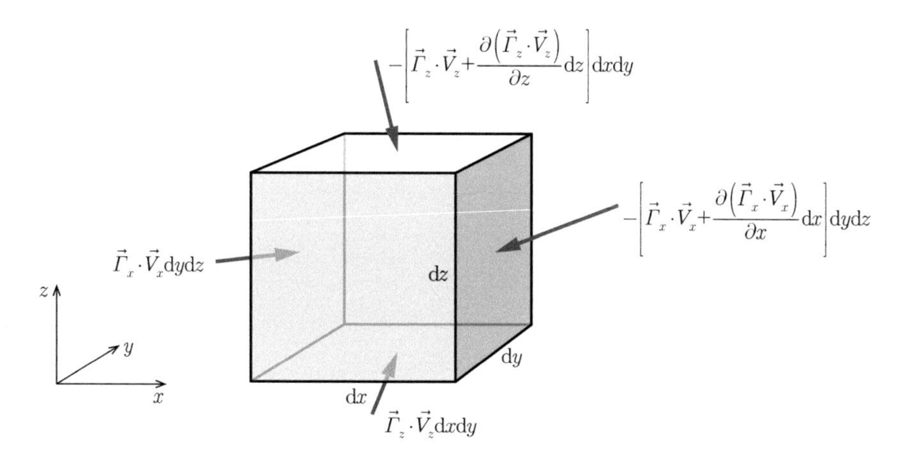

Figure 4.13 Rate of work done by surface forces on the surroundings.

The negative sign indicates that the body forces are actually the forces exerted on the fluid element by its surroundings.

As shown in Figure 4.13, the rate of work done by surface forces is slightly more complicated. On each side, the rate of work done by the fluid element on its surroundings is equal to the product of the local forces times the local velocity times the area of the face. On the left face, the rate of work done by surface forces on the surroundings is expressed as:

$$\dot{W}_{\text{left}} = \left(\vec{\Gamma}_x \mathrm{d}y\mathrm{d}z\right) \cdot \vec{V}_x = \vec{\Gamma}_x \cdot \vec{V}_x \mathrm{d}y\mathrm{d}z,$$

where the force $\vec{\Gamma}_x$ represents the force exerted *on the surroundings* by the fluid element. The tension is defined to be a positive force. Since this tension force is in the same direction as velocity, the rate of work done by the fluid element on the surroundings on the left is expressed as positive.

On the right face, the rate of work done by surface forces on the surroundings is expressed as:

$$\dot{W}_{\text{right}} = -\vec{\Gamma}_x \cdot \vec{V}_x \text{d}y\text{d}z - \frac{\partial\left(\vec{\Gamma}_x \cdot \vec{V}_x\right)}{\partial x}\text{d}x\text{d}y\text{d}z.$$

Remembering that tension is defined as positive, and that the force is in the opposite direction to velocity, the rate of work done by the fluid element on its surroundings on the right is negative.

On both left and right faces, the rate of total work done by surface forces on the surroundings is expressed as:

$$\dot{W}_{\text{surf},x} = \dot{W}_{\text{left}} + \dot{W}_{\text{right}} = -\frac{\partial\left(\vec{\Gamma}_x \cdot \vec{V}_x\right)}{\partial x}\text{d}x\text{d}y\text{d}z.$$

Using component form:

$$\dot{W}_{\text{surf},x} = -\frac{\partial}{\partial x}\left(\tau_{xx}u + \tau_{xy}v + \tau_{xz}w\right)\text{d}x\text{d}y\text{d}z.$$

Similarly, on the two y (back and front) faces and two z (bottom and top) faces, we have:

$$\dot{W}_{\text{surf},y} = -\frac{\partial}{\partial y}\left(\tau_{yx}u + \tau_{yy}v + \tau_{yz}w\right)\text{d}x\text{d}y\text{d}z.$$

$$\dot{W}_{\text{surf},z} = -\frac{\partial}{\partial z}\left(\tau_{zx}u + \tau_{zy}v + \tau_{zz}w\right)\text{d}x\text{d}y\text{d}z.$$

Thus, the rate of total work done by all the surface forces on the surroundings is:

$$\begin{aligned}\dot{W}_{\text{surface}} = &-\frac{\partial}{\partial x}\left(\tau_{xx}u + \tau_{xy}v + \tau_{xz}w\right)\text{d}x\text{d}y\text{d}z \\ &-\frac{\partial}{\partial y}\left(\tau_{yx}u + \tau_{yy}v + \tau_{yz}w\right)\text{d}x\text{d}y\text{d}z \\ &-\frac{\partial}{\partial z}\left(\tau_{zx}u + \tau_{zy}v + \tau_{zz}w\right)\text{d}x\text{d}y\text{d}z\,.\end{aligned} \tag{4.40}$$

The rate of change of total energy (consisting of internal and kinetic energy) of the fluid element is expressed as:

$$\frac{\text{D}e}{\text{D}t}\rho\text{d}x\text{d}y\text{d}z = \rho\frac{\text{D}}{\text{D}t}\left(\hat{u} + \frac{u^2 + v^2 + w^2}{2}\right)\text{d}x\text{d}y\text{d}z. \tag{4.41}$$

Substituting Equations (4.38)–(4.41) into the first law of thermodynamics, we have:

$$\begin{aligned}\rho\frac{\text{D}}{\text{D}t}\left(\hat{u} + \frac{u^2 + v^2 + w^2}{2}\right) = &\rho\left(f_{\text{b},x}u + f_{\text{b},y}v + f_{\text{b},z}w\right) \\ &+\frac{\partial\left(u\tau_{xx}\right)}{\partial x} + \frac{\partial\left(u\tau_{yx}\right)}{\partial y} + \frac{\partial\left(u\tau_{zx}\right)}{\partial z} \\ &+\frac{\partial\left(v\tau_{xy}\right)}{\partial x} + \frac{\partial\left(v\tau_{yy}\right)}{\partial y} + \frac{\partial\left(v\tau_{zy}\right)}{\partial z} \\ &+\frac{\partial\left(w\tau_{xz}\right)}{\partial x} + \frac{\partial\left(w\tau_{yz}\right)}{\partial y} + \frac{\partial\left(w\tau_{zz}\right)}{\partial z} \\ &+\frac{\partial}{\partial x}\left(\lambda\frac{\partial T}{\partial x}\right) + \frac{\partial}{\partial y}\left(\lambda\frac{\partial T}{\partial y}\right) + \frac{\partial}{\partial z}\left(\lambda\frac{\partial T}{\partial z}\right) + \rho\dot{q}.\end{aligned} \tag{4.42}$$

This formula can be rewritten in a simpler vector form:

$$\rho \frac{\mathrm{D}}{\mathrm{D}t}\left(\hat{u}+\frac{V^2}{2}\right)=\rho \vec{f}_{\mathrm{b}}\cdot\vec{V}+\nabla\cdot\left(\vec{V}\cdot\tau_{ij}\right)+\nabla\left(\lambda\nabla T\right)+\rho\dot{q}. \tag{4.42a}$$

or in tensor form:

$$\rho \frac{\mathrm{d}}{\mathrm{d}t}\left(\hat{u}+\frac{u_i u_i}{2}\right)=\rho f_{\mathrm{b},i}u_i+\frac{\partial}{\partial x_i}\left(\tau_{ij}u_j\right)+\frac{\partial}{\partial x_i}\left(\lambda\frac{\partial T}{\partial x_i}\right)+\rho\dot{q}. \tag{4.42b}$$

Equations (4.42), (4.42a), and (4.42b) are known as *differential energy equations*, and the physical meaning of each term from Equation (4.42b) is:

$\rho \frac{\mathrm{d}}{\mathrm{d}t}\left(\hat{u}+\frac{u_i u_i}{2}\right)$: rate of change of total energy (including internal and kinetic energy) of the fluid element;

$\rho f_{\mathrm{b},i}u_i$: rate of work done by body forces on the fluid element;

$\frac{\partial}{\partial x_i}\left(\tau_{ij}u_j\right)$: rate of work done by surface forces (including pressure and viscous forces) on the fluid element;

$\frac{\partial}{\partial x_i}\left(\lambda\frac{\partial T}{\partial x_i}\right)$: rate of heat transfer into the fluid element through heat conduction; and

$\rho\dot{q}$: rate of heat transfer into the fluid element through radiation.

Equation (4.42) describes the change of total energy, which can be classified into internal energy, reflected by temperature, and kinetic energy, reflected by flow velocity. For a better understanding of many actual flows, these two components of energy should be considered separately. For example, the basic form of Bernoulli's equation for steady and incompressible flows states that mechanical energy is conserved and would not be transformed into internal energy, while the extended Bernoulli's equation for compressible flow states that mechanical energy is not conserved and can be transformed into internal energy due to compression work.

It is important to understand the conversion between the internal energy and the kinetic energy for a flow. For example, let us consider the following questions:

(1) Does the internal energy of a fluid change when it is in free fall?
(2) When a fluid is heated, does its kinetic energy change?
(3) Can viscous forces increase the kinetic energy of a fluid without affecting its internal energy?

It is difficult to answer the above questions based only on Equation (4.42), so it is necessary to derive the kinetic and the internal energy equations separately. Next, we will derive these two equations to analyze the factors affecting the kinetic and the internal energy, respectively.

Referring to the derivation of Bernoulli's equation, the kinetic energy equation is an integral of momentum equation. Here we integrate the momentum equation to obtain the kinetic energy equation. The momentum equation in the x direction is rewritten as:

$$\frac{\mathrm{d}u}{\mathrm{d}t} = f_{\mathrm{b},x} + \frac{1}{\rho}\left(\frac{\partial \tau_{xx}}{\partial x} + \frac{\partial \tau_{yx}}{\partial y} + \frac{\partial \tau_{zx}}{\partial z}\right).$$

We multiply both sides of the above equation by the velocity component in the x direction, u:

$$u\frac{\mathrm{d}u}{\mathrm{d}t} = uf_{\mathrm{b},x} + \frac{1}{\rho}\left(u\frac{\partial \tau_{xx}}{\partial x} + u\frac{\partial \tau_{yx}}{\partial y} + u\frac{\partial \tau_{zx}}{\partial z}\right). \tag{4.43}$$

The left-hand side of Equation (4.43) represents the rate of change of kinetic energy per unit mass expressed in terms of the x component of the velocity:

$$u\frac{\mathrm{d}u}{\mathrm{d}t} = \frac{\mathrm{d}\left(u^2/2\right)}{\mathrm{d}t}.$$

The rate of change of total kinetic energy per unit mass expressed in terms of the three velocity components is:

$$\frac{\mathrm{d}\left(u_i u_i/2\right)}{\mathrm{d}t} = \frac{\mathrm{d}\left(V^2/2\right)}{\mathrm{d}t} = \frac{\mathrm{d}\left(u^2/2 + v^2/2 + w^2/2\right)}{\mathrm{d}t}.$$

The right-hand side of Equation (4.43) represents the rate of work done by the force component in the x direction. Adding up the rate of work done by the force components in the three coordinate directions gives:

$$\rho\frac{\mathrm{d}\left(u_i u_i/2\right)}{\mathrm{d}t} = \rho f_{\mathrm{b},i}u_i + u_j\frac{\partial \tau_{ij}}{x_i}. \tag{4.44}$$

This is the *differential kinetic energy equation*. As can be seen, there are two factors that cause the change of kinetic energy: one is the work done by body forces, and the other is the work done by surface forces.

Subtract Equation (4.44) from Equation (4.42b) to obtain:

$$\rho\frac{\mathrm{d}\hat{u}}{\mathrm{d}t} = \tau_{ij}\frac{\partial u_j}{x_i} + \frac{\partial}{\partial x_i}\left(\lambda\frac{\partial T}{\partial x_i}\right) + \rho\dot{q}. \tag{4.45}$$

This is the *differential internal energy equation*.

Next, we further analyze Equations (4.44) and (4.45) to understand how the kinetic and the internal energy of a fluid change.

First, all rates of heat transfer terms appear only in the internal energy equation:

$\frac{\partial}{\partial x_i}\left(\lambda\frac{\partial T}{\partial x_i}\right) + \rho\dot{q}$: rate of total heat transfer into the fluid element.

This means that the heat transfer between the fluid element and its surroundings affects only its internal energy, but not its kinetic energy (velocity). When a fluid is heated, it expands and internal energy is transformed into kinetic energy. The increase in kinetic energy in this case is directly reflected in the surface force terms, rather than the heat transfer terms.

Second, the body force terms appear only in the kinetic energy equation:

$\rho f_{b,i} u_i$: rate of work done by body forces on the fluid element

This indicates that the work done by body forces affects only the kinetic energy, but not its internal energy (temperature). As an object falls through the air, its temperature rise is a result of its friction and compression with the air. The conversion of kinetic energy into internal energy is due to the work done by surface forces, and has nothing to do with the work done by gravity.

Third, the surface force terms (pressure and viscous forces) appear in both the kinetic and the internal energy equations:

$u_j \dfrac{\partial \tau_{ij}}{\partial x_i}$: rate of work done by surface forces, which causes the kinetic energy of the fluid element to change.

$\tau_{ij} \dfrac{\partial u_j}{\partial x_i}$: rate of work done by surface forces, which causes the internal energy of the fluid element to change.

As a fluid element moves from a region of high pressure to another of lower pressure, it speeds up, which is due to differential pressure force. On the other hand, the friction and mixing produced by the viscous effects between adjacent fluid layers cause some of the kinetic energy to be transformed to internal energy, which accounts for the change in internal energy due to surface forces.

Since the work done by surface forces is rather complex, a thorough analysis is required.

Taken together with the total energy equation, it can be seen that the rate of work done by surface forces on the system consists of two parts:

$$\frac{\partial}{\partial x_i}\left(\tau_{ij} u_j\right) = u_j \frac{\partial \tau_{ij}}{\partial x_i} + \tau_{ij} \frac{\partial u_j}{\partial x_i}. \tag{4.46}$$

The term on the left-hand side of Equation (4.46) represents the rate of total work done by surface forces on the fluid element. The first term on the right-hand side appears only in the kinetic energy equation, and the second term appears only in the internal energy equation.

The first term on the right-hand side may be expanded as:

$$u_j \frac{\partial \tau_{ij}}{\partial x_i} = u\left(\frac{\partial \tau_{xx}}{\partial x} + \frac{\partial \tau_{yx}}{\partial y} + \frac{\partial \tau_{zx}}{\partial z}\right) + v\left(\frac{\partial \tau_{xy}}{\partial x} + \frac{\partial \tau_{yy}}{\partial y} + \frac{\partial \tau_{zy}}{\partial z}\right) + w\left(\frac{\partial \tau_{xz}}{\partial x} + \frac{\partial \tau_{yz}}{\partial y} + \frac{\partial \tau_{zz}}{\partial z}\right),$$

so that this term is divided into three parts which represent the rate of work done by surface forces on the fluid element to push it, moving translationally along one of the three coordinate directions, separately. The detailed descriptions are:

$u\left(\dfrac{\partial \tau_{xx}}{\partial x} + \dfrac{\partial \tau_{yx}}{\partial y} + \dfrac{\partial \tau_{zx}}{\partial z}\right)$: rate of work done by the x component of surface forces on the fluid element to push it, moving translationally along the x direction;

$v\left(\frac{\partial \tau_{xy}}{\partial x}+\frac{\partial \tau_{yy}}{\partial y}+\frac{\partial \tau_{zy}}{\partial z}\right)$: rate of work done by the y component of surface forces on the fluid element to push it, moving translationally along the y direction; and

$w\left(\frac{\partial \tau_{xz}}{\partial x}+\frac{\partial \tau_{yz}}{\partial y}+\frac{\partial \tau_{zz}}{\partial z}\right)$: rate of work done by the z component of surface forces on the fluid element to push it, moving translationally along the z direction.

Therefore, the sum of these three terms represents the rate of work done by surface forces on the fluid element in *translational motion*.

The second term on the right-hand side of Equation (4.46) can be expanded as:

$$\tau_{ij}\frac{\partial u_j}{\partial x_i}=\left(\tau_{xx}\frac{\partial u}{\partial x}+\tau_{yx}\frac{\partial u}{\partial y}+\tau_{zx}\frac{\partial u}{\partial z}\right)+\left(\tau_{xy}\frac{\partial v}{\partial x}+\tau_{yy}\frac{\partial v}{\partial y}+\tau_{zy}\frac{\partial v}{\partial z}\right)$$
$$+\left(\tau_{xz}\frac{\partial w}{\partial x}+\tau_{yz}\frac{\partial w}{\partial y}+\tau_{zz}\frac{\partial w}{\partial z}\right).$$

This formula represents the product of the surface stress times the strain rate.

As we saw in Chapter 3, the fluid deformation is classified into three categories: linear, rotational, and shear deformation. The work done by surface forces causes the internal energy to change when the fluid element experiences these three types of deformation.

According to the constitutive equation of Newtonian fluids, pressure and viscous terms in the surface stresses can be separated to obtain:

$$\begin{aligned}\tau_{ij}\frac{\partial u_j}{\partial x_i}&=-p\left(\nabla\cdot\vec{V}\right)-\frac{2}{3}\mu\left(\nabla\cdot\vec{V}\right)^2\\&+2\mu\left(\frac{\partial u}{\partial x}\right)^2+2\mu\left(\frac{\partial v}{\partial y}\right)^2+2\mu\left(\frac{\partial w}{\partial z}\right)^2\\&+\mu\left(\frac{\partial v}{\partial x}+\frac{\partial u}{\partial y}\right)^2+\mu\left(\frac{\partial w}{\partial y}+\frac{\partial v}{\partial z}\right)^2++\mu\left(\frac{\partial u}{\partial z}+\frac{\partial w}{\partial x}\right)^2.\end{aligned}\tag{4.47}$$

The second term on the right-hand side of Equation (4.47),

$$-\frac{2}{3}\mu\left(\nabla\cdot\vec{V}\right)^2.$$

represents the rate of volume work done by viscous normal stresses, which is negligible compared to the other terms. So, Equation (4.47) can be simplified as:

$$\tau_{ij}\frac{\partial u_j}{\partial x_i}=-p\left(\nabla\cdot\vec{V}\right)+\Phi_{\mathrm{v}}.\tag{4.48}$$

with

$$\begin{aligned}\Phi_{\mathrm{v}}&=2\mu\left(\frac{\partial u}{\partial x}\right)^2+2\mu\left(\frac{\partial v}{\partial y}\right)^2+2\mu\left(\frac{\partial w}{\partial z}\right)^2\\&+\mu\left(\frac{\partial v}{\partial x}+\frac{\partial u}{\partial y}\right)^2+\mu\left(\frac{\partial w}{\partial y}+\frac{\partial v}{\partial z}\right)^2+\mu\left(\frac{\partial u}{\partial z}+\frac{\partial w}{\partial x}\right)^2.\end{aligned}$$

The first term on the right-hand side of Equation (4.48),

$$-p\left(\nabla \cdot \vec{V}\right),$$

represents the rate of work done by pressure forces as the volume of the fluid element changes, which is the *volume work*.

When a fluid element is compressed, the work is done on the fluid element by its surroundings. Then, the term has a positive sign, $-p\left(\nabla \cdot \vec{V}\right) > 0$, indicating the conversion from mechanical to internal energy. When the fluid element expands, the work is done by the fluid element on its surroundings. In this case the term has a negative sign, $-p\left(\nabla \cdot \vec{V}\right) < 0$, indicating the conversion from internal into mechanical energy. In other words, this term represents the completely reversible conversion between internal and mechanical energy.

The second term on the right-hand side of Equation (4.48) is:

$$\begin{aligned}\Phi_{\mathrm{v}} &= 2\mu\left(\frac{\partial u}{\partial x}\right)^2 + 2\mu\left(\frac{\partial v}{\partial y}\right)^2 + 2\mu\left(\frac{\partial w}{\partial z}\right)^2 \\ &+ \mu\left(\frac{\partial v}{\partial x} + \frac{\partial u}{\partial y}\right)^2 + \mu\left(\frac{\partial w}{\partial y} + \frac{\partial v}{\partial z}\right)^2 + \mu\left(\frac{\partial u}{\partial z} + \frac{\partial w}{\partial x}\right)^2.\end{aligned}$$

This represents the rate of work done by viscous stresses as the fluid element deforms. Since all the terms in this equation are quadratic terms, Φ_{v} is always positive. In other words, this term only causes the internal energy of the fluid element to increase. Under adiabatic conditions, the mechanical energy of a flowing fluid will be irreversibly transformed into its internal energy. Φ_{v}, known as the *viscous dissipation term*, represents the mechanical energy loss due to viscous forces.

In addition, the expression for the viscous dissipation term indicates that here the deformation includes both linear and shear deformations, but not rotation. The derivation of the angular momentum differential equation has proven that the total torque acting on an infinitesimal fluid element is negligible. Therefore, even if the fluid element has an angular velocity, there is no work done due to its rotation.

4.6.3 Equations of Enthalpy, Entropy, Total Enthalpy, and Shaft Work

According to the relationship between enthalpy and internal energy, $h = \hat{u} + p/\rho$, the rate of change of enthalpy of a fluid is:

$$\frac{\mathrm{d}h}{\mathrm{d}t} = \frac{\mathrm{d}\hat{u}}{\mathrm{d}t} + \frac{\mathrm{d}}{\mathrm{d}t}\left(\frac{p}{\rho}\right). \tag{4.49}$$

Substituting Equation (4.48) into Equation (4.45) gives the internal energy equation with dissipation term:

$$\frac{\mathrm{d}\hat{u}}{\mathrm{d}t} = \frac{1}{\rho}\left[\Phi_{\mathrm{v}} - p\left(\nabla \cdot \vec{V}\right)\right] + \frac{1}{\rho}\frac{\partial}{\partial x_i}\left(\lambda \frac{\partial T}{\partial x_i}\right) + \dot{q}. \tag{4.50}$$

Substituting Equation (4.50) into Equation (4.49), we obtain the rate of change of enthalpy:

$$\begin{aligned}\frac{\mathrm{d}h}{\mathrm{d}t} &= \frac{\mathrm{d}}{\mathrm{d}t}\left(\frac{p}{\rho}\right)+\frac{1}{\rho}\left[\Phi_{\mathrm{v}}-p\left(\nabla\cdot\vec{V}\right)\right]+\frac{1}{\rho}\frac{\partial}{\partial x_i}\left(\lambda\frac{\partial T}{\partial x_i}\right)+\dot{q}\\ &= \frac{1}{\rho}\frac{\mathrm{d}p}{\mathrm{d}t}+p\frac{\mathrm{d}}{\mathrm{d}t}\left(\frac{1}{\rho}\right)-p\frac{1}{\rho}\left(\nabla\cdot\vec{V}\right)+\frac{1}{\rho}\Phi_{\mathrm{v}}+\frac{1}{\rho}\frac{\partial}{\partial x_i}\left(\lambda\frac{\partial T}{\partial x_i}\right)+\dot{q}\,.\end{aligned} \tag{4.51}$$

where the divergence of the velocity field represents the rate of change of volume of a fluid element, expressed as:

$$\nabla\cdot\vec{V}=\frac{1}{\delta B}\frac{\mathrm{d}\left(\delta B\right)}{\mathrm{d}t}=\rho\frac{\mathrm{d}}{\mathrm{d}t}\left(\frac{1}{\rho}\right). \tag{4.52}$$

Substituting Equation (4.52) into Equation (4.51), we obtain the *differential enthalpy equation for a flow*:

$$\frac{\mathrm{d}h}{\mathrm{d}t}=\frac{1}{\rho}\frac{\mathrm{d}p}{\mathrm{d}t}+\frac{1}{\rho}\Phi_{\mathrm{v}}+\frac{1}{\rho}\frac{\partial}{\partial x_i}(\lambda\frac{\partial T}{\partial x_i})+\dot{q}\,. \tag{4.53}$$

As can be seen, *there are three factors that cause the enthalpy of a flowing fluid to change: pressure change, viscous dissipation, and heat transfer with the surroundings.*

The relationship between entropy and enthalpy is:

$$T\frac{\mathrm{d}s}{\mathrm{d}t}=\frac{\mathrm{d}h}{\mathrm{d}t}-\frac{1}{\rho}\frac{\mathrm{d}p}{\mathrm{d}t}\,. \tag{4.54}$$

Substituting Equation (4.53) into Equation (4.54) gives:

$$T\frac{\mathrm{d}s}{\mathrm{d}t}=\frac{1}{\rho}\Phi_{\mathrm{v}}+\frac{1}{\rho}\frac{\partial}{\partial x_i}(\lambda\frac{\partial T}{\partial x_i})+\dot{q}\,. \tag{4.55}$$

This is the *differential entropy equation for a flow*. Notice that *there are two factors that cause the entropy of a flowing fluid to increase: viscous dissipation and heat transfer from the surroundings. For an adiabatic flow, entropy production is due solely to viscous dissipation.*

Often used in flow problems, the total enthalpy of a flowing fluid is the sum of its enthalpy and kinetic energy. In other words, it represents the total energy of a flow when neglecting the gravitational potential energy. Please refer to Equations (7.3) in Chapter 7 for the definition of total enthalpy. Dividing Equation (4.44) by density and adding it to Equation (4.53) gives:

$$\frac{\mathrm{d}h_{\mathrm{t}}}{\mathrm{d}t}=\left[f_{\mathrm{b},i}u_i+\frac{1}{\rho}u_j\frac{\partial\tau_{ij}}{\partial x_i}\right]+\left[\frac{1}{\rho}\frac{\mathrm{d}p}{\mathrm{d}t}+\frac{1}{\rho}\Phi_{\mathrm{v}}+\frac{1}{\rho}\frac{\partial}{\partial x_i}(\lambda\frac{\partial T}{\partial x_i})+\dot{q}\right]. \tag{4.56}$$

The total derivative of pressure with respect to time is the substantial derivative, expressed as:

$$\frac{\mathrm{d}p}{\mathrm{d}t}=\frac{\partial p}{\partial t}+u_i\frac{\partial p}{\partial x_i}\,. \tag{4.57}$$

The kinetic energy equation contains the work done by surface forces on the fluid element in translational motion. Rearrange the normal pressure and viscous forces separately to obtain the following relation:

$$u_j \frac{\partial \tau_{ij}}{\partial x_i} = u_j \frac{\partial\left(-p\delta_{ij}\right)}{\partial x_i} + u_j \frac{\partial \tau_{v,ij}}{\partial x_i}. \tag{4.58}$$

where $\tau_{v,ij}$ denotes the surface stresses without pressure, namely the viscous stresses.

Substituting Equations (4.57) and (4.58) into Equation (4.56) gives:

$$\frac{dh_t}{dt} = f_{b,i}u_i + \frac{1}{\rho}\frac{\partial p}{\partial t} + u_j \frac{\partial \tau_{v,ij}}{\partial x_i} + \frac{1}{\rho}\Phi_v + \frac{1}{\rho}\frac{\partial}{\partial x_i}(\lambda \frac{\partial T}{\partial x_i}) + \dot{q}. \tag{4.59}$$

This is the differential total enthalpy equation for a flow. There are four factors that cause the total enthalpy of a flowing fluid to increase: the work done by body forces, the work done by unsteady pressure forces, the work done by viscous forces, and heat transfer from the surroundings. Substituting Equation (4.55) into Equation (4.59) and neglecting the body forces (i.e., gravity) gives:

$$\frac{dh_t}{dt} = \frac{1}{\rho}\frac{\partial p}{\partial t} + u_j \frac{\partial \tau_{v,ij}}{\partial x_i} + T\frac{ds}{dt}. \tag{4.60}$$

To change the total enthalpy of a flowing fluid isentropically, we can only do work on it by unsteady pressure forces or drive it to move translationally by viscous forces. However, the translational motion generated by the viscous forces is always accompanied by deformation. Therefore, *unsteady pressure force work is the only way to increase the total enthalpy of a flowing fluid without loss*. In all kinds of machinery using fluid as the working medium, the total enthalpy of the fluid is always changed by devices that generate unsteady pressure forces, such as a piston or an impeller.

Equation (4.59) also states that the only way to increase the total enthalpy of an adiabatic, steady, and body force-ignored flow is to do work by viscous forces. Figure 4.14 schematically shows the work done by a fan impeller (such as the one in Figure 4.11) on the airflow passing through it. The increase in total enthalpy of the

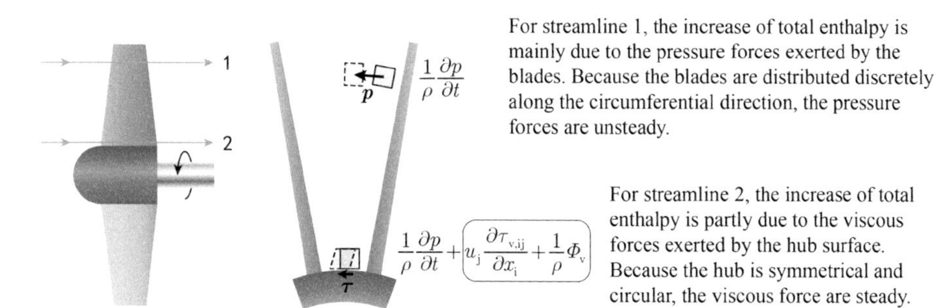

Figure 4.14 Work done by a fan impeller on airflow.

airflow is principally due to the unsteady pressure forces exerted by the fan blades on the airflow. Furthermore, the hub also accelerates the airflow along the circumferential direction by steady viscous forces, causing its total enthalpy to increase. This corresponds to the third and fourth terms on the right-hand side of Equation (4.59). In other words, the increase in total enthalpy of the airflow due to the viscous forces exerted by the hub is partly irreversible.

The shaft work was introduced earlier in the derivation of the one-dimensional energy integral equation (4.37). Here, let us see how the shaft work is reflected in the differential energy equation. Writing the rearranged Equation (4.37) and Equation (4.59) together gives:

$$h_{t2} - h_{t1} = -g(z_2 - z_1) - \dot{w}_s + \dot{q},$$

$$\frac{dh_t}{dt} = f_{b,i}u_i + \frac{1}{\rho}\frac{\partial p}{\partial t} + u_j\frac{\partial \tau_{v,ij}}{\partial x_i} + \frac{1}{\rho}\Phi_v + \frac{1}{\rho}\frac{\partial}{\partial x_i}(\lambda\frac{\partial T}{\partial x_i}) + \dot{q}.$$

As can be seen, the shaft work should be:

$$-\dot{w}_s = \int\left(\frac{1}{\rho}\frac{\partial p}{\partial t} + u_j\frac{\partial \tau_{v,ij}}{\partial x_i} + \frac{1}{\rho}\Phi_v\right)dt. \tag{4.61}$$

Notice that *the shaft work consists of three parts: the work done by unsteady pressure forces, the displacement work done by viscous forces, and the deforming work done by viscous forces*. Among them, the displacement work done by viscous forces is always accompanied by the deforming work that causes an increase in entropy. The efficiency of a fan is defined as the ratio of the effective work done by unsteady pressure forces to the total work done by unsteady pressure and viscous forces. To improve fan efficiency, the proportion of the work done by viscous forces should be reduced.

To sum up, the change in the energy of a fluid can be summarized as:

(1) When the potential energy of a flowing fluid is expressed in terms of work, its total energy consists of internal and kinetic energies.
(2) The work done by body forces only causes the kinetic energy of a flowing fluid to change, but does not affect its internal energy.
(3) The heat transfer with the surroundings only causes the internal energy of a flowing fluid to change, but does not affect its kinetic energy.
(4) The work done by the compression or expansion of a fluid, namely the volume work, causes a reversible transformation between kinetic and potential energy.
(5) The work done by viscous forces on a deforming fluid causes an irreversible transformation from kinetic to internal energy.
(6) Ignoring the body forces, the work done by unsteady pressure forces is the only way to increase the total enthalpy of a flowing fluid without loss.

4.7 Solution of the Governing Equations

4.7.1 Boundary Conditions

The three governing equations in differential form are:

continuity equation: $\dfrac{\partial \rho}{\partial t} + \nabla \cdot \left(\rho \vec{V}\right) = 0$;

momentum equation: $\dfrac{\mathrm{D}\vec{V}}{\mathrm{D}t} = \vec{f}_{\mathrm{b}} - \dfrac{1}{\rho}\nabla p + \dfrac{\mu}{\rho}\nabla^2 \vec{V} + \dfrac{1}{3}\dfrac{\mu}{\rho}\nabla\left(\nabla \cdot \vec{V}\right)$; and

energy equation: $\rho \dfrac{\mathrm{D}}{\mathrm{D}t}\left(\hat{u} + \dfrac{V^2}{2}\right) = \rho \vec{f}_{\mathrm{b}} \cdot \vec{V} + \nabla \cdot \left(\vec{V} \cdot \tau_{ij}\right) + \nabla\left(\lambda \nabla T\right) + \rho \dot{q}$.

Since the momentum (N-S) equation is the fundamental equation governing fluid motion and dynamics, the system of the three governing equations is generally called *N-S equations*. Basically, the flow of any Newtonian fluid satisfies the N-S equations.

For practical problems, the unknowns in the three governing equations are:

ρ, $\vec{V}$: the unknowns in the continuity equation;
ρ, $\vec{V}$, p: the unknowns in the momentum equation; and
ρ, $\vec{V}$, p, T: the unknowns in the energy equation.

Therefore, an additional relation is required to solve the system of three governing equations with four unknowns.

Actually, for an incompressible flow with known density there are only three unknowns, so the system can be solved. For an ideal compressible gas, its equation of state, $p = \rho RT$, is added, thus the system can be solved.

Any real flow is a specific solution to this system of differential equations. As described by Newton's second law, the speed of an object is only related to its speed at the previous moment and the force exerted over that time period. *Different flow types are determined by the initial state of a fluid and the combined contributions of forces and heat transfer to the process. In other words, they are determined by initial and boundary conditions.*

The initial conditions represent the state of a fluid at a certain time, specified in three-dimensional form as:

$$\Phi(t_0, x, y, z) = \Phi_0(x, y, z).$$

where Φ represents the unknowns of the system of governing equations, and Φ_0 the known state of the flow field. If the flow is steady, there are no initial conditions, which can be taken to mean that the influence of the initial state has vanished after a long enough period of time.

Fluids in motion have various types of boundary conditions. Consider the flow of water in a pipe, for instance: The boundary conditions include some of the flow characteristics such as pressure, temperature, and velocity at the inlet and outlet, as well as the friction and heat transfer between the water and the pipe wall. The boundary conditions between fluid and solid wall are relatively easy to specify. Basically, both no-slip and no-penetration boundary conditions assume that the fluid layer in immediate contact

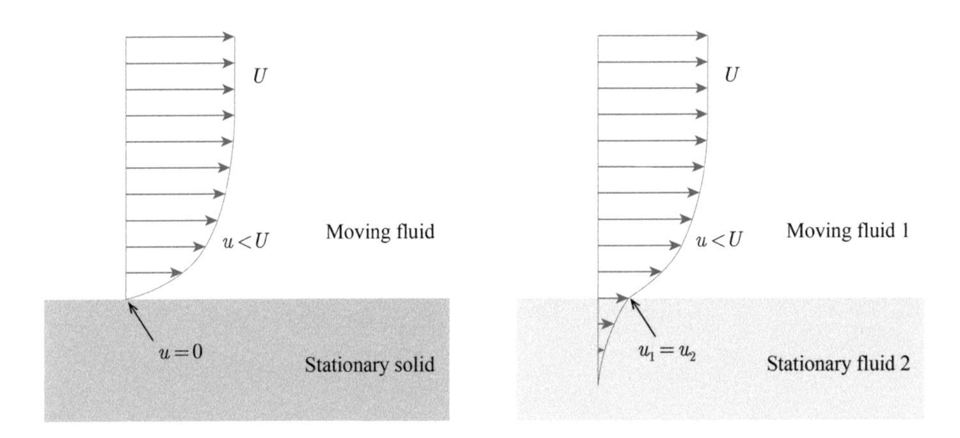

Figure 4.15 Velocity boundary conditions at the interfaces between a moving fluid and a stationary solid or fluid.

with a fixed wall remains at rest. That is to say, the tangential and normal components of the velocity of the fluid and the solid wall should be equal, expressed as:

$$\vec{V}_{\text{fluid}} = \vec{V}_{\text{solid}} .$$

This boundary condition is not readily accepted, and common sense dictates that the fluid layer in immediate contact with a solid wall can have a relatively high velocity component parallel to the wall surface. But on the molecular scale, both theory and experiments confirm that the fluid molecules in immediate contact with a solid are always attached to the solid surface, and the relative motion between adjacent fluid molecules is governed by the law of viscosity. However, the no-slip boundary condition at a wall is not always satisfied. It is usually accurate for liquids, but may cause large errors for some hypersonic rarefied-gas flows due to the large mean-free-path.

The no-slip boundary condition should also be satisfied at a fluid–fluid interface, expressed as:

$$\vec{V}_{\text{fluid1}} = \vec{V}_{\text{fluid2}} .$$

Figure 4.15 shows the flow pattern near the interfaces between a moving fluid and a stationary solid or a fluid.

When solving the energy equation, a commonly used boundary condition is to require that the solid and fluid temperatures be the same at the boundary walls:

$$T_{\text{fluid}} = T_{\text{solid}} .$$

Since it is affected by the fluid, the temperature at the boundary walls is often unknown and not easy to use. Another commonly used boundary condition is the heat flux through the wall:

$$-\left(\lambda \frac{\partial T}{\partial n}\right)_{\text{fluid}} = \dot{q}.$$

The flow is continuous, but generally only a part of it is investigated, involving the inlet and outlet boundary conditions. In unsteady-flow analysis, the boundary conditions are relatively difficult to determine; in steady-flow analysis, the inlet and outlet boundary conditions are easy to specify if some physical quantities (such as the mass flow rate of a fluid flowing through a pipe) are known.

The motion of a Newtonian fluid satisfying the continuum assumption is consistent with the system of N-S equations. If the boundary conditions for solutions are known, the total flow field can be obtained. Unfortunately, the analytical solutions of nonlinear N-S equations exist in theory but are not easy to obtain. Many unstable solutions of problems in mathematics correspond to some unstable type of flow in physics. So far, only a few analytical solutions of N-S equations have been obtained. We present two of them below.

4.7.2 Some Analytical Solutions of N-S Equations

4.7.2.1. Couette Flow

Maurice Couette (1858–1943) obtained an analytical solution of the N-S equations by investigating the viscous flow between two rotating, infinitely long, and concentric cylinders, as shown in Figure 4.16(a). If the annular gap between the cylinders is far smaller than the cylinders' diameters, the flow may be considered as a viscous flow between two infinitely large parallel plates, one of which is moving relative to the other with constant velocity, as shown in Figure 4.16(b). This viscous flow is affected by gravity, the pressure difference between upstream and downstream, the compressibility of the fluid, and the heat transfer with its surroundings. In the simplest case, the so-called Couette flow disregards these effects and considers only the viscous forces generated by the upper and lower plates. Here, let us consider only a steady, incompressible flow with no differential pressure or body forces.

For incompressible flows, the heat transfer does not affect the velocity or pressure. In other words, the energy equation has no influence on either velocity or pressure. To obtain the velocity and pressure distributions in the flow field, we need only to solve the continuity and momentum equations. The continuity equation for a two-dimensional, incompressible flow is:

$$\frac{\partial u}{\partial x}+\frac{\partial v}{\partial y}=0.$$

In this case, the fluid flows only in the x direction, that is $v = 0$. The above equation may be rearranged as: $\partial u/\partial x=0$. That is to say, u is only a function of y, expressed as:

$$u=f(y).$$

The momentum equations for a two-dimensional, steady, incompressible flow with no body forces are:

$$\begin{cases} u\dfrac{\partial u}{\partial x}+v\dfrac{\partial u}{\partial y}=-\dfrac{1}{\rho}\dfrac{\partial p}{\partial x}+\dfrac{\mu}{\rho}\left(\dfrac{\partial^2 u}{\partial x^2}+\dfrac{\partial^2 u}{\partial y^2}\right) \\ u\dfrac{\partial v}{\partial x}+v\dfrac{\partial v}{\partial y}=-\dfrac{1}{\rho}\dfrac{\partial p}{\partial y}+\dfrac{\mu}{\rho}\left(\dfrac{\partial^2 v}{\partial x^2}+\dfrac{\partial^2 v}{\partial y^2}\right). \end{cases}$$

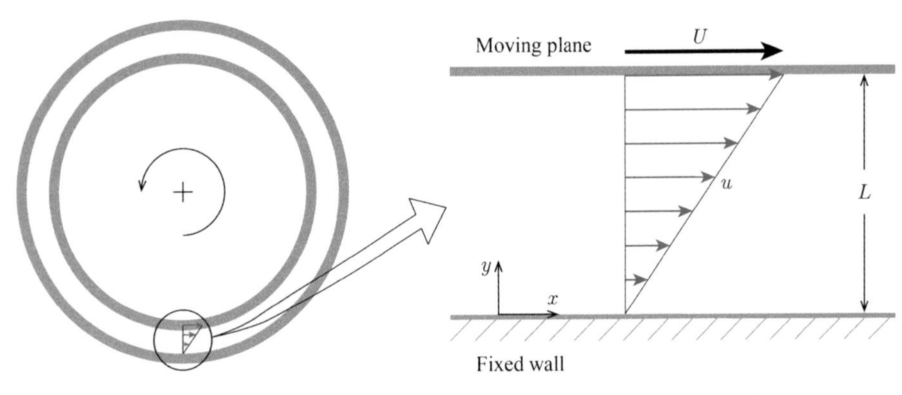

Figure 4.16 Flow studied by Maurice Couette and its simplified model. (a) Viscous flow between two infinitely long rotating concentric cylinders; (b) viscous flow between two infinitely large parallel plates (a moving plate, a fixed wall surface).

For the viscous flow between two rotating infinitely long concentric cylinders there is no change in pressure along the circumferential direction: $\partial p/\partial x = 0$.

Since $v = 0$ and $\partial u/\partial x = 0$, the above momentum equations along the x and y directions can be simplified as:

$$\frac{\partial^2 u}{\partial y^2} = 0, \quad \frac{\partial p}{\partial y} = 0 .$$

It results that the partial derivatives of pressure with respect to both x and y are equal to zero, corresponding to a uniform pressure field. Integrating the velocity with respect to y gives:

$$u = C_1 y + C_2 .$$

The boundary conditions at the upper and lower plates in Figure. 4.16(b) are:

$$\begin{cases} u = U, & y = L \\ u = 0, & y = 0 \end{cases} .$$

We can now solve for C_1 and C_2, and obtain an expression for velocity:

$$u = \frac{U}{L} y .$$

Therefore, the solutions of a simple Couette flow can be summarized as:

$$u = \frac{U}{L} y, \quad v = 0, \quad p = \text{const} . \tag{4.62}$$

Based on the solutions of the Couette flow, analytical solutions can be obtained in several more complex cases, such as a flow with streamwise pressure gradient or a compressible Couette flow with heat transfer. Interested readers are referred to textbooks treating these topics.

4.7.2.2. Hagen–Poiseuille Flow

The fully developed laminar flow in a circular pipe is another simple flow for which we can obtain an analytical solution. The flow model is shown in Figure 4.17. As the pipe wall produces drag on the fluid, the flow with constant velocity along the axis must also be acted upon by a driving force to balance the drag, which is the differential pressure. Let us consider a small element of flow with a length of $\mathrm{d}x$ as the control volume. Three control surfaces are: an inlet, an outlet and a cylindrical surface. The fluid is subjected to constant wall shear stress on the cylindrical surface and constant pressure at the inlet and outlet, respectively. Given the balance of forces, we have:

$$\tau_\mathrm{w} \cdot 2\pi R\mathrm{d}x = \left[p - \left(p + \frac{\partial p}{\partial x}\mathrm{d}x\right)\right] \cdot \pi R^2,$$

rearranged as:

$$\frac{\partial p}{\partial x} = -\frac{2\tau_\mathrm{w}}{R}.$$

As we can see, the streamwise pressure gradient must remain constant for a fully developed pipe flow. The momentum equation in cylindrical coordinates for steady, incompressible flow with no body forces is expressed as:

$$u_r \frac{\partial u_x}{\partial r} + \frac{u_\theta}{r}\frac{\partial u_x}{\partial \theta} + u_x \frac{\partial u_x}{\partial x} = -\frac{1}{\rho}\frac{\partial p}{\partial x} + \frac{\mu}{\rho}\left(\frac{\partial^2 u_x}{\partial r^2} + \frac{1}{r}\frac{\partial u_x}{\partial r} + \frac{1}{r^2}\frac{\partial^2 u_x}{\partial \theta^2} + \frac{\partial^2 u_x}{\partial x^2}\right).$$

We substitute $u_r = 0$, $u_\theta = 0$, $\partial u_x/\partial x = 0$, and $\partial u_x/\partial\theta = 0$ into the above equation to obtain:

$$\mu\left(\frac{\mathrm{d}^2 u_x}{\mathrm{d}r^2} + \frac{1}{r}\frac{\mathrm{d}u_x}{\mathrm{d}r}\right) = \frac{\mathrm{d}p}{\mathrm{d}x}.$$

The general solution of this equation is:

$$u = \frac{1}{\mu}\frac{\mathrm{d}p}{\mathrm{d}x}\frac{r^2}{4} + C_1 \ln r + C_2.$$

The boundary conditions at the symmetry at the center-line ($r = 0$) and pipe wall ($r = R$) are expressed as:

$$\begin{cases} \mathrm{d}u/\mathrm{d}r = 0, & r = 0 \\ u = 0, & r = R \end{cases}.$$

Substituting the boundary conditions into the general solution for velocity gives:

$$u = -\frac{1}{4\mu}\frac{\mathrm{d}p}{\mathrm{d}x}\left(R^2 - r^2\right).$$

Therefore, for a fully developed steady, incompressible flow in a circular pipe, the distribution of velocity is quadratic with radius. The average velocity across a cross-section of the pipe can be calculated:

$$U = -\frac{R^2}{8\mu}\frac{\mathrm{d}p}{\mathrm{d}x}.$$

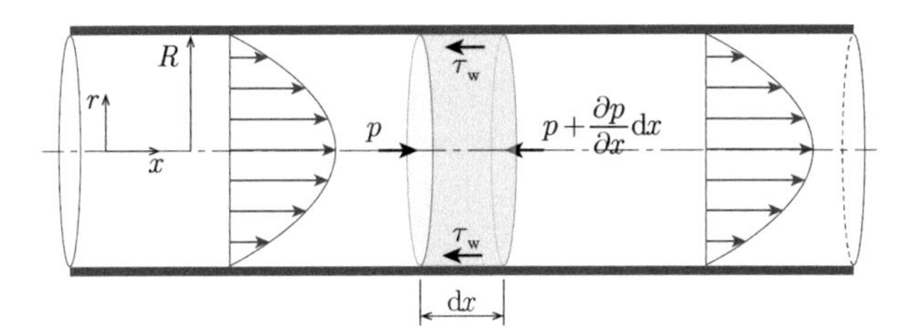

Figure 4.17 Hagen–Poiseuille flow in a circular pipe.

In practice, the flow rate or the average velocity is known. The velocity and pressure distributions in a Hagen–Poiseuille flow can be written as:

$$u = 2U\left[1-\left(\frac{r}{R}\right)^2\right], \quad \frac{dp}{dx} = -\frac{8\mu U}{R^2}.$$

Expanded Knowledge

A Comparison between Constitutive Equations for Fluids and Solids

Hooke's law, $F = kX$, is the earliest available constitutive equation for solids, which reflects the one-dimensional elastic force–deformation relationship. Correspondingly, the one-dimensional constitutive equation for fluids is Newton's law of viscosity (1.1). The three-dimensional constitutive equations for solids and fluids, developed by Navier, Cauchy, Poisson, St. Venen, Stokes, et al., are written as:

$$\tau_{xx} = \lambda\left(\frac{\partial u}{\partial x}+\frac{\partial v}{\partial y}+\frac{\partial w}{\partial z}\right)+2\mu\frac{\partial u}{\partial x}, \quad \tau_{xx} = -p+\lambda\left(\frac{\partial u}{\partial x}+\frac{\partial v}{\partial y}+\frac{\partial w}{\partial z}\right)+2\mu\frac{\partial u}{\partial x}$$

$$\tau_{yy} = \lambda\left(\frac{\partial u}{\partial x}+\frac{\partial v}{\partial y}+\frac{\partial w}{\partial z}\right)+2\mu\frac{\partial v}{\partial y}, \quad \tau_{yy} = -p+\lambda\left(\frac{\partial u}{\partial x}+\frac{\partial v}{\partial y}+\frac{\partial w}{\partial z}\right)+2\mu\frac{\partial v}{\partial y}$$

$$\tau_{zz} = \lambda\left(\frac{\partial u}{\partial x}+\frac{\partial v}{\partial y}+\frac{\partial w}{\partial z}\right)+2\mu\frac{\partial w}{\partial z}, \quad \tau_{zz} = -p+\lambda\left(\frac{\partial u}{\partial x}+\frac{\partial v}{\partial y}+\frac{\partial w}{\partial z}\right)+2\mu\frac{\partial w}{\partial z}$$

$$\tau_{xy} = \mu\left(\frac{\partial u}{\partial y}+\frac{\partial v}{\partial x}\right), \quad \tau_{xy} = \mu\left(\frac{\partial u}{\partial y}+\frac{\partial v}{\partial x}\right)$$

$$\tau_{yz} = \mu\left(\frac{\partial v}{\partial z}+\frac{\partial w}{\partial y}\right), \quad \tau_{yz} = \mu\left(\frac{\partial v}{\partial z}+\frac{\partial w}{\partial y}\right),$$

$$\tau_{zx} = \mu\left(\frac{\partial w}{\partial x}+\frac{\partial u}{\partial z}\right), \quad \tau_{zx} = \mu\left(\frac{\partial w}{\partial x}+\frac{\partial u}{\partial z}\right)$$

where the left-hand column lists the constitutive equations for elastic solids, and the right-hand column those for Newtonian fluids. Notice that the corresponding terms between the two columns are almost identical, but actually different in three ways.

First, the u, v, and w in the constitutive equations for solids represent the small displacement along the three coordinate directions, respectively, while the u, v, and w in the constitutive equations for fluids represent the velocities along the three coordinate directions, respectively. This means that the stress of a solid produces strain, while the stress of a fluid produces flow (strain rate).

Second, there is an additional pressure term, p, in the normal stress term of the constitutive equations for fluids. This represents the stress within a fluid at rest. In an atmospheric environment there is also such a term in the constitutive equations for solids, which is usually ignored.

Third, λ in the constitutive equations for fluids is generally expressed as $\lambda = -2\mu/3$. Therefore, one parameter – μ, the coefficient of viscosity – is sufficient to accurately describe the force–flow relationship for fluids, while two parameters – Young's modulus and Poisson's ratio – are required to describe the force–deformation relationship for solids.

For the derivation of the constitutive equations for Newtonian fluids, we refer the reader to some textbooks on *viscous fluid dynamics*.

Mathematical Properties of N-S Equations

In mathematics, partial differential equations are classified into three categories, depending on boundary conditions and solution methods. The classification is based on the following quasi-linear second-order partial differential equation:

$$A\frac{\partial^2 \Phi}{\partial x^2} + B\frac{\partial^2 \Phi}{\partial x \partial y} + C\frac{\partial^2 \Phi}{\partial y^2} + D = 0,$$

where the symbols A, B, C, and D can be any nonlinear function consisting of x, y, Φ, $\partial\Phi/\partial x$, and $\partial\Phi/\partial y$, except for the second derivative of Φ.

The properties involved in this equation can be determined by the following criteria:

$$B^2 - 4AC \begin{cases} < 0 & \text{elliptic type} \\ = 0 & \text{parabolic type}\,. \\ > 0 & \text{hyperbolic type} \end{cases}$$

Elliptic equations have no characteristic curves; parabolic equations have one characteristic curve; and hyperbolic equations have two. In mathematics, the characteristic curve represents the domain of dependence and region of influence for the solution, and it has a similar meaning in physics. However, the N-S equations are more complex than the above standard equations, and cannot be simply classified as any one of them. In other words, the N-S equations have the characteristics of the three types of differential equations at the same time, and display different characteristics in different situations. For example, a steady, inviscid, subsonic flow has the characteristics of an elliptic equation, while a steady, inviscid, supersonic flow has that of a hyperbolic equation. The characteristic curves in a steady flow are reflected as discontinuity. They represent the expansion and compression waves (see Chapter 7), which occur only in supersonic flow.

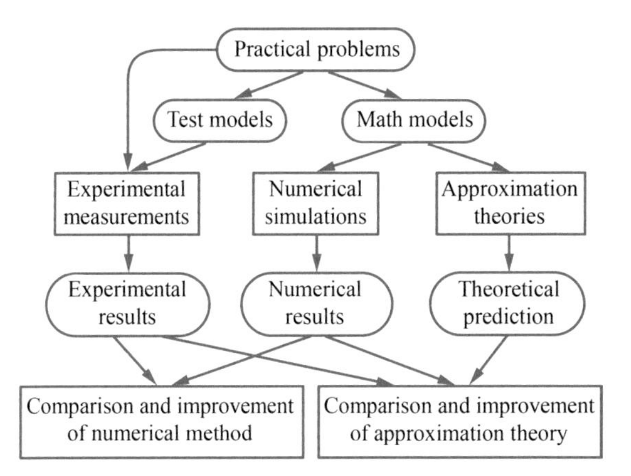

Figure 4.18 Relationship between theoretical solutions, numerical simulations, and experimental measurements.

The complexity of the N-S equations has attracted considerable attention, and they have been the object of in-depth study by many mathematicians, but their mathematical properties have not yet been figured out. In 2000, the Clay Mathematics Institute (CMI) of Cambridge, MA, USA, announced seven Millennium Problems, with a $1 million reward for the solution of each of them. One of the problems is to prove the existence and smoothness of the N-S equations' solutions.

The N-S equations represent a family of equations that are used to describe a wide variety of physical phenomena in nature. Any small progress in research on this topic may bring unexpected applications. However, these problems are of interest to mathematicians, while fluid mechanics researchers are more concerned with the solution of specific flow problems, and have developed various simplification methods. Fluid mechanics books often contain more than a dozen chapters, most of which are devoted to these simplification methods, such as potential flow theory, boundary layer theory, turbulence flow theory, and so on. In recent years, computers became increasingly important for solving complex flow problems.

Solving Flow Problems

Due to the complexity of solving N-S equations, the analytical solutions such as those in Section 4.7.2 are not readily available for general flow problems. Therefore, there are no theoretical solutions to slightly more complex flow problems, and simplified theoretical models or experimental measurements were commonly used in the early stages. Modern *computational fluid dynamics* has developed rapidly and extensively since the 1960s. By the end of the last century, numerical simulation techniques had largely substituted experimental measurements and become the main tool for the solution of flow problems.

Nowadays, to solve a specific flow problem, researchers can choose theoretical solutions, numerical simulations, or experimental measurements. Figure 4.18 shows the application of these methods in scientific research. The number of flow problems that can be solved by theoretical solutions is very limited. However, numerical simulations and experimental measurements also have their shortcomings, such as uncontrollable error, high cost, lack of information, and so on. It is significantly important for researchers to understand the advantages and disadvantages of numerical simulations and experimental measurements correctly. Since the theory is the basis of both experimental and numerical methods, it is crucial for researchers to master and understand fluid theory.

Questions

4.1 There is a hole in the window glass of a moving car, with all the other doors and windows well sealed. Analyze whether the air flows in or out of the car through the hole. Discuss the cases in which the hole is in the front, side, or rear window.

4.2 Consider the physical meanings of $D\rho/Dt = 0$ and $\partial\rho/\partial t = 0$. Can the continuity equation for incompressible flow (4.6) be used for unsteady flow? Why?

4.3 There is no tension forces between gas molecules, so why does the normal stress term in the constitutive equations for Newtonian fluids contain a positive term (such as $2\mu(\partial u/\partial x)$)?

4.4 What happens to the still air if a unpowered glider flies horizontally through it at a constant speed? What happens to the still air if that is a powered plane?

4.5 Where does the symbol 1/2 in Bernoulli's equation come from? What kind of physical meaning or process does it represent?

4.6 The restrictions on Bernoulli's equation do not include the adiabatic flow, which means that the heat transfer from or into the airflow does not affect its total pressure. Is that correct?

4.7 Why is the specific heat of air at constant pressure larger than that at constant volume? What is the specific heat of liquid water at constant pressure or constant volume?

5 Inviscid Flow and Potential Flow Method

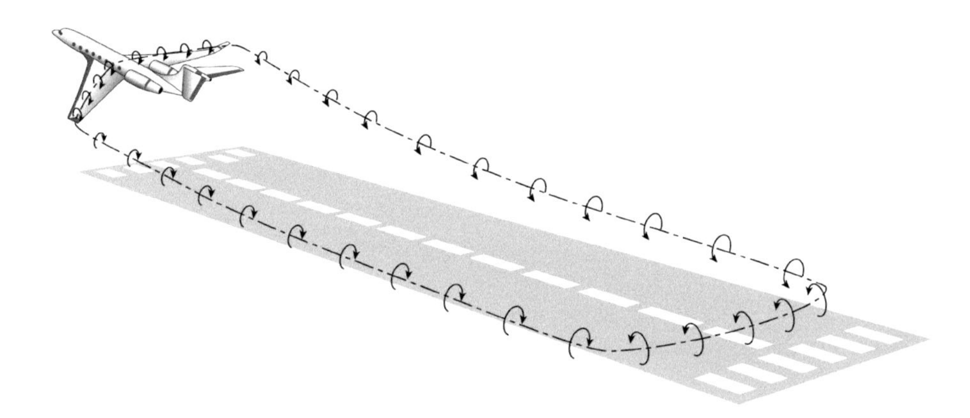

The attached vortex, tip vortex, and starting vortex form a closed vortex ring.

5.1 Characteristics of Inviscid Flow

All real fluid flows involve viscous effects to some degree (superfluids are not discussed here), but we do not always feel them. For the two most common fluids (air and water), the viscosity is very low. In many flow problems, the viscous forces are much smaller than the differential pressure and the gravitational and inertial forces. In this case, the results obtained by neglecting viscosity are close to the actual values. When viscosity is neglected, the N-S equations reduce to the Eulerian equations, which makes it possible to obtain analytical solutions. There is a special branch of fluid mechanics, called *inviscid flow*, which studies the motion law of inviscid flow.

A flow in which the viscous forces do not exist is known as inviscid flow. It can be classified into two categories: the flow of an inviscid fluid (generally called ideal fluid) in which the viscosity of the fluid is zero, and the flow of a viscous fluid in which the viscous forces are completely ignored under certain circumstances. Observe the constitutive equations for viscous fluids (4.22) carefully. When all the terms multiplied by the coefficient of viscosity are equal to zero, the total viscous stress is zero. Since the viscous normal stresses are negligible, only the viscous shear stresses are considered. To make the viscous shear stresses of a fluid be equal to zero, we must have:

$$\frac{\partial u}{\partial y}+\frac{\partial v}{\partial x}=0, \quad \frac{\partial v}{\partial z}+\frac{\partial w}{\partial y}=0, \quad \frac{\partial w}{\partial x}+\frac{\partial u}{\partial z}=0. \tag{5.1}$$

The three terms actually represent the shear rates in the three coordinate directions. In other words, there is no viscous shear stress in a fluid flowing without shear deformation. It is impossible for general flows to satisfy the rather rigorous restriction of having no shear deformation. Strictly speaking, inviscid flow refers only to the flow of an inviscid fluid.

As long as the shearing motion of a real fluid is not too strong, the viscous forces are much smaller than the differential pressure and inertial forces, and the flow can be considered approximately inviscid. Even for a strong vortex flow such as a tornado, the centrifugal forces from the rotation of airflow are balanced by the differential pressure forces inside and outside the vortex. The shear effects inside the tornado are actually negligible, and the flow can be treated as inviscid. For the airflow around an airfoil, viscous forces only need to be taken into account in the thin layer near the wall surface, whereas the flow outside this viscous layer can be treated as inviscid flow. In reality, many flows can be treated as inviscid, which simplifies these problems to a large extent.

Since the viscous forces in an inviscid flow are zero, pressure forces are the only surface forces. Since a pressure force cannot generate torque for a barotropic fluid (as defined in Tip 5.1), there is no torque in inviscid flow. The angular momentum equation states that *for inviscid flow, a fluid element initially free of rotation will never rotate, while a rotational fluid element will rotate forever.*

5.2 Inviscid Rotational Flow

Inviscid theory states that for an irrotational flow, the governing equations can be reduced to a simpler form, which is convenient for solving flow problems. Therefore, it is important to determine what is an irrotational flow. In this section, we shall discuss what is rotational and what is irrotational. A flow is known as irrotational when there is no rotation of fluid elements. Equation (3.12a) states that a flow is irrotational when it satisfies the following relations:

$$\frac{\partial w}{\partial y}-\frac{\partial v}{\partial z}=0, \quad \frac{\partial u}{\partial z}-\frac{\partial w}{\partial x}=0, \quad \frac{\partial v}{\partial x}-\frac{\partial u}{\partial y}=0. \tag{5.2}$$

Equation (5.2) is used to determine whether a given flow is rotational or not. Let us consider a parallel flow with a transverse velocity gradient, as shown in Figure 5.1. This is a simple Couette flow, and is characterized by:

$$u \neq 0, \quad v=0, \quad w=0, \quad \frac{\partial u}{\partial y} \neq 0.$$

The angular velocity at any point in the x–y plane is given by:

$$\Omega_z=\frac{1}{2}\left(\frac{\partial v}{\partial x}-\frac{\partial u}{\partial y}\right)=-\frac{1}{2}\frac{\partial u}{\partial y} \neq 0.$$

This result states that the Couette flow is a rotational flow in which all the fluid elements spin about their own centers at the same rate.

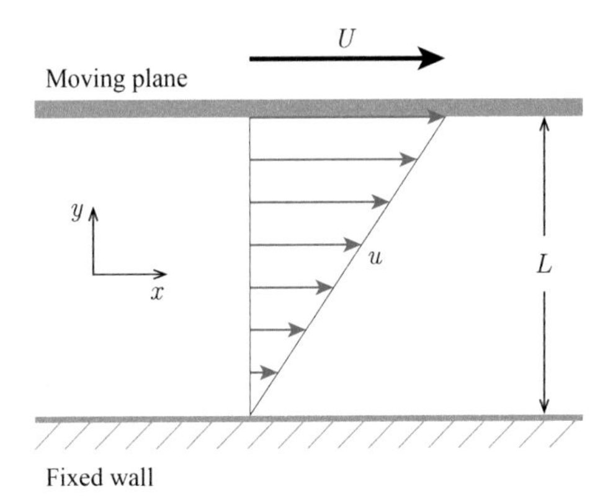

Figure 5.1 Viscous flow between two infinitely large parallel plates.

Tip 5.1: A Barotropic Fluid

A barotropic fluid is one whose density is a function of pressure only. In a barotropic fluid, the surfaces of constant pressure and constant density are parallel. By contrast, a baroclinic fluid is a fluid whose density is not only a function of pressure, but also depends on its temperature and/or other properties. In a baroclinic fluid, the surfaces of constant pressure and constant density can intersect. In a broad sense, the mechanical properties of a barotropic fluid are independent of its thermal characteristics.

Common fluids are actually baroclinic – for example, seawater, whose density depends on salinity; and air, whose density is a function of both temperature and humidity. Therefore, the barotropic fluid is an idealized model of fluid behavior, just like the concept of ideal gas or incompressible flow.

The equation of state for an ideal gas is $p = \rho RT$. In some particular cases, this equation can be reduced to $p = f(\rho)$, that is:

(1) Fluid of constant density: $\rho = C$.
(2) Fluid of constant temperature: $p = C\rho$.
(3) Fluid of adiabatic and isentropic: $p = C\rho^k$.

Many gas flows are approximately adiabatic and isentropic, and therefore can be treated as barotropic. For liquids, when temperature changes slightly they behave approximately as a fluid of constant density, and therefore can be treated as barotropic.

Why does a parallel flow have the characteristics of rotation? Figure 5.2 shows the microscopic motion of all the deformed fluid elements, which rotate at the same rate due to the difference in velocity between the upper and lower surfaces. This illustration considers only the rotation and neglects shear motion for clarity purposes.

The vortex flow shown in Figure 5.3 is generally known as a *free vortex*. Unlike the rotation of water in a vessel in Chapter 2, the existence of such a free vortex does not

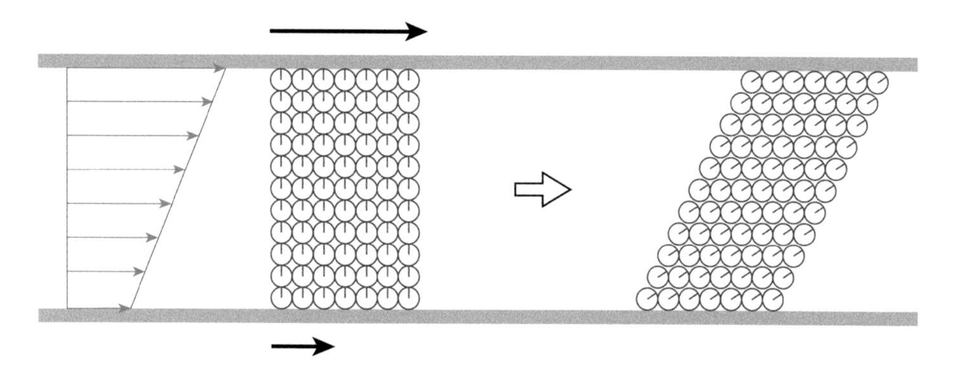

Figure 5.2 Rotation of all the fluid elements between two parallel plates.

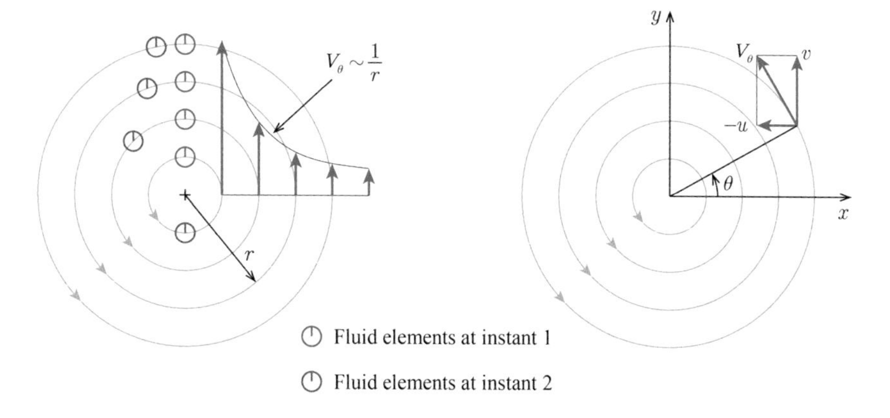

Figure 5.3 Velocity distribution and decomposition of a free vortex.

require any external torque. Sometimes we even feel that a free vortex can be created out of nothing, with examples such as the whirlpool at a drain port and giant vortices such as tornadoes and hurricanes. The angular momentum equation states that the tangential velocity of a free vortex is inversely proportional to the radius of rotation, that is:

$$rV_\theta = C, \tag{5.3}$$

where C is a constant.

The tangential velocity in Cartesian coordinates can be decomposed as:

$$u = -V_\theta \sin\theta, \quad v = V_\theta \cos\theta.$$

The angular velocity is:

$$\Omega_z = \frac{1}{2}\left(\frac{\partial v}{\partial x} - \frac{\partial u}{\partial y}\right) = \frac{1}{2}\left[\frac{\partial\left(V_\theta \cos\theta\right)}{\partial x} - \frac{\partial\left(-V_\theta \sin\theta\right)}{\partial y}\right].$$

Substitute Equation (5.3) into the above equation to get:

$$\Omega_z = \frac{C}{2}\left[\frac{\partial(\cos\theta/r)}{\partial x} + \frac{\partial(\sin\theta/r)}{\partial y}\right].$$

Substituting the relations $\sin\theta = y/r$ and $\cos\theta = x/r$:

$$\Omega_z = \frac{C}{2}\left[\frac{\partial\left(x/r^2\right)}{\partial x} + \frac{\partial\left(y/r^2\right)}{\partial y}\right].$$

Finally, since $r^2 = x^2 + y^2$:

$$\Omega_z = \frac{C}{2}\left[\frac{\partial\left[x/\left(x^2+y^2\right)\right]}{\partial x} + \frac{\partial\left[y/\left(x^2+y^2\right)\right]}{\partial y}\right] = 0.$$

The above analysis is applicable to all the spatial points except the vortex center. As can be seen, the fluid elements at these points have no angular velocity. Figure 5.3 shows the positions and directions of several fluid elements distributed along a radius at two instants of time. After some interval of time, the inner and outer fluid elements rotate around the vortex center at different angles, but they do not rotate around their own center and remain in the original direction. Therefore, the free vortex flow is irrotational except at the center ($r = 0$). The vortex center is treated as a singular point in theory since the angular velocity there is infinite. Near the center of an actual free vertex flow, viscous effects cannot be ignored due to the very large velocity gradient, and a small forced vortex rotates as a whole there.

To sum up, a seemingly parallel flow without vortex may actually be rotational, while a vortex flow that appears to be spinning may be irrotational. A flow is said to be rotational if fluid elements rotate about their own center; otherwise the flow is irrotational. Whether the flow is rotational or irrotational is determined by the self-rotation of each fluid element rather than its translational trajectory. Vorticity is commonly used to quantitatively express the rotation of a fluid element, defined as the curl of velocity:

$$\vec{\omega} = \nabla\times\vec{V} = \begin{bmatrix} \vec{i} & \vec{j} & \vec{k} \\ \dfrac{\partial}{\partial x} & \dfrac{\partial}{\partial y} & \dfrac{\partial}{\partial z} \\ u & v & w \end{bmatrix}. \tag{5.4}$$

This is a vector quantity, and its components in the three coordinate directions are:

$$\omega_x = \frac{\partial w}{\partial y} - \frac{\partial v}{\partial z}, \quad \omega_y = \frac{\partial u}{\partial z} - \frac{\partial w}{\partial x}, \quad \omega_z = \frac{\partial v}{\partial x} - \frac{\partial u}{\partial y}.$$

Notice that the *vorticity is twice the angular velocity*:

$$\vec{\omega} = 2\vec{\Omega}.$$

As we can see, the vorticity directly measures the magnitude of angular velocity of a fluid element.

For inviscid, barotropic flows with conservative body forces, Hermann Ludwig Ferdinand von Helmholtz (1821–1894) proposed three theorems about vortex motion:

(1) Fluid elements initially free of vorticity remain free of vorticity.
(2) Fluid elements lying on a vortex line at some instant continue to lie on that vortex line.
(3) The strength of a vortex tube is constant along its length.

Next, let us consider the physical meaning and restrictions of these three theorems.

The first one can be understood as stating that a fluid element initially free of rotation cannot be made to rotate. Angular momentum theorem states that a torque is required to produce an angular acceleration of an object. Therefore, the first theorem is valid only if the fluid elements are not subjected to external torques. The assumptions of inviscid, baroclinic, and conservative body forces actually guarantee that no net external torque acts on the fluid elements; this will be discussed later.

The second theorem can be understood as stating that a rotating fluid element will rotate forever. The first two theorems can be summarized in one statement: an initially irrotational flow will remain irrotational, and an initially rotational flow will remain rotational. In other words, vorticity cannot be created or destroyed in the interior of a fluid.

The third theorem is a type of quantitative description. The strength of a vortex tube can be expressed as:

$$\Gamma = \oint_L \vec{V} \cdot \mathrm{d}l = \iint_A \vec{\omega} \cdot \vec{n} \mathrm{d}A .$$

If the strength of a vortex tube is constant along its length, the average magnitude of the vorticity inside it is inversely proportional to its cross-sectional area. As the vortex tube stretches, the cross-sectional area decreases; thus, the vorticity must increase proportionally. This can explain the increase in angular velocity of the water flowing out of a drain. In fact, the third theorem can be understood as another expression of the conservation of angular momentum.

If we observe carefully, we will find that real-life situations are not in complete accord with the above three theorems. For example, small whirlwinds and whirlpools will stop rotating after a while, and originally static air will be full of vortices after passing through a fan. These phenomena are caused by the violation of at least one of the three restrictions on vortex maintenance. Let us discuss next how a torque is generated in the fluid and causes the rotation of a fluid element to change.

5.2.1 Vorticity Generated by Viscous Force

The shear stresses in a fluid are generated by viscous forces. Figure 5.4 shows the viscous flow around an airfoil. The vorticity at any point in the thin viscous flow region near the solid wall is expressed as:

$$\omega_z = \frac{\partial v}{\partial x} - \frac{\partial u}{\partial y} = -\frac{\partial u}{\partial y} \neq 0 .$$

As can be seen, the viscous flow around the airfoil is rotational. Since the incoming is irrotational, the vorticity is generated by the viscous shear stresses exerted by the wing surface.

Consider the irrotational translational motion of an ice skater. If the skater on the ice accidentally slides onto the cement floor, he will suddenly fall forward and roll over, as

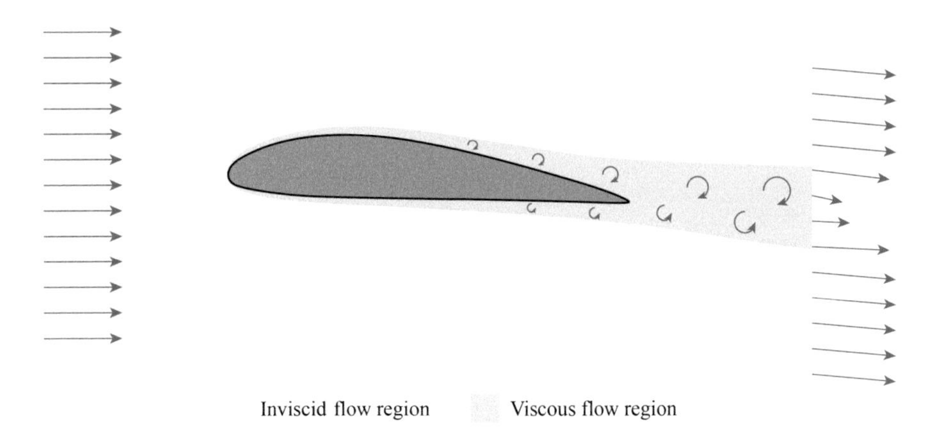

Figure 5.4 Viscous and inviscid flows around an airfoil.

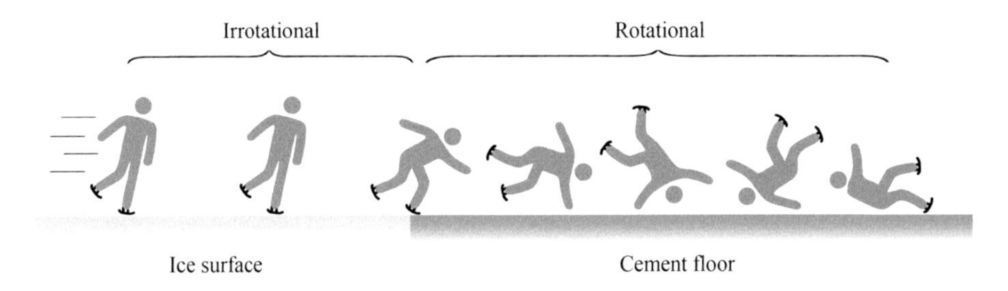

Figure 5.5 A skater rolls over due to friction force exerted by the ground.

shown in Figure 5.5. The transition of the skater from high-speed sliding to tumbling is similar to that of a fluid element from irrotational motion to rotational motion as it hits the wall. The skater is caused to roll over by a sudden torque. As a viscous fluid flows over a plate wall, the friction force from the wall produces a torque, and the fluid elements then deform and rotate. In other words, the wall is a source of vorticity. As the generated vorticity diffuses deeper into the interior of the fluid, more irrotational fluid elements begin to rotate, reflected as a thicker boundary layer.

In some cases, vorticity exists even if there is no wall in the flow field. For example, there is a strong shearing action between two parallel flows at different speeds, which is the source of vorticity.

As an airfoil is accelerated from rest in a fluid, the faster fluids over the lower surfaces meet the slower ones over the upper surfaces at the trailing edge and the so-called *starting vortex* forms (Figure 5.6)

5.2.2 Vorticity Generation in Baroclinic Flow

In an inviscid flow with conservative body force, if the fluid is baroclinic, the unbalanced pressure forces could produce torque, which could generate or destroy vorticity. Figure 5.7 shows the natural convection through an open window in a building.

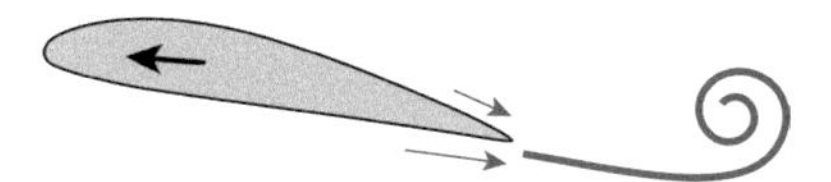

Figure 5.6 Starting vortex of an airfoil.

With the window closed, the indoor and outdoor air pressures change with height respectively

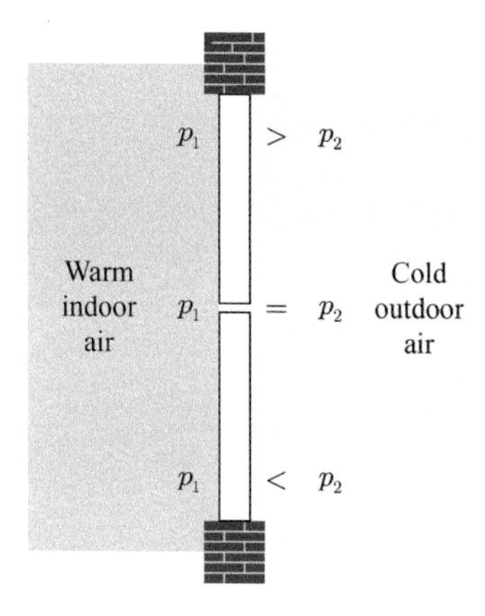

With the window opened, vorticity is generated by differential pressure between indoor and outdoor air

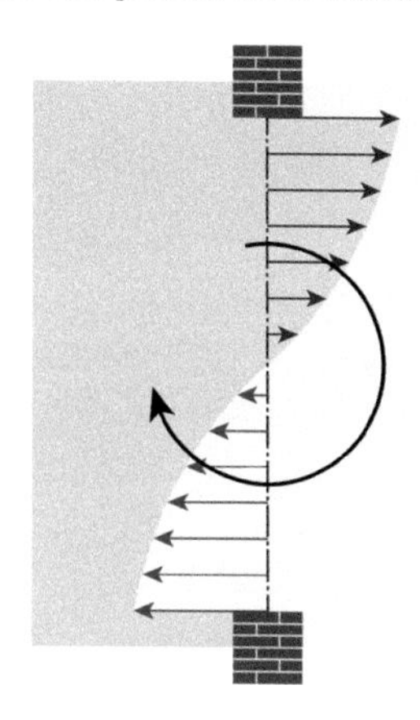

Figure 5.7 A rotational, baroclinic flow caused by the difference between indoor and outdoor air temperatures.

When the window is closed, the indoor and outdoor air are at rest, and the indoor air temperature is higher than that of the outdoors. Suppose there is a vent hole in the middle of the window where the indoor and outdoor air pressures are equal. Due to the lower temperature and higher density of the outdoor air, the outdoor air pressure in the lower part of the window is higher than that indoors, while the outdoor air pressure in the upper part of the window is lower than that indoors.

Suppose that the pressure in the middle of the window is p_0, the indoor air density is ρ_1, and the outdoor air density is ρ_2. Then, the vertical pressure distributions of indoor and outdoor air are expressed as:

$$p_1 = p_0 + \rho_1 gz, \quad p_2 = p_0 + \rho_2 gz.$$

The vertical distribution of the differential pressure between indoor and outdoor air is expressed as:

$$\Delta p = p_1 - p_2 = (\rho_1 - \rho_2) gz.$$

If the window is suddenly opened, the air flows in through the lower part of the window, and flows out through its upper part. According to Bernoulli's equation, the relationship between velocity and differential pressure is approximately expressed as:

$$V \sim (\Delta p)^{1/2}.$$

So, the change of velocity with height is:

$$V \sim z^{1/2},$$

and the vertical vorticity distribution is expressed as:

$$\omega = -\frac{\partial V}{\partial z} = -\frac{\partial\left(Cz^{1/2}\right)}{\partial z} = -\frac{C}{2}z^{-1/2} \neq 0.$$

As we have seen, the vorticity in baroclinic flow can be created from zero.

5.2.3 Vorticity Generation with Nonconservative Body Forces

A force acting on an object is said to be conservative if the work done by this force depends only on the initial and final positions of the object, independent of the moving path. Then, the potential function can be defined to express the work done by body forces as potential energy. It can be proven that the conservative body forces do not generate torques that cause fluid elements to rotate. Therefore, they have no effect on the vorticity within the flow. In many flows, gravity is the only body force, which is a conservative force.

Nonconservative body forces will generate torques. Coriolis force is a common example of a nonconservative force in a flowing fluid, which should be taken into account in rotating reference frames. For example, many vortices are generated and dissipated continuously in the oceanic and atmospheric circulations over the Earth's surface. Due to the Earth's rotation on its axis, the Coriolis force will affect the distribution of vorticity in the vortex across different latitudes. For researchers studying atmospheric dynamics, the Coriolis effect cannot be ignored.

Tip 5.2: Coriolis Force

The Coriolis force is a fictitious force that acts on objects that are in motion within a rotating reference frame. Rotating reference frames are not inertial. When a person stands on a rotating system, the observed object does not move in uniform straight motion even if there is no external force acting on it. The Coriolis force is expressed as:

$$\vec{F}_{\text{Coriolis}} = -2m\left(\vec{\Omega} \times \vec{V}\right),$$

where $\vec{\Omega}$ is the angular velocity of the rotating system and $\vec{V}$ is the velocity of the object relative to this system.

To make it easier to understand, three special cases are considered: (1) If the object moves along the axis of rotation, $\vec{\Omega}$ and $\vec{V}$ are in the same direction and the Coriolis force is zero. (2) If the object moves along the circumferential direction, the Coriolis and centrifugal forces are in the same straight line. The Coriolis force changes the magnitude of the centrifugal force, but not the trajectory of the object. Therefore, the Coriolis effect is not often perceived. (3) Only when the object has a radial velocity the Coriolis force will cause a noticeable change in its trajectory.

5.3 Irrotational Flow and Velocity Potential

Velocity potential function is a scalar function, with its gradient equal to the velocity vector of an irrotational flow. A flow governed by the velocity potential function is called a *potential flow*.

In an irrotational flow, we have:

$$\nabla \times \vec{V} = 0 .$$

The vector algorithm states that for any scalar ϕ, there is always the equation:

$$\nabla \times \nabla \phi = 0 .$$

We can define a relation between a scalar ϕ and velocity as:

$$\vec{V} = \nabla \phi , \tag{5.5}$$

or written as components:

$$u = \frac{\partial \phi}{\partial x}, \quad v = \frac{\partial \phi}{\partial y}, \quad w = \frac{\partial \phi}{\partial z}, \tag{5.5a}$$

where ϕ is called velocity potential.

By defining the velocity potential, the three components of velocity in an irrotational flow can be expressed as the gradients of a scalar function of position with respect to each of the three directions. The theory of potential flow is analogous to that of potentials in electricity and magnetism; the velocity and velocity potential are analogous to the electric current and electric potential, respectively. Just as the potential difference and resistance between two points in an electric field determine the electric current, the potential difference and distance between two points in a flow field determine the velocity.

The continuity equation for an incompressible flow is expressed as:

$$\nabla \cdot \vec{V} = \frac{\partial u}{\partial x} + \frac{\partial v}{\partial y} + \frac{\partial w}{\partial z} = 0 .$$

Substituting Equation (5.5a) into the continuity equation, we will obtain the governing equations in terms of velocity potential:

$$\nabla \cdot \nabla \phi = \nabla^2 \phi = \frac{\partial^2 \phi}{\partial x^2} + \frac{\partial^2 \phi}{\partial y^2} + \frac{\partial^2 \phi}{\partial z^2} = 0.$$

This is the *Laplace equation* for three-dimensional flow. In a two-dimensional flow, Laplace's equation reduces to:

$$\frac{\partial^2 \phi}{\partial x^2} + \frac{\partial^2 \phi}{\partial y^2} = 0 .$$

Based on this equation, the *stream function*, usually denoted by ψ, is defined as:

$$u = \frac{\partial \psi}{\partial y}, \quad v = -\frac{\partial \psi}{\partial x}. \tag{5.6}$$

Substituting the stream functions into the condition of irrotationality for a two-dimensional flow, $\omega_z = 0$, we obtain:

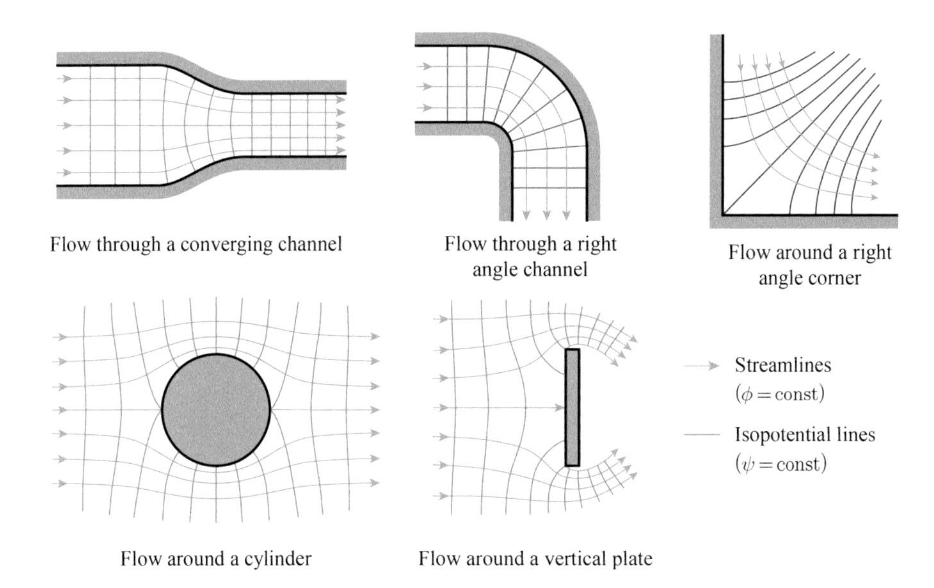

Figure 5.8 Streamlines and isopotential lines given by several typical potential flow solutions.

$$\frac{\partial^2 \psi}{\partial x^2} + \frac{\partial^2 \psi}{\partial y^2} = 0.$$

Thus, the stream function also conforms to Laplace's equation.

For two-dimensional, incompressible, irrotational flows, the velocity field may be calculated by solving both the potential and the stream functions – that is, by solving the Laplace equation mathematically. In a flow field, *lines of constant stream function are streamlines, while lines of constant potential functions are everywhere orthogonal to streamlines*. The orthogonal grid of streamlines and isopotential lines is called a *flow net*. The flow net is an intuitive graphical representation method to calculate two-dimensional potential flows. Numerous engineering problems may be quickly and easily solved by sketching flow nets. Figure 5.8 shows several types of flow nets. The flow net can directly give the direction of velocity at any position in the flow field, which is very useful for understanding the characteristics of flows. However, whether a specific flow can be solved by potential functions depends on the effect of viscosity and compressibility. For flow around a cylinder, as shown in Figure 5.8, the viscous flow field behind the cylinder is quite different from the potential flow field, while the viscous flow is similar to the potential flow upstream of the cylinder.

5.4 Planar Potential Flow

A planar potential flow describes the velocity field of a two-dimensional, incompressible, irrotational flow as the potential and stream functions. Because the governing equations obey the linear Laplace equation, the superposition of solutions is also a

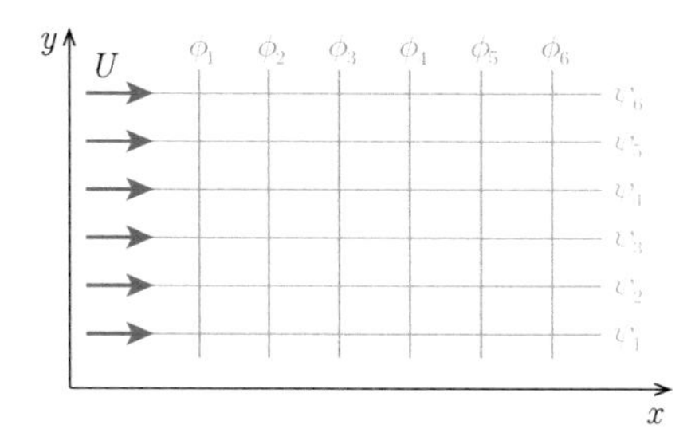

Figure 5.9 Streamlines and isopotential lines of a uniform flow.

solution. A complex planar potential flow can thus be decomposed into several simple ones, which can be solved separately and then superposed. Some elementary flows are taken as an example to illustrate the computational methods for planar potential flows below.

5.4.1 Uniform Flow

This is a flow with a constant velocity U along the x direction. The velocity distribution is expressed as:

$$u = U, \quad v = 0.$$

According to Equations (5.5) and (5.6), the potential function and stream function can be expressed as:

$$\phi = Ux, \quad \psi = Uy.$$

The streamlines and isopotential lines are straight lines parallel to the x axis and y axis, respectively, as shown in Figure 5.9.

5.4.2 Point Source and Point Sink

A point source or a point sink is a single point in the flow field at which a constant magnitude of flux emerges or disappears. Viewed directly from above, a fountain or drain is analogous to a point source or sink, respectively. It is often convenient to describe these types of flow using polar coordinates. If the point source or sink is placed at the origin of the coordinate system and the volume flow rate per unit length is Q, then the mathematical expressions for radial and tangential velocities are:

$$V_r = \frac{Q}{2\pi r}, \quad V_\theta = 0\,.$$

The potential and stream functions are:

$$\phi = \frac{Q}{2\pi}\ln r, \quad \psi = \frac{Q}{2\pi}\theta.$$

As we can see, the isopotential lines are lines of equal radius (i.e., concentric circles) and the streamlines are lines of equal angle (i.e., radial lines). A point source and

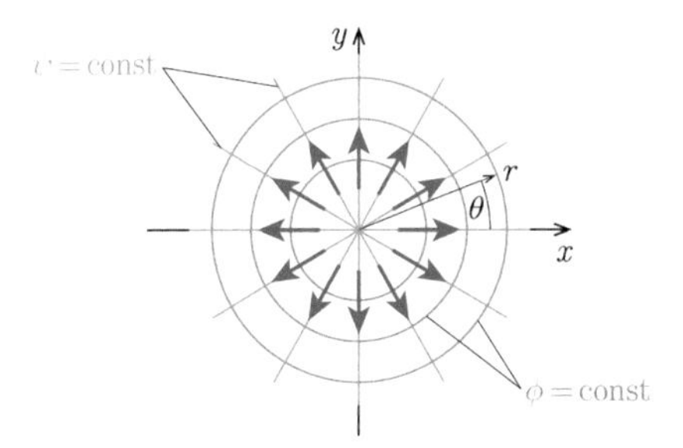

Figure 5.10 Streamlines and isopotential lines of a point source.

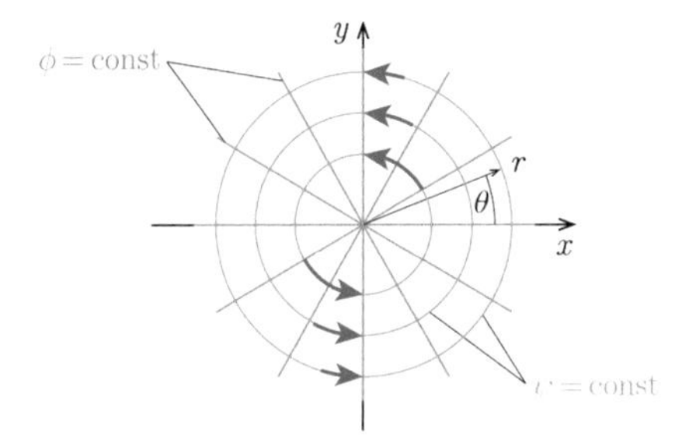

Figure 5.11 Streamlines and isopotential lines of a point vortex.

a point sink of equal strength have exactly the same streamlines and isopotential lines, but with opposite velocity direction. Figure 5.10 shows the streamlines and isopotential lines of a point source.

5.4.3 Point Vortex

A point vortex is the idealization of a two-dimensional free vortex with circular flow paths around a central point, such that the velocity distribution satisfies the irrotational condition (i.e., there is no radial velocity, the tangential velocity is inversely proportional to the radius, and the individual particles do not rotate about their own centers). If the intensity of a vortex is Γ, then the velocity distribution is expressed as:

$$V_r = 0, \quad V_\theta = \frac{\Gamma}{2\pi r}.$$

The potential and stream functions are:

$$\phi = \frac{\Gamma}{2\pi}\theta, \quad \psi = -\frac{\Gamma}{2\pi}\ln r.$$

As shown in Figure 5.11, the isopotential lines are radial lines and the streamlines are concentric circles.

5.4.4 Dipole

A dipole is not the simplest flow, but the superposition of a point source and a point sink. If a point source and a point sink of equal strength are placed at the

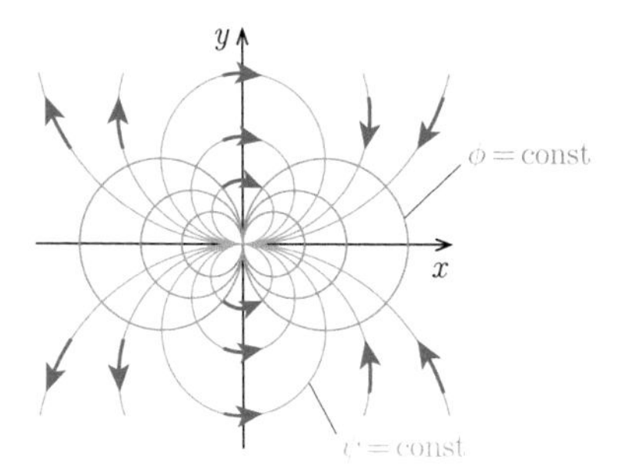

Figure 5.12 Streamlines and isopotential lines of a dipole.

same point, they will cancel each other out. Suppose the two points are separated by a certain distance and a plate is placed between them so that the fluids from the point source cannot directly flow into the point sink. As the distance between the two points and the size of the plate becomes infinitesimal, the fluids from the point source will flow into the point sink along a large circle. Then, a dipole is produced. The infinitesimal dipole model is more commonly used in electromagnetics. The magnetic field lines emerge out of the north pole of an infinitesimal magnetic dipole and enter its south pole.

A point source and a point sink cannot be superposed directly to produce a dipole. The concept of limit is needed for the derivation of the dipole. Suppose that a point source (left) and a point sink (right) are symmetrically distributed about the origin of the x axis at a distance of δ, and the volume flow rates per unit length of both points are Q. Consequently, the potential and stream functions are expressed as:

$$\phi = \frac{Q\delta}{2\pi}\frac{\cos\theta}{r}, \quad \psi = -\frac{Q\delta}{2\pi}\frac{\sin\theta}{r}.$$

The velocity distributions are expressed as:

$$V_r = -\frac{Q\delta}{2\pi}\frac{\cos\theta}{r^2}, \quad V_\theta = -\frac{Q\delta}{2\pi}\frac{\sin\theta}{r^2}.$$

With a constant stream function value ($\psi = C$), a single streamline can be calculated:

$$x^2 + \left(y - \frac{1}{2C}\right)^2 = \frac{1}{4C^2}.$$

The streamlines of a dipole are a cluster of circles tangential to the x axis at the origin, and with centers on the y axis, as shown in Figure 5.12.

5.4.5 Uniform Flow Around a Circular Cylinder

The uniform flow of an inviscid, incompressible fluid around a circular cylinder is a classic problem in fluid dynamics, and the potential flow theory provides an exact analytical solution. Consider a uniform flow around a circular cylinder with a freestream velocity of U. Place the cylinder of radius R with the center of its base at the origin. This flow can be obtained by combining a uniform flow with a dipole. The freestream

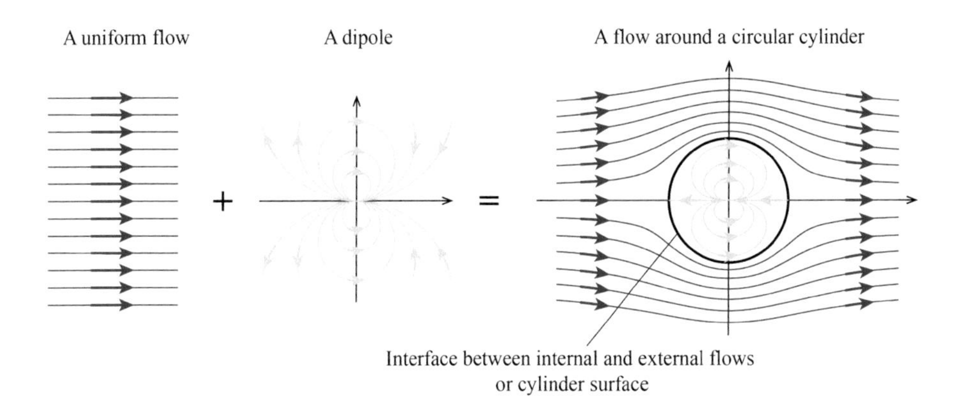

Figure 5.13 A flow around a circular cylinder is equivalent to the superposition of a uniform flow and a dipole.

is blocked by the fluids emerging from the point source, and drawn to the point sink, as shown in Figure 5.13.

The superposed solutions of a uniform flow and a dipole gives the solution of a flow around a circular cylinder, provided the dipole is of appropriate strength to make the radius of the innermost streamline of the freestream equal to R. Then, the potential and stream functions can be calculated:

$$\phi = U\left(1+\frac{R^2}{r^2}\right)r\cos\theta,\quad \psi = U\left(1-\frac{R^2}{r^2}\right)r\sin\theta.$$

The velocity distributions are expressed as:

$$V_r = U\left(1-\frac{R^2}{r^2}\right)\cos\theta,\quad V_\theta = -U\left(1+\frac{R^2}{r^2}\right)\sin\theta.$$

Notice that the flow field is symmetric, not only up and down but also front and back. A potential flow is characterized by an irrotational velocity field in which the velocity depends only on the spatial coordinates. If the geometry is symmetric, the flow field will also be symmetric.

On the cylinder surface where $r = R$, the velocity distribution can be calculated:

$$V_r = 0,\quad V_\theta = -2U\sin\theta\,.$$

The zero radial velocity indicates that the nonpenetration boundary condition is satisfied. However, since the inviscid flow does not satisfy the no-slip boundary condition, the tangential velocity is not zero.

As expected, the tangential velocity of the fluid on the cylinder surface increases from zero at the front stagnation point ($\theta = 0°$) to a maximum of $2U$ at the "equator" ($\theta = 90°$), and then decreases again to zero at the rear stagnation point ($\theta = 180°$). Due to viscosity within a flowing fluid, the actual distribution of the streamlines is quite different from the potential flow, but the potential flow can approximately represent the actual flow for the front half of the cylinder. Therefore, for airflow around a blunt object such as a cylinder, the maximum velocity on the two sides of the object is of the order of twice the freestream velocity.

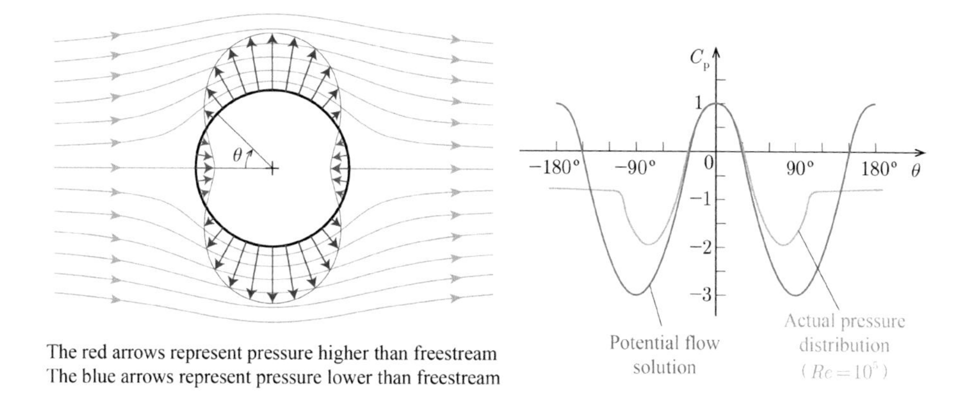

Figure 5.14 Pressure distributions on the cylinder surface in actual viscous and potential flows.

In engineering applications the pressure distribution on the surface of an object is often of interest. By Bernoulli's equation, the pressure distribution on the cylinder surface can be derived from the velocity distribution, expressed in terms of nondimensional pressure coefficient as:

$$C_{\mathrm{p}} = \frac{p - p_\infty}{\rho U^2/2} = 1 - 4\sin^2\theta\,.$$

Figure 5.14 compares the pressure distributions on the cylinder surface in the actual viscous and potential flows. The actual pressure distribution is similar to the potential flows value up to about 40°, but departs distinctly from it thereafter. Viscous effects cause the thickness of the boundary layer over the front portion of the cylinder to increase, so that the equivalent solid surface that influences the freestream is no longer circular. The streamlines of the separated flow over the rear portion of the cylinder are completely different from the potential flow solutions. The boundary layer separation will be further discussed in the next chapter.

5.5 Complex Potential

In the previous section, the flow around a circular cylinder was calculated by combining the potential and stream functions for both a uniform flow and a dipole. However, this method may be difficult to apply to flows with complex boundary conditions such as that over an airfoil. Therefore, to solve a potential flow, the complex velocity potential has been proposed, which has the following advantages.

5.5.1 A More Concise Expression

The definition of potential function ϕ and stream function ψ can lead to the following relations:

$$u = \frac{\partial \phi}{\partial x} = \frac{\partial \psi}{\partial y}, \quad v = \frac{\partial \phi}{\partial y} = -\frac{\partial \psi}{\partial x},$$

where ϕ and ψ satisfy the relation:

$$\frac{\partial \phi}{\partial x} = \frac{\partial \psi}{\partial y}, \quad \frac{\partial \phi}{\partial y} = -\frac{\partial \psi}{\partial x}.$$

This are the Cauchy–Riemann conditions. A complex function consisting of the two real functions ϕ and ψ can be defined as:

$$F(z) = \phi + i\psi,$$

where $z = x + iy$ is a complex variable.

$F(z)$ is often called the *complex potential*. When we replace ϕ and ψ with it, the flow governing equations are transformed into equations of a single variable, which are easier to solve. For example, the complex potential for a point source, a point vortex, and a dipole can be expressed as:

$$F(z) = \frac{Q}{2\pi} \ln z, \quad F(z) = \frac{\Gamma}{2\pi i} \ln z, \quad F(z) = \frac{Q\delta}{2\pi} \ln \frac{1}{z}.$$

5.5.2 Conformal Transformations

The *conformal transformation/mapping* is a mathematical concept from complex analysis, which can transform a coordinate from complicated planar domains to simpler ones while preserving the proportion of line segments and corresponding angles in the two domains. In fluid mechanics, one of the most useful applications of the conformal transformation is the design of simple airfoil shapes. We first superimpose a uniform flow with rotating cylinders of different radii, and then convert the circular cylinder into a family of airfoil shapes through a conformal transformation. Thus, both the airfoil shape and the analytical solution for the flow around it can be obtained directly, as shown in Figure 5.15. However, such a simple method is not sufficient to design the

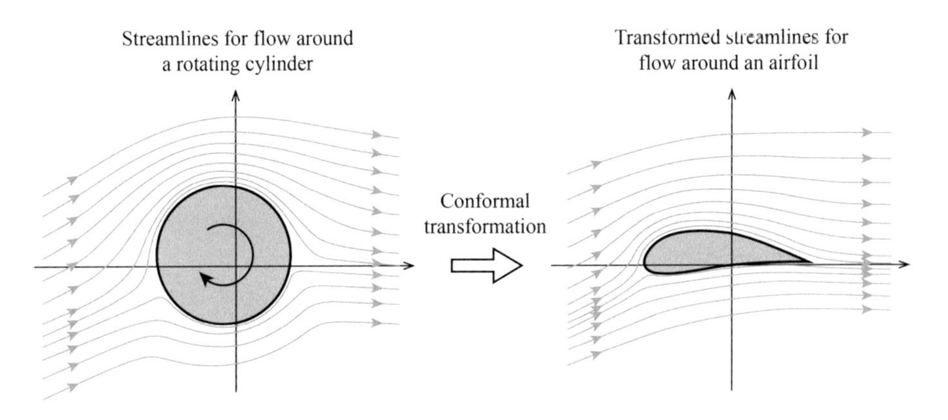

Figure 5.15 Conformal transformation from a rotating cylinder to an airfoil and the corresponding flow solutions.

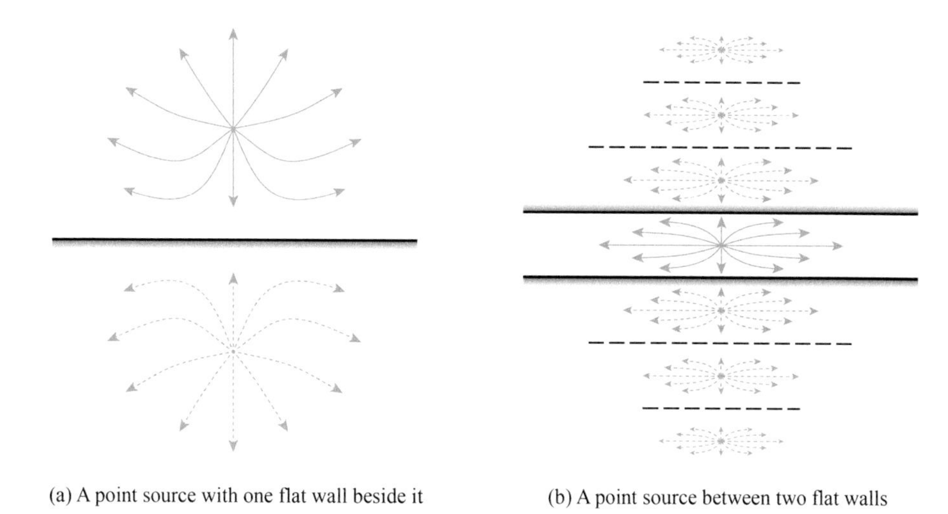

(a) A point source with one flat wall beside it

(b) A point source between two flat walls

Figure 5.16 Simulation of wind tunnel walls by the image method.

airfoil with superior aerodynamic performance. A row of point vortices, rather than a single point vortex, is required to simulate the flow around a better designed airfoil.

5.5.3 The Method of Images

Being subjected to the condition of infinite freestream, the potential flow cannot deal with flow problems with external boundary conditions. However, in actual flows there are always boundaries. Consider the airfoil mounted in a wind tunnel of limited size, for example. The pressure distribution over the airfoil surface cannot be directly compared with the solution for potential flow. In this case, the method of images can be used to simulate a wind tunnel wall by adding a symmetric flow.

Figure 5.16(a) shows a point source with one flat wall beside it. The fluid adjacent to the wall surface moves in a straight line. Subsequently, a fictitious point source is symmetrically added relative to the real one. In the resulting solution for the potential flow, the line in the middle between the two point sources represents the wind tunnel wall. It's a little more complicated if there are two flat walls on both sides of the point source. This is because the real point source is limited by two walls, while the two added fictitious point sources are not limited on the outside. Then, the line between the real and fictitious point sources is no longer a straight line and cannot represent the wind tunnel wall. An infinite number of fictitious point sources of smaller strength should be placed on the outside in sequence, as shown in Figure 5.16(b), to truly simulate the two wind tunnel walls.

5.6 Engineering Applications of Potential Flow and Its Current Status

In engineering, two-dimensional potential flows are quite different from three-dimensional ones. A number of researchers have been attempting to develop new methods to calculate some simple three-dimensional viscous flows based on potential flow theory. For example, the aforementioned airfoil generated by conformal transformation is limited to a two-dimensional airfoil, or an infinitely long airfoil. An actual airfoil has a finite length, so additional methods are needed to evaluate its lift and drag characteristics. One of the most useful is called the *lifting line theory*. It replaces a three-dimensional airfoil with a single vortex filament, or "lifting line," to predict the spanwise distribution of lift over the airfoil.

At the tip of the airfoil, the high-pressure area on the bottom of the airfoil pushes the air around the tip to the low-pressure area on the top. This action generates a vortex flow called a tip vortex. From the perspective of vortex dynamics, the tip vortex can also be interpreted as the extension of each vortex filament over the airfoil. It can be simply understood as follows: An attached vortex is generated by the velocity difference on the top and bottom surfaces of the airfoil. This vortex is the source of lift. Since a vortex cannot have endpoints in inviscid flow, the attached vortex is shed from the wingtip to create a tip vortex. The two tip vortices will be carried downstream away from the airfoil until they reach the starting vortex, forming a closed vortex ring, as shown on the title page of this chapter.

Potential flow theory and vortex dynamics based on inviscid flows have played a historically important role. Although it is now common practice to solve the N-S equations for three-dimensional flow numerically, potential flow theory still has its place. On the one hand, the analytical solution of potential flow can judge the mainstream structure theoretically. On the other hand, this method combined with the boundary layer theory can yield more reliable results, which in many cases are more accurate than solving the N-S equations numerically. In addition, potential flow theory provides solutions much faster than solving the N-S equations, which is convenient for optimization problems. Therefore, potential flow theory remains a useful method, worth studying and mastering in depth.

Expanded Knowledge

Complex Variable Functions and Fluid Mechanics

The theory of complex variable functions comes in naturally in the study of fluid mechanics. In the eighteenth century, d'Alembert put forward this concept in his paper on fluid mechanics, and in the same period Euler also carried out some investigations. By the nineteenth century, Cauchy and Riemann had further developed the theory of functions of a complex variable in the study of fluid mechanics. These functions are principally applied to the solution for inviscid flow, which has been introduced in this chapter. They are also used in the Kutta–Zhukowski theorem to derive the airfoil lift formula based on inviscid flow theory. Nowadays, complex variable functions have been exploited in a wide range of physical, mathematical, and other applications.

Questions

5.1 Use the inviscid vortex theory to explain the formation of starting vortex at the trailing edge of an airfoil.

5.2 The pressure distribution along the cylinder surface, as shown in Figure 5.14, may be similar to the surface pressure distribution of a ball traveling through the air. Thus, analyze the shape of a falling raindrop.

6 Viscous Shear Flow

A skydiver adjusts his posture to change the air drag and control the falling speed.

6.1 Shearing Motion and Flow Patterns of Viscous Fluids

Although the Navier–Stokes (N-S) equations describing viscous flows were established as early as the First Industrial Revolution, their practical applications could not be exploited at the time due to the complexity and difficulty in solving them. Some useful results can only be obtained based on the Eulerian equation for inviscid flow, and potential flow theory for irrotational flow. However, potential flow theory is unable to solve those problems in which viscosity plays a key role, such as the flow drag exerted on a particular object and the hydraulic losses through fluid machinery. Only when Ludwig Prandtl (1875–1953) put forward the famous *boundary layer flow theory* in 1904 could researchers finally quantitatively evaluate the effect of viscosity by mathematical methods.

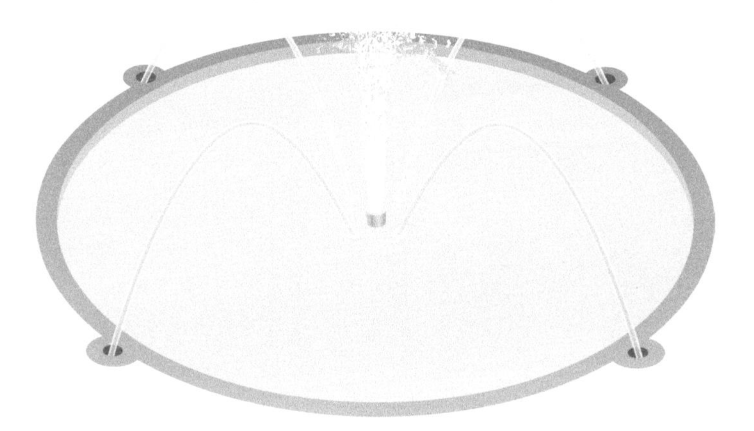

Figure 6.1 Two kinds of flow in a fountain.

The viscosity of a fluid is reflected by the shearing motion of fluid elements, which occurs in a variety of different situations, such as the flow near a wall, the mixing of two fluids with different velocities, and the jet and wake. Among these, two types of flow have been extensively studied: the flow over a flat plate, and the flow through a long circular pipe of constant diameter. The former is the simplest form of external flow, while the latter is the simplest form of internal flow.

Flow problems are not too difficult to solve if the viscous effect produces only the viscous shear and normal stresses given by Equations (4.22) and (4.22a), respectively. In most actual flows, the fluid does not seem to follow fixed laws of motion but behaves in a chaotic and irregular fashion. Figure 6.1 shows an artificial fountain. The water spouted from the four side nozzles moves regularly in stratified and parallel layers; this is called *laminar flow*. The water spouted from the middle nozzle is characterized by the irregular and unsteady movement; this is called *turbulent flow*. As a complex phenomenon in fluid dynamics, turbulent flow is caused by the combined action of inertia forces and viscous forces. Its unpredictability is one of the major obstacles to solving fluid mechanics problems.

Turbulent flow is common in everyday life, and research efforts to characterize it have a long history. Leonardo Da Vinci (1452–1519) might be the first to describe and tackle turbulence in great detail. However, it was Osborne Reynolds (1842–1912) who conducted the famous Reynolds experiment to carefully study the behavior of flow in a pipe by varying different flow conditions. This experiment was the first to demonstrate that turbulent or laminar flow is determined by the magnitude of a dimensionless number, later named the Reynolds number and defined as:

$$Re = \frac{\rho VL}{\mu}, \tag{6.1}$$

where ρ is the fluid density; V is the fluid velocity; L is the characteristic length scale within a flow field; and μ is the dynamic viscosity of the fluid. In the expression of the Reynolds number, the variable most difficult to understand is the characteristic length

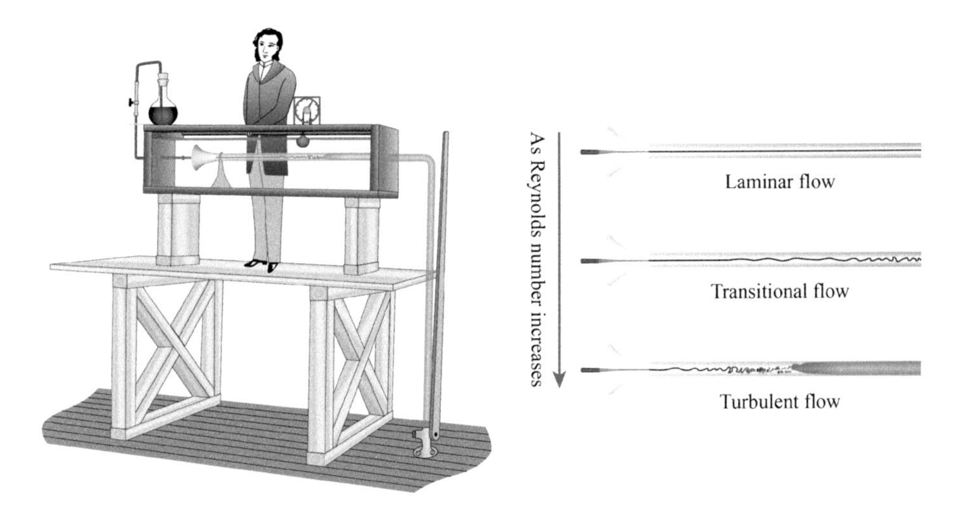

Figure 6.2 The Reynolds test (redrawn from the original drawing).

scale L. In many cases, there is no apparent length scale to choose as characteristic. Reynolds conducted his experiments in a circular pipe, and thus chose the diameter of the pipe as the characteristic length scale.

As shown in Figure 6.2, Reynold's apparatus enables water to enter the circular pipe as undisturbed as possible. A dye is injected at the inlet of the pipe to display the state of flow. Over the years, the same experiment has been carried out by a number of researchers with different results. Reynolds was the first to discover a critical Reynolds number, $Re_{cr} = 2{,}100$, determined as a limit where laminar flow changes to turbulent flow. Furthermore, he pointed out that Re_{cr} is related to the disturbance, environmental vibration, and noise at the inlet of the pipe. If the Reynolds number is larger than Re_{cr}, the flow regime is turbulent; otherwise, it is laminar.

Further clarifications of the critical Reynolds number have been attempted. Reynolds himself increased Re_{cr} to 12,000 later by minimizing disturbances. In 1910, Ekman increased it to 40,000, Pfenniger pushed up that number to 10^5 in 1961, and Salwen even gave it a value of 10^7 in 1980. It is reasonable to assume that *the transition from a laminar to turbulent regime requires two conditions: that the Reynolds number be high enough and also that the disturbance be large enough.* The Reynolds number only determines the degree of instability of the flowing fluid; whether a flow becomes turbulent or not needs to be triggered by disturbance. When the Reynolds number is high enough, the slightest disturbance will turn the flow turbulent. On the other hand, for a low enough Reynolds number, the flow will be laminar even if it is strongly disturbed. For actual flows in a pipe, if the Reynolds number is less than 2,000, the flow is laminar; if it is greater than 10^5, the flow is turbulent; Reynolds numbers between 2000 and 10^5 cover a critical zone between laminar and turbulent flow, and in this case the state of the flow is not easy to predict.

Turbulent flow does not only happen in pipe flow. The smoke coming out of a chimney and the water in a rocky river are also turbulent flows. Turbulence is only generated and sustained by the shearing motion within the flowing fluid. If there is no shearing motion, a turbulent flow normally will gradually diffuse and dissipate, and eventually return to laminar flow. Therefore, turbulence usually refers to the flow in the near-wall region (pipe flow and boundary layer flow) or the flow in which the layers of fluid mix together via shearing motion (mixing layer, jet and wake).

6.2 Laminar Boundary Layer

When the Eulerian equations were established, the importance of fluid viscosity was not fully realized and they were thought to be accurate enough. However, it can be derived directly from the Eulerian equations that the drag force exerted on any three-dimensional object in a flow is zero. This conclusion, known as *d'Alembert's paradox*, is obviously inconsistent with reality. It is a contradiction raised by the French mathematician Jean le Rond d'Alembert in 1752. It was not until 1904 that Prandtl put forward the concept of the boundary layer and provided the appropriate theoretical basis which fundamentally explained the d'Alembert's paradox and made fluid mechanics a useful subject. For this reason, *boundary layer theory* has been regarded as the most important developmental milestone in modern fluid mechanics.

A boundary layer is a thin layer of viscous fluid in the immediate vicinity of a solid surface, which is also known as a frictional layer. The no-slip condition at the interface between the fluid and the solid surface creates a sharp velocity gradient normal to the wall. Therefore, viscous forces cannot be disregarded inside a boundary layer. Outside the thin boundary region, the viscous forces are usually ignored due to the small velocity gradient. Figure 6.3 shows the airflow around an airfoil and the velocity profiles within the near-wall region where the viscous effects are significant. Unlike the viscous region, magnified for clarity in Figure 5.4 (most of the drawings in books

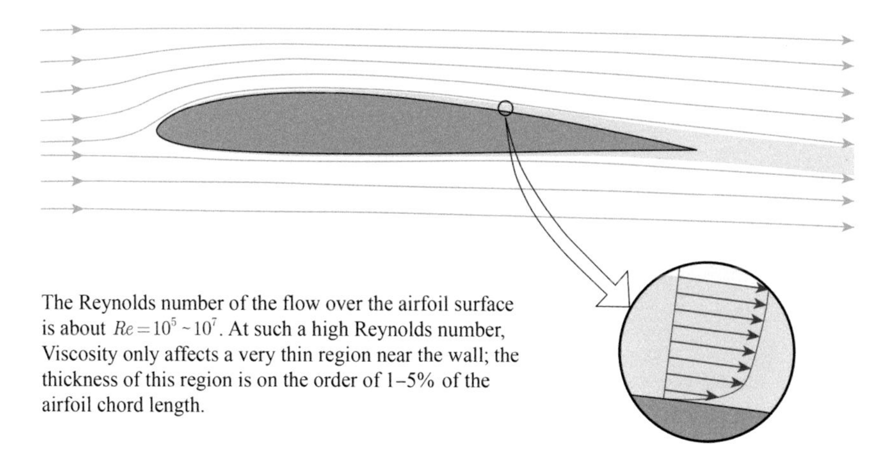

Figure 6.3 The flow around an airfoil and the near-wall region affected by viscosity.

on the subject magnify the boundary layer thickness), this region is drawn according to the real situation to give readers an idea of its true size.

Prandtl's breakthrough was built on the fact that, in general flows, viscosity is only significant within the very thin near-wall layer called the boundary layer, outside of which the flow (generally called *mainstream*) is essentially inviscid. Within the very thin boundary layer, the N-S equations reduce to the boundary layer equations whose accuracy satisfies the needs of engineering projects. For general flow problems, the unsolvable N-S equations are therefore broken up into solvable boundary layer equations and potential functions, which gives Prandtl's boundary layer theory its particular significance.

Since the flow near the wall slows down due to the resistance of wall shear stress, a part of the fluid is pushed outwards and no longer flows parallel to the wall. However, since the thickness of the very thin boundary layer increases negligibly in the streamwise direction, all fluid particles within the boundary layer are deemed to move parallel to the wall surface with no velocity component perpendicular to it. Therefore, the critical condition for simplifying N-S equations is that the boundary layer must be thin enough, expressed as:

$$\delta \ll L,$$

where δ is the boundary layer thickness and L is the length in the streamwise direction.

Why is the boundary layer so thin? It is because the action range of viscosity is limited. Since the Reynolds number measures the magnitude of viscous forces, the viscous effects are negligible when this number is large. A requirement for the boundary layer to be thin is that the Reynolds number be large. In other words, *boundary layer theory is valid only if the Reynolds number is large enough.*

6.2.1 Prandtl's Boundary Layer Equations for Two-Dimensional Flows

The N-S equations can be reduced to the boundary layer equations through dimensional analysis. We will derive the boundary layer equations for a two-dimensional, steady, incompressible flow with no body forces.

$$\text{Continuity equation: } \frac{\partial u}{\partial x} + \frac{\partial v}{\partial y} = 0. \tag{6.2}$$

$$\text{Momentum equation: } \begin{cases} u\dfrac{\partial u}{\partial x} + v\dfrac{\partial u}{\partial y} = -\dfrac{1}{\rho}\dfrac{\partial p}{\partial x} + \dfrac{\mu}{\rho}\left(\dfrac{\partial^2 u}{\partial x^2} + \dfrac{\partial^2 u}{\partial y^2}\right) \\ u\dfrac{\partial v}{\partial x} + v\dfrac{\partial v}{\partial y} = -\dfrac{1}{\rho}\dfrac{\partial p}{\partial y} + \dfrac{\mu}{\rho}\left(\dfrac{\partial^2 v}{\partial x^2} + \dfrac{\partial^2 v}{\partial y^2}\right) \end{cases}. \tag{6.3}$$

Figure 6.4 shows the velocity profile inside the boundary layer over a flat plate. Three physical quantities, namely the uniform freestream velocity U, the distance from the leading edge of the flat plate L, and the boundary layer thickness δ, can be used to measure the magnitude of relevant quantities:

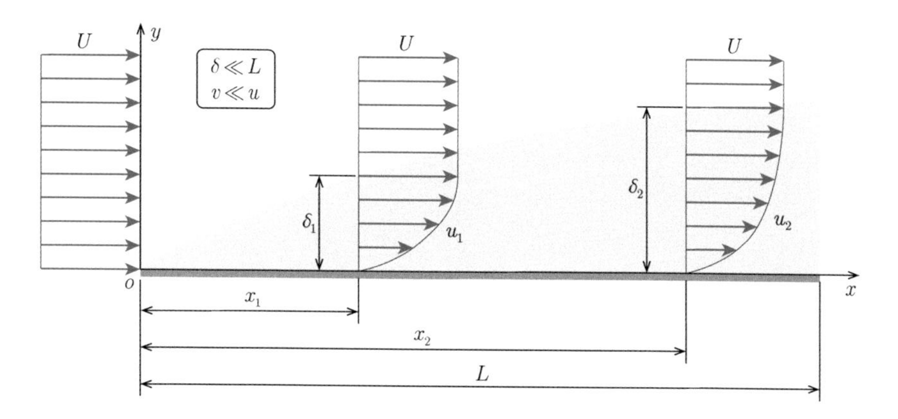

Figure 6.4 Flow in the boundary layer over a flat plate.

$$u \sim U,\ \ x \sim L,\ \ y \sim \delta .$$

Substituting the above reference quantities into the continuity equation,

$$\frac{U}{L} + \frac{v}{\delta} \sim 0 .$$

Subsequently, the magnitude of the normal velocity component v can be measured as:

$$v \sim \frac{\delta}{L} U .$$

Since $\delta \ll L$, the relation $v \ll u$ can be obtained from the above equation. In other words, the normal component of the fluid velocity in the boundary layer is negligible compared to its streamwise component. Therefore, the flow in the boundary layer may be regarded as being parallel to the plate.

The terms in Equation (6.3) are expressed by the reference quantities that measure their magnitudes:

$$u\frac{\partial u}{\partial x} \longrightarrow U\frac{U}{L} = \frac{U^2}{L},$$

$$v\frac{\partial u}{\partial y} \longrightarrow \frac{\delta}{L}U \cdot \frac{U}{\delta} = \frac{U^2}{L},$$

$$u\frac{\partial v}{\partial x} \longrightarrow U\frac{\frac{\delta}{L}U}{L} = \frac{\delta}{L} \cdot \frac{U^2}{L},$$

$$v\frac{\partial v}{\partial y} \longrightarrow \frac{\delta}{L}U \cdot \frac{\frac{\delta}{L}U}{\delta} = \frac{\delta}{L} \cdot \frac{U^2}{L}.$$

$$-\frac{1}{\rho}\frac{\partial p}{\partial x} \text{(this term is generally a known quantity)},$$

$$-\frac{1}{\rho}\frac{\partial p}{\partial y}(\text{this term will be discussed later}),$$

$$\frac{\mu}{\rho}\left(\frac{\partial^2 u}{\partial x^2}\right) \longrightarrow \frac{\mu}{\rho}\left(\frac{\partial^2 u}{\partial x^2}\right)=\frac{\mu}{\rho UL}UL\left(\frac{\partial^2 u}{\partial x^2}\right)=\frac{1}{Re}UL\left(\frac{\partial^2 u}{\partial x^2}\right),$$
$$\longrightarrow \frac{1}{Re}UL\frac{U}{L^2}=\frac{1}{Re}\cdot\frac{U^2}{L}$$

$$\frac{\mu}{\rho}\left(\frac{\partial^2 u}{\partial y^2}\right) \longrightarrow \frac{\mu}{\rho}\left(\frac{\partial^2 u}{\partial y^2}\right)=\frac{\mu}{\rho UL}UL\left(\frac{\partial^2 u}{\partial y^2}\right)=\frac{1}{Re}UL\left(\frac{\partial^2 u}{\partial y^2}\right),$$
$$\longrightarrow \frac{1}{Re}UL\frac{U}{\delta^2}=\frac{1}{Re}\cdot\frac{L^2}{\delta^2}\cdot\frac{U^2}{L}$$

$$\frac{\mu}{\rho}\left(\frac{\partial^2 v}{\partial x^2}\right) \longrightarrow \frac{\mu}{\rho}\left(\frac{\partial^2 v}{\partial x^2}\right)=\frac{\mu}{\rho UL}UL\left(\frac{\partial^2 v}{\partial x^2}\right)=\frac{1}{Re}UL\left(\frac{\partial^2 v}{\partial x^2}\right),$$
$$\longrightarrow \frac{1}{Re}UL\frac{\frac{\delta}{L}U}{L^2}=\frac{1}{Re}\cdot\frac{\delta}{L}\cdot\frac{U^2}{L}$$

$$\frac{\mu}{\rho}\left(\frac{\partial^2 v}{\partial y^2}\right) \longrightarrow \frac{\mu}{\rho}\left(\frac{\partial^2 v}{\partial y^2}\right)=\frac{\mu}{\rho UL}UL\left(\frac{\partial^2 v}{\partial y^2}\right)=\frac{1}{Re}UL\left(\frac{\partial^2 v}{\partial y^2}\right).$$
$$\longrightarrow \frac{1}{Re}UL\frac{\frac{\delta}{L}U}{\delta^2}=\frac{1}{Re}\cdot\frac{L}{\delta}\cdot\frac{U^2}{L}$$

Previous qualitative analyses have concluded that the Reynolds number is high if the boundary layer is very thin. Now let us quantitatively estimate the magnitude of this number. In the boundary layer, the magnitudes of viscous and inertia forces near the wall should be of the same order to ensure that the flow is balanced. According to the dimensional analysis of the momentum equation in the x direction, the magnitudes of inertial and viscous forces can be expressed as:

$$\text{inertial force: } u\frac{\partial u}{\partial x}+v\frac{\partial v}{\partial y}\longrightarrow\frac{U^2}{L},$$

$$\text{viscous force: } \frac{\mu}{\rho}\left(\frac{\partial^2 u}{\partial y^2}\right)\longrightarrow\frac{1}{Re}\cdot\frac{L^2}{\delta^2}\cdot\frac{U^2}{L}.$$

Setting these two terms equal, we obtain:

$$\frac{\delta}{L}\sim\frac{1}{\sqrt{Re}}.$$

Now, if $\delta \ll L$, we must have $Re \gg 1$.

By substituting this formula into the above dimensional equations and eliminating the Reynolds number, we obtain the magnitude of each term and write it next to the corresponding terms in Equation (6.3):

$$\begin{cases} \overbrace{u\dfrac{\partial u}{\partial x}}^{\frac{U^2}{L}} + \overbrace{v\dfrac{\partial u}{\partial y}}^{\frac{U^2}{L}} = -\dfrac{1}{\rho}\dfrac{\partial p}{\partial x} + \overbrace{\dfrac{\mu}{\rho}\left(\dfrac{\partial^2 u}{\partial x^2}\right)}^{\frac{\delta^2}{L^2}\cdot\frac{U^2}{L}} + \overbrace{\dfrac{\mu}{\rho}\left(\dfrac{\partial^2 u}{\partial y^2}\right)}^{\frac{U^2}{L}} \\ \underbrace{u\dfrac{\partial v}{\partial x}}_{\frac{\delta}{L}\cdot\frac{U^2}{L}} + \underbrace{v\dfrac{\partial v}{\partial y}}_{\frac{\delta}{L}\cdot\frac{U^2}{L}} = -\dfrac{1}{\rho}\dfrac{\partial p}{\partial y} + \underbrace{\dfrac{\mu}{\rho}\left(\dfrac{\partial^2 v}{\partial x^2}\right)}_{\frac{\delta^3}{L^3}\cdot\frac{U^2}{L}} + \underbrace{\dfrac{\mu}{\rho}\left(\dfrac{\partial^2 v}{\partial y^2}\right)}_{\frac{\delta}{L}\cdot\frac{U^2}{L}} \end{cases}$$

If the magnitude of the inertia force in the x direction, U^2/L, is taken as the reference quantity, then any quantity multiplied by δ/L is negligible, and the original equation reduces to:

$$\begin{cases} u\dfrac{\partial u}{\partial x}+v\dfrac{\partial u}{\partial y}=-\dfrac{1}{\rho}\dfrac{\partial p}{\partial x}+\dfrac{\mu}{\rho}\left(\dfrac{\partial^2 u}{\partial y^2}\right) \\ \dfrac{\partial p}{\partial y}=0 \end{cases}.$$

In the boundary layer, the momentum equation in the y direction reduces to $\partial p/\partial y = 0$. In other words, *the pressure in the boundary layer remains constant along the normal direction*. So, the pressure depends only on the x coordinate, and the pressure gradient term in the momentum equation in the x direction can be written in ordinary differential form:

$$u\frac{\partial u}{\partial x}+v\frac{\partial u}{\partial y}=-\frac{1}{\rho}\frac{\mathrm{d}P}{\mathrm{d}x}+\frac{\mu}{\rho}\left(\frac{\partial^2 u}{\partial y^2}\right), \tag{6.4}$$

where P is the fluid pressure outside the boundary layer that might change in the streamwise direction.

Equation (6.4) is the boundary layer equation for steady, incompressible, laminar flow. First proposed by Prandtl, it is also called *Prandtl's boundary layer equation*. Since only $\delta \ll L$ is used in its derivation and it is not necessary to have a wall surface, this equation is also applicable to other thin shear layers, sometimes called two-dimensional laminar thin-shear-layer equations.

The fluid pressure in Equation (6.4) refers to the pressure outside the boundary layer where the inviscid flow satisfies the Eulerian equations. The change of pressure can be represented by that of velocity. Since the mainstream is regarded as a good approximation to a one-dimensional flow, the Eulerian equation for a one-dimensional flow with no body forces reduces to:

$$U\frac{\partial U}{\partial x}=-\frac{1}{\rho}\frac{\mathrm{d}P}{\mathrm{d}x}.$$

So, Equation (6.4) can be rearranged as:

$$u\frac{\partial u}{\partial x}+v\frac{\partial u}{\partial y}=U\frac{\partial U}{\partial x}+\frac{\mu}{\rho}\left(\frac{\partial^2 u}{\partial y^2}\right). \tag{6.4a}$$

Although the boundary layer equations are much simpler than the N-S equations, they still require a substantial computation effort due to the existence of a nonlinear term (inertial force terms). Just as the N-S equations, the boundary layer equations can be analytically solved only under limited and specific conditions. Heinrich Blasius (1883–1970), one of Prandtl's students, obtained a quasi-analytical solution for the laminar boundary layer over a flat plate in 1908. Blasius's expression, based on the above boundary layer equations, has been experimentally verified.

Although known as the exact solution for laminar boundary layer, the Blasius solution is only accurate for high Reynolds number flows. For low Reynolds number flows where viscous forces dominate, the boundary layer becomes so thick that the boundary layer equations are no longer valid, and the Blasius solution is not applicable. Even for low-viscosity, high-speed air or water flow, the boundary layer equations are no longer valid in the small region near the leading edge of the plate, where both $\delta \sim L$ and $Re \gg 1$ are not satisfied. In reality, the boundary layer thickness near the leading edge is not zero, but has a certain value at the front stagnation point of the plate surface.

6.2.2 Boundary Layer Thickness

For practical engineering problems, we can sometimes ignore the flow within the boundary layer and consider only its influence on the mainstream. Taking the wind tunnel experiments as an example, the boundary layer over the tunnel walls is not the object of study. However, the low-speed fluids inside the boundary layer occupy a certain space, so that the effective flow area of the wind tunnel is reduced, corresponding to a certain flow blockage. In a low-speed wind tunnel of constant cross-section, the airflow accelerates slightly along the central line. It is necessary to quantify the effects of airflow acceleration, or slightly enlarge the area of the test section along streamwise direction by considering the existence of a near-wall boundary layer in the prior design to compensate for the blockage.

For an experimental model in the wind tunnel, the consideration is different. The drag on the model may be an important test objective, which is directly related to the frictional forces between flow and model surface. We might need to study the flow inside the boundary layer, and calculate the skin-frictional drag acting on the model by integrating the product of the viscous shear stress and the local surface area at all points of the model.

Sometimes, we are more concerned with how much mechanical energy has been converted into internal energy (or how much mechanical energy has been lost) during a flow process. The strong shear flow inside the boundary layer results in mechanical energy losses whose magnitude depends directly on the magnitudes of shear force and shear rate (see the energy equation in Chapter 4) – that is to say, directly on the flow patterns inside the boundary layer. The blockage, drag, and mechanical energy losses mentioned above are the macroscopic effects generated by the boundary layer. Although essentially determined by the flow inside the boundary layer, these macroscopic effects can be measured by certain macroscopic parameters. Traditionally, three types of boundary layer thickness are introduced to aid in the solution of these problems: blockage corresponds to *displacement thickness*, drag corresponds to *momentum thickness*, and mechanical energy loss corresponds to *energy thickness*. Next, let us look at the definitions and meaning of these three boundary layer thicknesses.

Tip 6.1: Blasius solution

The Blasius problem deals with the laminar boundary layer flow over a flat plate. By using the similarity principle in mathematics, the boundary layer system of partial differential equations reduces to an ordinary differential equation. The basic method is the following:

The dimensionless variables of the original equation are defined as:

$$u^* = u/U, \quad \eta = y\Big/\sqrt{\frac{\mu x}{\rho U}}, \quad f = \int u^* \mathrm{d}\eta\,.$$

Then, the original problem becomes one with η as the independent variable and f as the dependent variable, expressed as:

$$\frac{\mathrm{d}^3 f}{\mathrm{d}\eta^3} + \frac{1}{2} f \frac{\mathrm{d}^2 f}{\mathrm{d}\eta^2} = 0 \qquad \begin{cases} \text{at wall } (\eta = 0): & f = 0,\ \mathrm{d}f/\mathrm{d}\eta = 0 \\ \text{at infinity } (\eta = \infty): & \mathrm{d}f/\mathrm{d}\eta = 1 \end{cases}.$$

The Blasius equation is a well-known third-order nonlinear ordinary differential equation. Blasius solved it by using a series-expansion method:

$$f = \sum_{n=0}^{\infty} \left(-\frac{1}{2}\right)^n \frac{0.332^{n+1} C_n}{(3n+2)!} \eta^{(3n+2)}, \quad C_0 = 1,\ C_1 = 1,\ C_2 = 11,\ C_3 = 375,\ \ldots$$

At present, researchers prefer to use numerical methods to solve the Blasius equation. Figure T6.1 shows the comparison between the Blasius solution and the experimental data given by Liepmann.

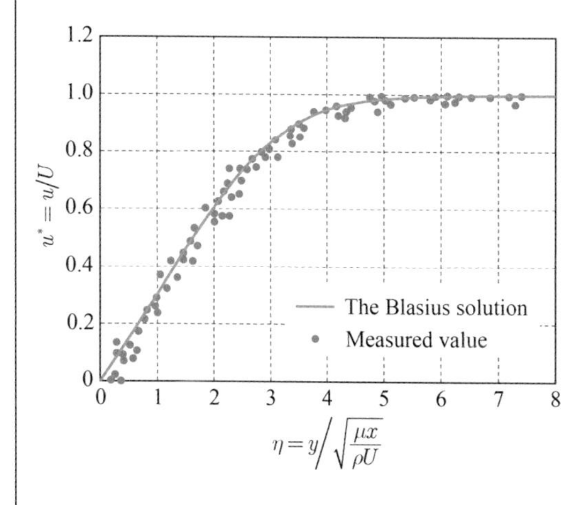

Figure T6.1.

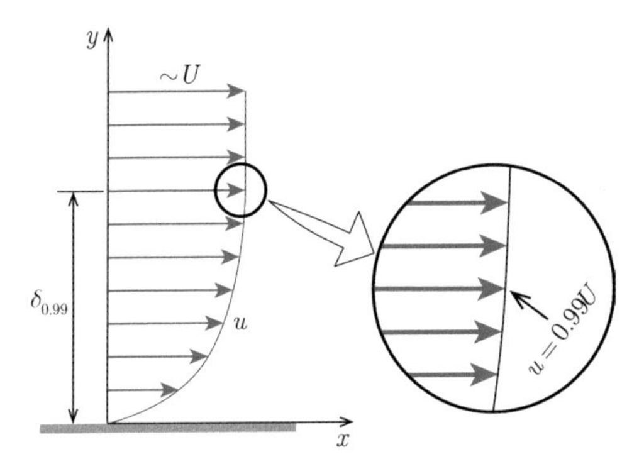

Figure 6.5 Nominal thickness of the boundary layer.

The fluid velocity at a stationary wall is zero, and the velocity at the outer edge of the boundary layer is equal to the mainstream velocity. Moreover, the velocity gradient at the outer edge of the boundary layer should be zero. Therefore, the boundary layer satisfies the following boundary conditions:

At the wall: $u = 0, \quad \partial u/\partial y = \tau_{\mathrm{w}}/\alpha.$

At the outer edge of the boundary layer: $u = U, \quad \partial u/\partial y = 0.$

Based on these boundary conditions, we can assume an approximate velocity profile inside the boundary layer. As shown in Figure 6.5, the streamwise velocity component varies greatly with y in the normal direction, from zero at the wall surface to nearly the mainstream value across the relatively thin boundary layer. The velocity gradient reaches its maximum value at the wall surface, and decreases with the distance away from it. Accordingly, the viscous effects follow the same trend.

In theory, the viscous force will not equal zero until the distance approaches infinity, hence the boundary layer does not have a definite outer edge, and it is problematic to define just what the thickness is. The outer edge of the boundary layer is based on the fact that, beyond this boundary, the effect of wall viscous stresses can be neglected. The commonly used engineering *nominal thickness* of the boundary layer, δ or $\delta_{0.99}$, is defined as the thickness of the zone extending from the wall surface to a point where the fluid velocity is 99% of the freestream velocity and the local velocity gradient is negligible, as shown in Figure 6.5. The physical quantity defined in this way is not strict, because the velocity change takes place gradually. Of course, other definitions of boundary layer thickness can be used according to the specific type of problem. For example, the outer edge of the boundary layer, $\delta_{0.95}$, can also be defined as the distance from the wall surface to the point where the fluid velocity is 95% of the freestream velocity.

The nominal thickness of the boundary layer does not have much practical significance, and some other types of boundary layer thickness are more meaningful in engineering. Now let us look at the *displacement thickness* of the boundary layer. As shown in Figure 6.6, the near-wall flow decelerates due to viscosity and causes a reduction in flow rate; subsequently, the fluid will be forced upwards. The mainstream

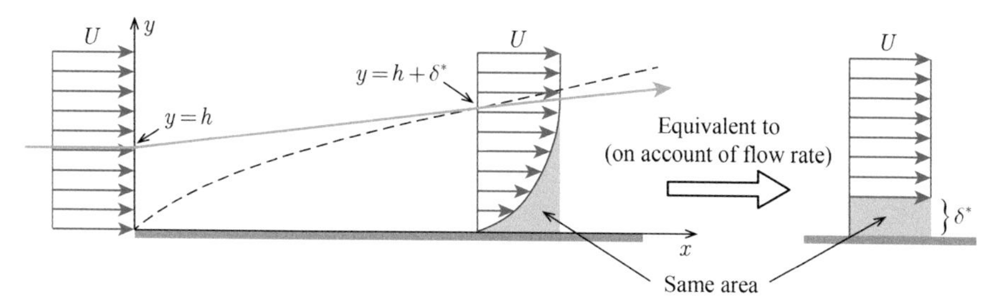

Figure 6.6 Derivation of displacement thickness.

above the boundary layer will be slightly displaced away from the wall by a height, which is defined as the displacement thickness of the boundary layer and generally expressed by δ^*. Now let's derive the formula for the displacement thickness in detail.

The total flow rate inside the boundary layer is

$$\dot{m}_{\text{real}} = \int_0^{\delta} \rho u b \mathrm{d}y,$$

where b is the crosswise width of the plate (the scale perpendicular to the page).

For an incompressible flow, the above equation can be rewritten as

$$\dot{m}_{\text{real}} = \rho b \int_0^{\delta} u \mathrm{d}y.$$

If the fluid inside the boundary layer does not decelerate, the mass flow rate passing through the same height should be

$$\dot{m}_{\text{ideal}} = \rho U b \delta.$$

The reduction in flow rate on account of boundary layer is expressed as

$$\Delta \dot{m} = \dot{m}_{\text{ideal}} - \dot{m}_{\text{real}} = \rho U b \delta - \rho b \int_0^{\delta} u \mathrm{d}y = \rho b \int_0^{\delta} (U - u) \mathrm{d}y. \tag{6.5}$$

The distance by which the external flow is displaced outwards is δ^*, which is caused by the flow rate reduction inside the boundary layer. The flow with velocity of U is displaced a distance δ^* to compensate this flow rate reduction, that is

$$\Delta \dot{m} = \rho U b \delta^*. \tag{6.6}$$

By combining Equations (6.5) and (6.6), the displacement thickness can be derived as

$$\delta^* = \int_0^{\delta} \left(1 - \frac{u}{U}\right) \mathrm{d}y. \tag{6.7}$$

The displacement thickness seems to depend on both the velocity profile inside the boundary layer and the boundary layer thickness. In fact, the integral term of Equation (6.7) is zero outside the boundary layer where $u = U$. Therefore, the upper limit of integration needs only to be greater than the boundary layer thickness δ, but independent of δ itself. That is to say, the displacement thickness depends solely on the velocity profile inside the boundary layer, and its rigorous definition is expressed as:

$$\delta^* = \int_0^{\infty} \left(1 - \frac{u}{U}\right) \mathrm{d}y. \tag{6.8}$$

We define the momentum thickness and energy thickness in a similar way. The corresponding derivations are not carried out here, and their expressions for incompressible flow are directly given by:

$$\theta = \int_0^\infty \frac{u}{U}\left(1 - \frac{u}{U}\right)\mathrm{d}y \tag{6.9}$$

$$\delta_3 = \int_0^\infty \frac{u}{U}\left(1 - \frac{u^2}{U^2}\right)\mathrm{d}y \tag{6.10}$$

The displacement thickness δ^* represents the outward displacement of the streamlines of the free flow caused by the boundary layer. The momentum thickness θ is the distance by which the boundary should be displaced to compensate for the deficit in momentum of the free flow caused by the boundary layer. The energy thickness δ_3 is the distance by which the boundary should be displaced to compensate for the deficit in kinetic energy of the free flow caused by the boundary layer.

An additional parameter, known as *shape factor* and usually denoted by H, helps describe the shape of the boundary layer velocity profile. It is defined as the ratio of displacement thickness to momentum thickness:

$$H = \frac{\delta^*}{\theta}. \tag{6.11}$$

All these parameters depend only on the velocity profile inside the boundary layer. We can simply evaluate their magnitudes by assuming that the velocity profile inside the laminar boundary layer satisfies the quadratic curve:

$$\frac{u}{U} = a + b\left(\frac{y}{\delta}\right) + c\left(\frac{y}{\delta}\right)^2.$$

The values of the coefficients are determined by the boundary conditions at the wall and outer edge of the boundary layer:

At the wall $(y=0)$: $u = 0$.

At the outer edge of the boundary layer $(y=\delta)$: $u = U,\ \ \partial u/\partial y = 0$.

Then, the velocity profile inside the boundary layer is given by:

$$\frac{u}{U} = 2\left(\frac{y}{\delta}\right) - \left(\frac{y}{\delta}\right)^2, \quad 0 \le y \le \delta.$$

Substituting the above formula into Equations (6.8), (6.9) and (6.11), and setting the upper limit to be δ, we have:

$$\delta^* = \frac{1}{3}\delta, \quad \theta = \frac{1}{7.5}\delta, \quad H = 2.5.$$

Figure 6.7 illustrates a comparison of the thicknesses of the three types of boundary layers.

6.2.3 Integral Approach for Solving Boundary Layer Problems

In engineering practice, plenty of attention has been paid to the effects of boundary layer on the overall flow field, such as blockage, drag, mechanical energy loss, and so

Figure 6.7 Comparison of three boundary layer thicknesses.

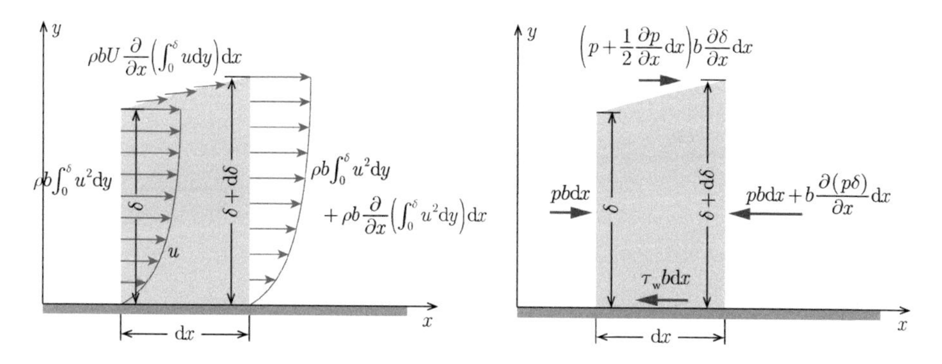

Figure 6.8 Force and momentum analysis for a control volume inside the boundary layer.

on. For general solutions, the integral approach is sufficient and easy to implement. In 1921, von Karman (1881–1963) derived the momentum integral equation for a two-dimensional boundary layer, which can be obtained by integrating the boundary layer equation in the normal direction, or by performing a control volume analysis on the boundary layer. The second approach is more useful for understanding the flow characteristics and it is introduced below.

As shown in Figure 6.8, the crosswise width of the wall is b and the flow is assumed to be incompressible. A small element with a length of $\mathrm{d}x$ inside the boundary layer is chosen as the control volume. Its lower and upper surfaces are the wall surface and outer edge of the boundary layer, respectively. Based on the control volume analysis, we can obtain the boundary layer momentum integral equation.

6.2.3.1 Momentum into and out of the Control Volume

The mass flow rate into the control volume across the left control surface is expressed as:

$$\dot{m}_{\text{left}} = \int_0^\delta \rho u b \mathrm{d}y = \rho b \int_0^\delta u \mathrm{d}y \,.$$

The mass flow rate out of the control volume across the right control surface is:

$$\dot{m}_{\text{right}} = \rho b\left[\int_0^\delta u\mathrm{d}y + \frac{\partial}{\partial x}\left(\int_0^\delta u\mathrm{d}y\right)\mathrm{d}x\right].$$

The net mass flow rate out of the control volume across the two control surfaces is expressed as:

$$\delta \dot{m} = \dot{m}_{\text{right}} - \dot{m}_{\text{left}} = \rho b \frac{\partial}{\partial x}\left(\int_0^\delta u \mathrm{d}y\right)\mathrm{d}x \, .$$

By the principle of conservation of mass, $\delta \dot{m}$ must be equal to the mass flow rate into the control volume across the upper control surface.

The momentum flow rate into the control volume across the left control surface is:

$$\dot{M}_{\text{left}} = \rho b \int_0^\delta u^2 \mathrm{d}y \, .$$

The momentum flow rate out of the control volume across the right control surface is expressed as:

$$\dot{M}_{\text{right}} = \rho b \left[\int_0^\delta u^2 \mathrm{d}y + \frac{\partial}{\partial x}\left(\int_0^\delta u^2 \mathrm{d}y\right)\mathrm{d}x\right].$$

The velocity of the fluid flowing into the control volume across the upper control surface is equal to the mainstream velocity U. The mass flow rate into the control volume across the upper control surface is already obtained. Therefore, the x component of momentum flow rate into the control volume across the upper control surface is expressed as:

$$\dot{M}_{\text{upper}} = \delta \dot{m} \cdot U = \rho b U \frac{\partial}{\partial x}\left(\int_0^\delta u \mathrm{d}y\right)\mathrm{d}x \, .$$

The net momentum flow rate out of the control volume is:

$$\begin{aligned} \Delta \dot{M} &= \dot{M}_{\text{right}} - \dot{M}_{\text{left}} - \dot{M}_{\text{upper}} \\ &= \rho b \left[\int_0^\delta u^2 \mathrm{d}y + \frac{\partial}{\partial x}\left(\int_0^\delta u^2 \mathrm{d}y\right)\mathrm{d}x\right] - \rho b \int_0^\delta u^2 \mathrm{d}y - \rho b U \frac{\partial}{\partial x}\left(\int_0^\delta u \mathrm{d}y\right)\mathrm{d}x \\ &= \rho b \left[\frac{\partial}{\partial x}\left(\int_0^\delta u^2 \mathrm{d}y\right)\mathrm{d}x - U \frac{\partial}{\partial x}\left(\int_0^\delta u \mathrm{d}y\right)\mathrm{d}x\right]. \end{aligned} \tag{6.12}$$

6.2.3.2 Forces Acting on the Control Volume

We now analyze the forces on the control volume. Neglecting the body forces and the shear force at the outer edge, only surface forces need to be considered. The surface forces acting on the control volume include: the pressure forces acting inward on the upper, left, and right control surfaces, and the shear force applied by the wall.

Since the pressure remains unchanged across the boundary layer in the y direction, the force acting on the left control surface is equal to the product of the pressure times the surface area:

$$F_{\text{left}} = p b \delta \, .$$

The forces acting on the right control surface are toward the left, and can be expressed as the Taylor expansion of the force acting on the left control surface:

$$F_{\text{right}} = -\left(p b \delta + b \frac{\partial (p\delta)}{\partial x}\mathrm{d}x\right).$$

The pressure on the upper control surface varies in the x direction, and its average value can be expressed as:

$$p_{\text{upper}} = p + \frac{1}{2}\frac{\partial p}{\partial x}\mathrm{d}x \, .$$

So, the x component of the force acting on the upper control surface is:

$$F_{\text{upper}} = \left(p + \frac{1}{2}\frac{\partial p}{\partial x}\mathrm{d}x\right)b\frac{\partial \delta}{\partial x}\mathrm{d}x\,.$$

Neglecting the small quantity of the second order, the above formula reduces to:

$$F_{\text{upper}} = pb\frac{\partial \delta}{\partial x}\mathrm{d}x\,.$$

The force acting on the lower control surface is equal to the shear force acted by the wall:

$$F_{\text{lower}} = -\tau_{\text{w}} b\mathrm{d}x\,.$$

Thus, the x component of the resultant force acting on the control volume is expressed as:

$$\begin{aligned}\sum F &= F_{\text{left}} + F_{\text{right}} + F_{\text{upper}} + F_{\text{lower}}\\ &= pb\delta - pb\delta - \frac{b\partial(p\delta)}{\partial x}\mathrm{d}x + pb\frac{\partial \delta}{\partial x}\mathrm{d}x - \tau_{\text{w}} b\mathrm{d}x\\ &= -\frac{\partial p}{\partial x}b\delta\mathrm{d}x - \tau_{\text{w}} b\mathrm{d}x.\end{aligned} \tag{6.13}$$

Substituting Equations (6.12) and (6.13) into the momentum equation:

$$-\frac{\partial p}{\partial x}b\delta\mathrm{d}x - \tau_{\text{w}} b\mathrm{d}x = \rho b\left[\frac{\partial}{\partial x}\left(\int_0^\delta u^2\mathrm{d}y\right)\mathrm{d}x - U\frac{\partial}{\partial x}\left(\int_0^\delta u\mathrm{d}y\right)\mathrm{d}x\right].$$

Rearranged, this is:

$$\frac{\partial}{\partial x}\left(\int_0^\delta u^2\mathrm{d}y\right) - U\frac{\partial}{\partial x}\left(\int_0^\delta u\mathrm{d}y\right) = -\frac{\delta}{\rho}\frac{\partial p}{\partial x} - \frac{\tau_{\text{w}}}{\rho}. \tag{6.14}$$

This is the integral relationship equation for the boundary layer.

It is troublesome to use the above formula directly, and it is simpler to express it by using the displacement and momentum thickness. The derivation is given below.

The streamwise pressure gradient inside the boundary layer is equal to that outside the boundary layer. The external flow outside the boundary layer satisfies the one-dimensional Eulerian equation:

$$\frac{\partial p}{\partial x} = -\rho U\frac{\mathrm{d}U}{\mathrm{d}x}.$$

The first term on the right-hand side of Equation (6.14) can be rewritten as:

$$-\frac{\delta}{\rho}\frac{\partial p}{\partial x} = U\frac{\mathrm{d}U}{\mathrm{d}x}\delta = U\frac{\mathrm{d}U}{\mathrm{d}x}\int_0^\delta \mathrm{d}y. \tag{6.15}$$

By partial integration, the second term on the left-hand side of Equation (6.14) can be rewritten as:

$$-U\frac{\mathrm{d}}{\mathrm{d}x}\left(\int_0^\delta u\mathrm{d}y\right) = -\frac{\mathrm{d}}{\mathrm{d}x}\left(\int_0^\delta Uu\mathrm{d}y\right) + \int_0^\delta \frac{\mathrm{d}U}{\mathrm{d}x}u\mathrm{d}y \tag{6.16}$$

Substituting Equations (6.15) and (6.16) into Equation (6.14):

$$\frac{\mathrm{d}U}{\mathrm{d}x}\int_0^\delta (U-u)\mathrm{d}y + \frac{\mathrm{d}}{\mathrm{d}x}\int_0^\delta u(U-u)\mathrm{d}y = \frac{\tau_{\text{w}}}{\rho}.$$

By applying the definitions of displacement thickness and momentum thickness, the above equation can be rewritten as:

$$U\frac{\mathrm{d}U}{\mathrm{d}x}\delta^* + \frac{\mathrm{d}}{\mathrm{d}x}\left(U^2\theta\right) = \frac{\tau_\mathrm{w}}{\rho}.$$

This equation, rewritten in a more general form, becomes:

$$\frac{\mathrm{d}\theta}{\mathrm{d}x} + (2+H)\frac{\theta}{U}\frac{\mathrm{dU}}{\mathrm{d}x} = \frac{\tau_\mathrm{w}}{\rho U^2}. \tag{6.17}$$

The shear force term in the formula is sometimes replaced by the skin friction coefficient, which is defined as the ratio of shear stress at the plate to the dynamic pressure of the freestream:

$$C_\mathrm{f} = \frac{\tau_\mathrm{w}}{\rho U^2/2}.$$

Thus, Equation (6.17) can be rearranged as:

$$\frac{\mathrm{d}\theta}{\mathrm{d}x} + (2+H)\frac{\theta}{U}\frac{\mathrm{dU}}{\mathrm{d}x} = \frac{C_\mathrm{f}}{2}. \tag{6.17a}$$

Equations (6.17) and (6.17a), called *von Karman boundary layer momentum integral equations*, were obtained by von Karman in 1921. Subsequently, an iteration method has been proposed to solve a general flow with boundary layer, and the calculation steps for performing this kind of analysis are summarized below:

(1) The flow is assumed to be inviscid, and the theory of potential flow is used to calculate the velocity along the wall, $U(x)$. We can then obtain the velocity gradient in the x direction, $\mathrm{d}U/\mathrm{d}x$.
(2) We assume an approximate velocity profile inside the boundary layer, $u = f(y)$, and substitute it into the Newtonian viscosity model for the wall shear stress, $\tau_\mathrm{w} = \mu(\partial u/\partial y)$, to solve Equation (6.17).
(3) The obtained displacement thickness is used to correct the wall shape, and $U(x)$ is solved by the theory of potential flow again.
(4) Repeat the above calculation steps several times until the desired accuracy is reached.

For the boundary layer over a flat plate, we have $\mathrm{d}U/\mathrm{d}x = 0$. Equation (6.17) is then simplified as:

$$\frac{\mathrm{d}\theta}{\mathrm{d}x} = \frac{\tau_\mathrm{w}}{\rho U^2}. \tag{6.18}$$

Then, we need only the additional relationship $u = f(y)$ to solve the above equation. Von Karman assumed that the velocity profile inside the laminar boundary layer over a flat plate was a quadratic curve, and the approximate solution was very close to the Blasius solution. Figure 6.9 compares the quadratic velocity profile and the Blasius solution.

We can see that the quadratic velocity profile for the laminar boundary layer over a flat plate is close to the Blasius solution. The streamwise change in the boundary layer thickness is expressed as:

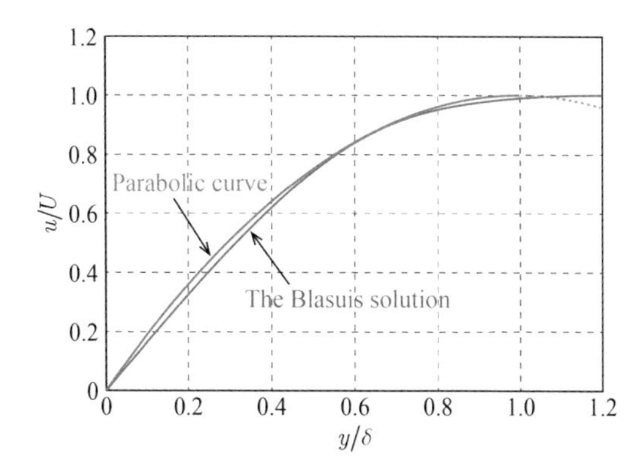

Figure 6.9 A comparison of quadratic velocity profile and the Blasius solution.

$$\text{Quadratic curve integral solution: } \frac{\delta}{x} = \frac{5.5}{\sqrt{Re_x}}, \quad \frac{\delta^*}{x} = \frac{1.83}{\sqrt{Re_x}}, \quad \frac{\theta}{x} = \frac{0.74}{\sqrt{Re_x}}.$$

$$\text{Blasius solution: } \frac{\delta}{x} = \frac{5.0}{\sqrt{Re_x}}, \quad \frac{\delta^*}{x} = \frac{1.72}{\sqrt{Re_x}}, \quad \frac{\theta}{x} = \frac{0.66}{\sqrt{Re_x}}.$$

Since the Blasius solution is almost an accurate solution, the above-mentioned integral approach is apparently not useful for a flat plate. However, this approach is helpful for some other boundary layer problems, such as on the surface of an airfoil, where the Blasius solution is not applicable.

6.3 Turbulent Boundary Layer

Turbulent boundary layers are very different from laminar ones. As shown in Figure 6.10, the turbulent boundary layer is characterized by chaotic and irregular velocity fluctuations, and there is almost no general law for the velocity profile. Moreover, the outer edge of the boundary layer does not remain steady, so the boundary layer thickness at the same streamwise position may vary significantly with time. The flow through a point at a distance from the wall may be either turbulent or laminar at different instants of time. The alternation of turbulent and laminar flows is called the intermittent motion of the turbulent boundary layer, and the *intermittent factor* can be defined according to the proportion of time occupied by turbulent and laminar flows.

$$\gamma = t_{\text{turb}} / (t_{\text{lam}} + t_{\text{turb}}).$$

The flow near the wall is always turbulent, or $\gamma = 1$, and the flow far enough away from the wall is always laminar, or $\gamma = 0$. For the turbulent boundary layer over a flat plate, the distribution of intermittent factor normal to the wall is roughly shown in Figure 6.11. The flow below $0.4\gamma$is always turbulent, or $\gamma = 1$, and the flow above 1.2δ is always laminar, or $\gamma = 0$.

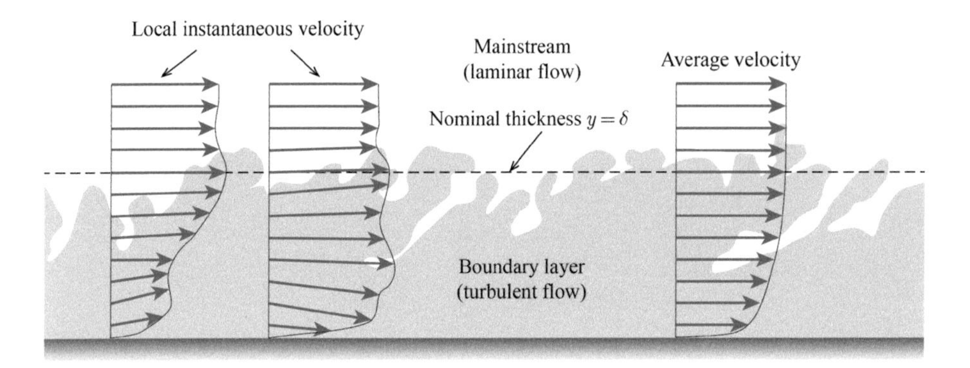

Figure 6.10 A schematic diagram of the velocity profile inside the turbulent boundary layer.

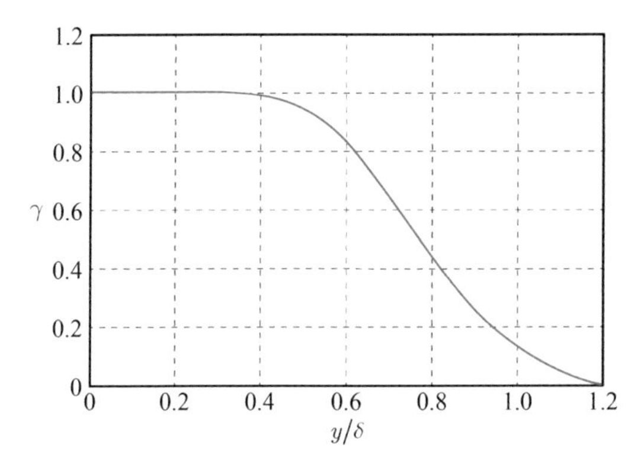

Figure 6.11 Intermittent factor of a typical turbulent boundary layer over a flat plate.

Since the turbulent boundary layer has no definite thickness δ, where do the 0.4δ and 1.2δ presented here come from? Actually, the δ is an averaged value. Averaging is an important method in turbulence theory.

Turbulent flow is characterized by irregular and unsteady movement of fluid particles. However, if the instantaneous velocity of the fluid at each point is averaged over a sufficiently long period of time, the average flow field can be found to have a regular pattern. The flow properties in the turbulent boundary layer defined by the average velocity have characteristics similar to that in the laminar boundary layer. The turbulent boundary layer thickness also increases along the streamwise direction. At the wall, the fluid velocity is assumed to be zero, and the velocity gradient normal to the wall reaches its maximum value at the wall. Figure 6.12 compares the velocity profiles of laminar boundary layers and time-averaged velocity profile of turbulent boundary layers over a flat plate. Under the same boundary conditions, the thickness of a turbulent boundary layer is larger than that of a laminar boundary layer. The two types of boundary layer thickness are converted into their dimensionless forms to compare the difference in the shape of the velocity profiles. Notice that the time-averaged velocity profile for turbulent boundary layer is "fuller" than the laminar one.

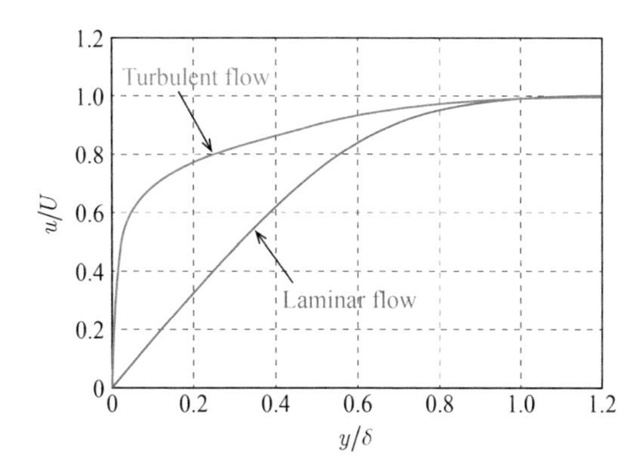

Figure 6.12 Velocity profiles of laminar and turbulent boundary layers over a flat plate.

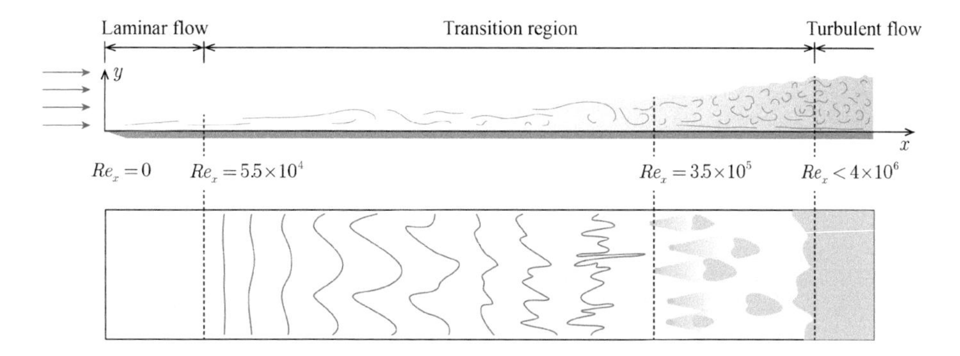

Figure 6.13 A complete schematic diagram of the boundary layer over a flat plate (laminar flow region–transition region–turbulent flow region).

For laminar boundary layers over a flat plate, the shape factor is about 2.5, while for turbulent boundary layers this factor is about 1.3.

Turbulent flow is supposed to occur at high Reynolds numbers. For the boundary layer, the mainstream velocity can be used as the characteristic velocity. There are two common types of characteristic length scales: one is the distance from the leading edge of the plate to the current section, and the other is the boundary layer thickness at the current streamwise station. No matter which scale is selected, the Reynolds number at the leading edge should be equal to zero in theory, and it gradually increases along the streamwise direction. Therefore, *as the fluid flows over a long flat plate, the boundary layer will always be laminar at the entrance region, and change from laminar to turbulent (known as laminar-to-turbulent transition) at a critical distance where the flow cannot remain stable*. The entire boundary layer over a flat plate is illustrated in Figure 6.13.

Just downstream of the leading edge of the plate, the Reynolds number is very low, and the boundary layer flow is dominated by viscous forces. Therefore, there will be a low-velocity layer which grows sharply in thickness. As the Reynolds number increases, the viscous effects decrease. Then, the characteristics of the laminar boundary layer conforms to Prandtl's statement that the boundary layer thickness is far less than the streamwise distance. When the Reynolds number at some location x reaches a certain number (usually in the order of $Re_x \sim 5.5 \times 10^4$), unstable fluctuations start to occur inside the boundary layer, which are generated by external disturbances, but the fluid can still remain laminar. When the Reynolds number downstream at x reaches a certain number, usually on the order that exceeds $Re_x \sim 3.5 \times 10^5$, there is not enough viscous damping to remove those small disturbances in the flow. More small-scale vortices are generated, and turbulent spots appear locally and sporadically. Further downstream, the transition from laminar to fully turbulent flow occurs when the Reynolds number at x exceeds $Re_x \sim 4.0 \times 10^6$. The laminar-to-turbulent transition in the flat plate boundary layer is a long process, and the length of the transition zone may reach ten or even tens of times the length of laminar flow.

The turbulent boundary layer flow is strongly unsteady. Although the average velocity profile is quite regular, the velocity fluctuations still have an important influence on the averaged flow. Subsequently, the averaged turbulent boundary layer equation is different from the laminar one, with the terms of fluctuating quantities in the equation.

Applying Newton's law of viscosity: $\tau = \mu(\partial u/\partial y)$, the boundary layer equation (6.4a) can be rewritten as:

$$u\frac{\partial u}{\partial x} + v\frac{\partial u}{\partial y} = U\frac{\partial U}{\partial x} + \frac{1}{\rho}\frac{\partial \tau}{\partial y}.$$

This equation is applicable to both laminar and averaged turbulent boundary layers. For the turbulent boundary layer equation, the velocity and shear force terms are substituted by time-averaged values, expressed as:

$$\bar{u}\frac{\partial \bar{u}}{\partial x} + \bar{v}\frac{\partial \bar{u}}{\partial y} = U\frac{\partial U}{\partial x} + \frac{1}{\rho}\frac{\partial \bar{\tau}}{\partial y}. \tag{6.19}$$

The averaged shear stress is not equal to the gradient of the averaged velocity:

$$\bar{\tau} \neq \mu\frac{\partial \bar{u}}{\partial y}.$$

It can be proved that the average shear stress for turbulent flows is:

$$\bar{\tau} = \mu\frac{\partial \bar{u}}{\partial y} - \rho\overline{u'v'}, \tag{6.20}$$

where u' and v' represent the streamwise and normal components of fluctuating velocity (i.e., the difference between instantaneous velocity and average velocity), respectively. First proposed by Reynolds, this decomposition, called *Reynolds decomposition* or *Reynolds averaging*, has become the leading method to simulate turbulent flows. The instantaneous velocity components in all three coordinate directions are decomposed into average and fluctuating velocity components:

$$u = \bar{u} + u', \quad v = \bar{v} + v', \quad w = \bar{w} + w'. \tag{6.21}$$

For the boundary layer equation (6.19) expressed by average quantities, the inertia force term is exactly the same as that expressed by instantaneous velocities. However, the shear stress term satisfies Equation (6.20). It must be emphasized here that for both laminar and turbulent flows, the viscous shear stress at any moment satisfies the constitutive equation (which reduces to Newton's law of viscosity, $\tau = \mu(\partial u/\partial y)$, for

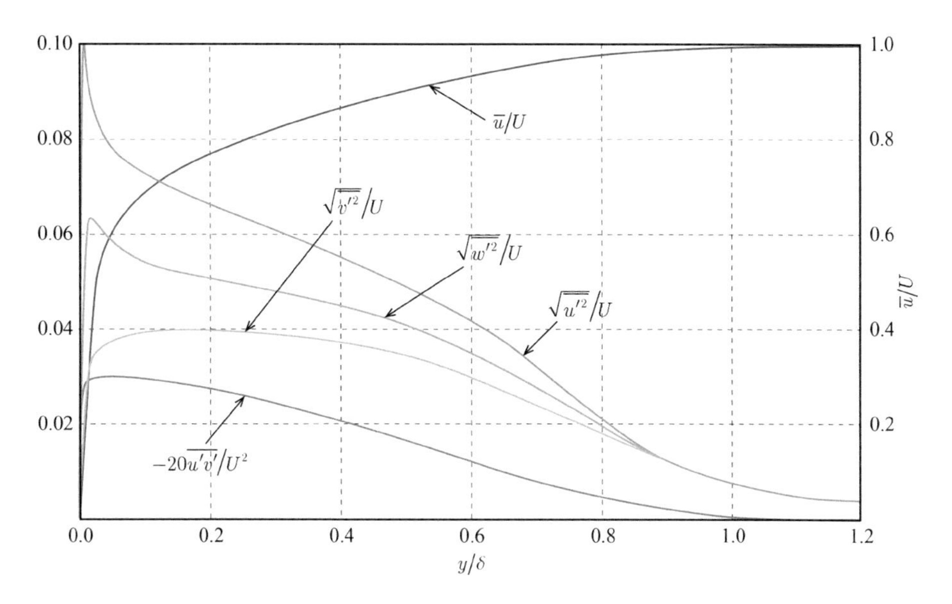

Figure 6.14 Average and fluctuating velocity profiles of the turbulent boundary layer over a flat plate.

parallel flows). The difference between turbulent shear stress and laminar shear stress is caused by the time-averaging of governing equations. The shear stress for turbulent flow defined by average velocities is not equal to its average shear stress. The time-averaged equation induces an extra shear stress term, $-\rho\overline{u'v'}$, known as *Reynolds stress*. In other words, *the average shear stress for turbulent flow is decomposed into two components: the component related to average velocity gradient is called laminar viscous force (or molecular viscous force), and the other component, related to fluctuating velocity, is called turbulent viscous force (or eddy viscous force)*. Equation (6.20) can be further interpreted as:

$$\overline{\tau} = \tau_{\text{lam}} + \tau_{\text{turb}}, \quad \tau_{\text{lam}} = \mu\frac{\partial \overline{u}}{\partial y}, \quad \tau_{\text{turb}} = -\rho\overline{u'v'}. \tag{6.22}$$

In the turbulent boundary layer, the average velocity profile has a very regular pattern, so does the magnitude of the fluctuating velocity. The Reynolds stress represents a measure of fluctuating velocity. Traditionally, the root mean square of the turbulent velocity fluctuation is used to represent the magnitudes of the fluctuating velocities in the three coordinate directions. Figure 6.14 shows the experimentally measured average and fluctuating velocity profiles inside the turbulent boundary layer over a flat plate. The average velocity at the wall is zero, and it rapidly increases in the direction normal to the wall. At a distance of 0.2δ from the wall, the average velocity reaches 0.8 times the mainstream velocity. The fluctuating velocity components in all three coordinate directions (streamwise direction: $\sqrt{\overline{u'^2}}/U$; normal direction: $\sqrt{\overline{v'^2}}/U$; spanwise direction: $\sqrt{\overline{w'^2}}/U$) are different in magnitude, but their changing laws are similar. The streamwise fluctuating velocity component is the largest, followed by the spanwise fluctuating velocity component; the normal fluctuating velocity component is the smallest, since the wall offers the largest constraints in the normal direction. All three fluctuating velocity components tend to zero outside 1.2δ. This is consistent with the previous conclusion that the intermittent factor tends to zero outside 1.2δ. Turbulent flow is an unstable phenomenon generated by shear. The stronger the shear, the larger the fluctuating velocity component. Therefore, the fluctuating velocity components inside the boundary layer are found to change in magnitude as a function of distance to wall, being larger where it is closer to the wall. However, all three fluctuating velocity components adjacent to the wall tend to zero due to the no-penetration and no-slip boundary conditions. Some studies have claimed that there is a near-wall laminar sublayer inside the turbulent boundary layer.

As an important parameter for the boundary layer, the Reynolds stress represents the magnitude of turbulent viscous forces. Figure 6.14 also shows that the Reynolds stress, $-\rho\overline{u'v'}$, remains almost unchanged within the range of $0 \sim 0.2\delta$, gradually diminishes outside 0.2δ, and at 1.2δ eventually vanishes.

At the wall, the Reynolds stress ($\tau_{\text{turb}} = -\rho\overline{u'v'}$) is assumed to be zero, while the laminar viscous force ($\tau_{\text{lam}} = \mu(\partial\overline{u}/\partial y)$) is dominant due to the very large velocity gradient. Further away from the wall, the laminar viscous force decreases sharply, while the turbulent viscous force rapidly becomes dominant. Except for the thin layer adjacent to the wall, inside a turbulent boundary layer the shear stress is mostly provided by the Reynolds stress. Inside the near-wall region within the range of $0 \sim 0.2\delta$, the total shear stress $\overline{\tau} = \tau_{\text{lam}} + \tau_{\text{turb}}$ remains virtually unchanged. The turbulent boundary layer therefore has a two-layer structure. Inside the near-wall sublayer, the Reynolds stress remains unchanged, implying that there is a dynamic equilibrium between the turbulent fluctuations generated by wall shear stress and the dissipation of turbulent energy to internal energy. Inside the outer sublayer, where the viscous dissipation dominates, the turbulent fluctuations gradually diminish in the direction normal to the wall.

Tip 6.2: Reynolds Stress

After averaging, an additional shear stress, known as Reynolds stress, is added into the governing equations:

$$\overline{\tau} = \tau_{\text{lam}} + \tau_{\text{turb}}, \quad \tau_{\text{lam}} = \mu\frac{\partial\overline{u}}{\partial y}, \quad \tau_{\text{turb}} = -\rho\overline{u'v'},$$

where u' and v' represent the streamwise and normal fluctuating velocity components, respectively.

By definition, the average value of a single fluctuating velocity component is zero (i.e., $\overline{u'} = 0,\ \overline{v'} = 0$). However, the average value of the product of two fluctuating velocity components, $\overline{u'v'}$, is not zero, and is opposite in sign to the average velocity gradient.

We suppose that there is a positive average velocity gradient inside the boundary layer, as shown in Figure T6.2 (i.e., $\mu(\partial\overline{u}/\partial y) > 0$). A detailed analysis follows.

Case 1: When $v' > 0$, the fluid element A will move upwards from the low-velocity layer into the high-velocity one, equivalently generating a negative disturbance to the local streamwise velocity component (i.e., $u' < 0$). So, u' and v' will have opposite signs (i.e., $u'v' < 0$).

Case 2: When $v' < 0$, the fluid element A will move downwards from the high-velocity layer into the low-velocity layer, equivalently generating a positive disturbance to the local streamwise velocity (i.e., $u' > 0$). So, u' and v' will still have opposite signs (i.e., $u'v' < 0$).

When the average shear stress, $\mu\left(\partial\overline{u}/\partial y\right)$, is positive or negative, $-\rho u'v'$ is also positive or negative. In other words, $-\rho u'v'$ is always in the same direction as the average shear stress.

It is worth noticing that the fluid element here refers to an element in the turbulence, and its speed corresponds to the statistical average of the turbulence. Therefore, $-\rho u'v'$ should rather be written as $-\rho\overline{u'v'}$ (i.e., the Reynolds stress).

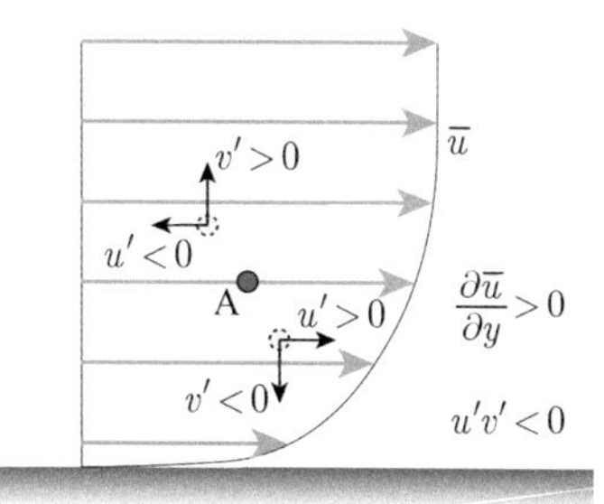

Figure T6.2.

Numerous attempts have been undertaken to study the theory of the turbulent boundary layer, and many of them are still under development. Unlike those for the laminar boundary layer, the theories describing the turbulent boundary layer have not been well established. Early-stage research often relies on experiments and, in recent years, numerical simulations have been widely used. Presently, most of the "theories" about the turbulent boundary layer applied in engineering are actually based on experiments and numerical simulations.

6.4 Pipe Flow

Pipe flow is very common in everyday life and production, so it has been studied for a long time. For example, the water supply and drainage system of ancient Rome clearly required some knowledge of pipe flow. Long before Prandtl published the boundary layer theory, engineers working on pipe flow were able to deal with some practical problems on the basis of experiments.

Pipe flow research is principally concerned with performance parameters, such as flow rate, head loss, and required pumping power. The laws for these parameters can be derived from a large number of experiments. However, in order to have universal

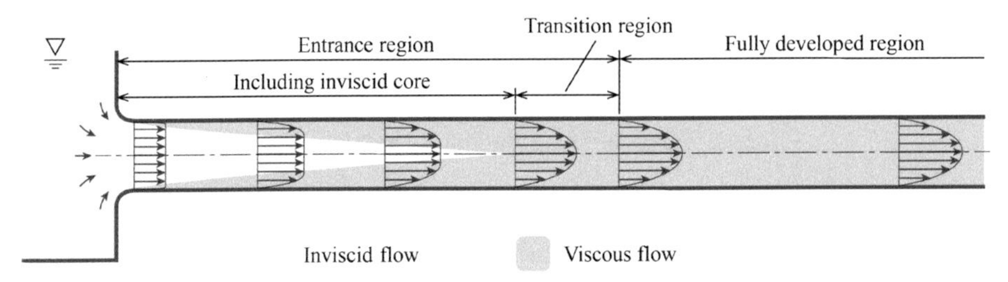

Figure. 6.15 Entrance region and fully developed region of the laminar flow in a pipe.

laws and sound predictions for new problems, it is necessary to develop the corresponding theories. According to Prandtl's boundary layer theory, it is valid to split any flow over a body into two regions: a thin boundary layer adjacent to the wall where the viscous effects are significant, and a mainstream outside the boundary layer where such effects are negligible. However, this description is not suitable for a pipe flow confined within a long, closed conduit, where the viscosity may influence the entire flow field.

As shown in Figure 6.15, the concepts of boundary layer and mainstream are only applicable in the entrance region. When the fluid travels along the pipe over a long enough distance, the boundary layer from the wall will exceed the radius of the pipe and merge at the center. Then, there is no mainstream free from viscous effects, and further increase in boundary layer thickness becomes impossible. Viscous forces provide the resisting force, while the favorable streamwise pressure gradient provides the driving force. With the driving force precisely balancing the resisting force, the fluid flows with constant velocity. Generally, *the region where the velocity profile still changes in the streamwise direction is called the entrance region, and the far downstream region where the velocity profile remains unchanged in the streamwise direction is called the fully developed region.* The flow principles and analysis methods applicable to the two regions are described in more detail below.

6.4.1 Entrance Region

If fluid enters the pipe with uniform velocity, a boundary layer begins to develop at the circumference wall, forming an entrance region. The boundary layer causes the effective area to decrease, accompanied by the loss of momentum and kinetic energy. In the case of flow over a flat plate, the boundary layer causes the mainstream to shift away from the wall by a certain distance. For the pipe, since both mass flow rate and cross-section remain the same along the streamwise direction in the entrance region, the inviscid core velocity needs to increase to compensate the velocity decrease near the wall. The inviscid core in the entrance region satisfies Bernoulli's equation: an increase in velocity is accompanied by a decrease in pressure.

Figure 6.16 shows the profiles of velocity for different cross-sections in the entrance region. As the flow proceeds downstream, the fluid near the wall continuously decelerates under the action of viscous forces, while the fluid in the inviscid core continuously accelerates under the action of differential pressure forces. The wall

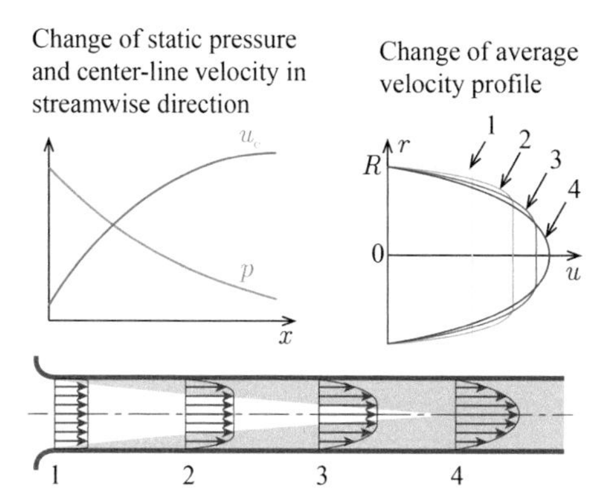

Figure 6.16 Streamwise velocity profiles across the cross-sections in the entrance region.

shear stress determines the magnitude of pressure drop, while the differential pressure force determines the acceleration of the fluid in the inviscid core.

Evidently, the larger the viscosity of a fluid, the faster the boundary layer thickness grows, and the shorter the inviscid core. Calculating the entrance length remains a difficult problem, and there is no exact theoretical solution, especially for turbulent flow. However, the relationship between entrance length, pipe diameter, and flow parameters has been experimentally determined. The relative length of the entrance region (the ratio of entrance length to pipe diameter) depends only on the Reynolds number. For a laminar flow, we have:

$$\frac{L_e}{D} = 0.06 Re_D, \tag{6.23}$$

where L_e is the entrance length; D is the pipe diameter; and Re_D is the Reynolds number defined by the pipe inner diameter and the average fluid velocity.

At low Reynolds numbers, the entrance region can be very short. For example, the entrance length is $0.6D$ for $Re_D = 10$. However, common flows are generally characterized by much higher Reynolds numbers. According to Equation (6.23), at $Re_D = 10^5$, the entrance length will reach $6{,}000D$, which is actually not realistic because when the Reynolds number exceeds a critical value transition from laminar to turbulent occurs, and the laminar formula (6.23) is not applicable any more. For a turbulent flow, the relationship between entrance length and Reynolds number is:

$$\frac{L_e}{D} = 4.4 {Re_D}^{1/6} . \tag{6.24}$$

Figure 6.17 depicts the entrance length as a function of the Reynolds number. At the same Reynolds number, the entrance length for turbulent flow is much smaller than that for laminar flow. Taking $Re_D = 2{,}300$ as the critical Reynolds number, the corresponding entrance lengths for laminar and turbulent flows are:

$$\left(\frac{L_e}{D}\right)_{\text{lam}} = 0.06 \times 2{,}300 = 138 .$$

$$\left(\frac{L_e}{D}\right)_{\text{turb}} = 4.4 \times 2{,}300^{1/6} = 16 .$$

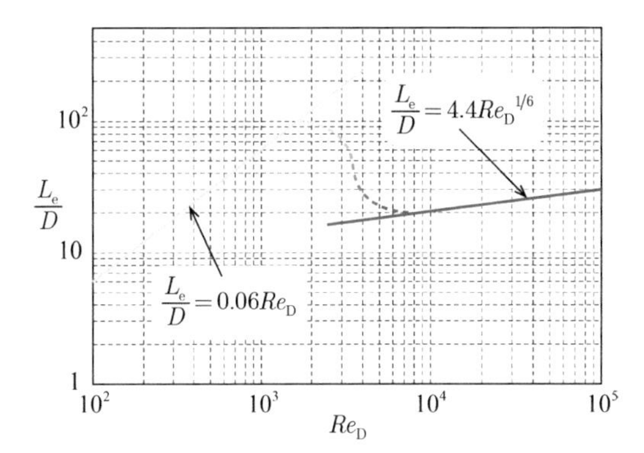

Figure 6.17 Entrance length as a function of Reynolds number.

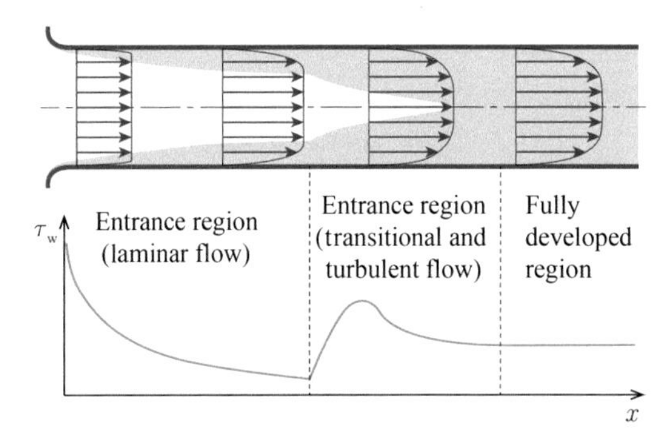

Figure 6.18 Entrance region with transition and the corresponding wall shear stress.

In a turbulent flow the intense turbulent diffusion makes the boundary layer grow faster, so the entrance length is relatively shorter. For $Re_D = 2{,}300$, the entrance length for a laminar flow is 138 times the pipe diameter, which might be the maximum entrance length encountered in reality. For a turbulent flow, such a long distance can only be achieved at rare high Reynolds numbers, typically at $Re_D = 10^8$.

In pipe flows, Reynolds numbers are generally much higher than 2,300. The flow phenomenon would be like this: As the flowing fluid enters a pipe, the boundary layer will initially be of laminar form. Before the velocity profile reaches a fully developed distribution, the boundary layer flow will have become turbulent. A short distance further downstream, the flow becomes fully developed. Therefore, the actual flow in the entrance region of a pipe is more like that shown in Figure 6.18 rather than that in Figure 6.15. The circumferential laminar boundary layer becomes turbulent at a critical distance downstream from the entrance of the pipe. The turbulent boundary layer thickens faster, and the flow rapidly becomes fully developed. In the laminar entrance region, the wall shear stress decreases rapidly in the streamwise direction with increased boundary layer thickness. The laminar-to-turbulent transitional region experiences relatively high wall shear stress, which also decreases with increased boundary layer thickness. After the flow becomes fully developed, the wall shear stress no longer changes in the streamwise direction.

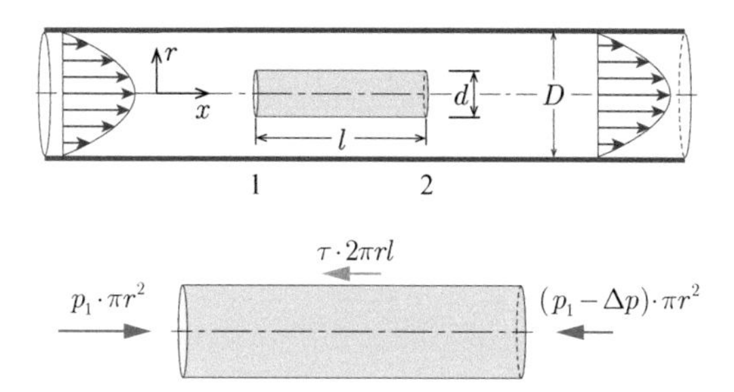

Figure 6.19 Control volume analysis in the fully developed region.

6.4.2 Fully Developed Region

As the circumferential boundary layers merge at the center of the pipe, the inviscid core disappears. A short distance further downstream, the velocity profiles no longer change in the streamwise direction, and then the flow in the pipe is fully developed. Consequently, there must be a driving force acting on the fluid to balance the resisting force caused by friction between the pipe wall and the fluid. In the flow through a horizontal pipe, the streamwise differential pressure is the only driving force. That is to say, the pressure continuously drops in the streamwise direction. *In the fully developed region, static pressure decreases in the streamwise direction, and total pressure also decreases. The total pressure drop is equal to the static pressure drop.*

Since the velocity profile remains unchanged, the velocity gradient normal to the wall, $\partial u/\partial y$, should remain constant. For a laminar flow, the wall shear stress, $\tau_w = \mu(\partial u/\partial y)$, stays the same everywhere. Subsequently, the wall frictional force per streamwise unit length remains constant. Therefore, the pressure declines linearly in the streamwise direction. Next, let us derive the streamwise variation of pressure in the fully developed region.

As shown in Figure 6.19, a cylindrical control volume of diameter d is chosen from a pipe with diameter D ($0 < d < D$). The forces acting on the left, right, and cylindrical surfaces are expressed as:

$$\begin{aligned} F_{\text{left}} &= p_1 \cdot \pi r^2 \\ F_{\text{right}} &= -(p_1 - \Delta p) \cdot \pi r^2, \\ F_{\text{side}} &= -\tau \cdot 2\pi rl \end{aligned}$$

where p_1 is the inlet pressure; Δp is the pressure difference between inlet and outlet; τ is the shear stress on the cylindrical surfaces; r is the radius of the cylindrical control volume; and l is the length of the control volume.

Since the vector sum of these three forces is equal to zero, it follows:

$$\sum F = F_{\text{left}} + F_{\text{right}} + F_{\text{side}} = p_1 \cdot \pi r^2 - (p_1 - \Delta p) \cdot \pi r^2 - \tau \cdot 2\pi rl = 0.$$

Substitute the shear stress, $\tau = \mu(\partial u/\partial y)$, into the above equation, rearranged as:

$$\frac{\mathrm{d}u}{\mathrm{d}r} = -\frac{\Delta pr}{2\mu l}.$$

By integrating the above formula and substituting the no-slip boundary condition of the wall surface, we obtain:

$$u(r) = \frac{\Delta p D^2}{16\mu l}\left[1 - \left(\frac{r}{R}\right)^2\right], \tag{6.25}$$

where D is the pipe diameter and R is the radius. Comparing this solution and the Hagen–Poiseuille solution in Section 4.7.2 shows that, if $\Delta p/l$ is replaced by $\mathrm{d}p/\mathrm{d}x$ that is applicable to the elemental control volume, these two formulas are exactly the same.

The maximum velocity occurs at the center-line where $r = 0$, expressed as:

$$u_c = \frac{\Delta p D^2}{16\mu l}.$$

Thus, the velocity profiles across the cross-sections can be written as:

$$u(r) = u_c\left[1 - \left(\frac{r}{R}\right)^2\right].$$

Previous discussion about the laminar boundary layer over a flat plate established that the velocity profile is nearly parabolic in shape. Here, we prove that the velocity profile for laminar flow in a circular pipe is exactly parabolic in shape.

It can be proved that the average velocity in a cross-section is exactly one-half of the maximum velocity. If the average velocity is denoted by V, then:

$$V = \frac{1}{2}u_c = \frac{\Delta p D^2}{32\mu l}.$$

Rearranging the above formula, we obtain the pressure drop per unit length of pipe:

$$\frac{\Delta p}{l} = \frac{32\mu V}{D^2}. \tag{6.26}$$

As can be seen, the pressure in the fully developed region declines linearly in the streamwise direction.

Tip 6.3: There Is a Mixing Region between the Inviscid Core Region and the Fully Developed Region

Some books regard the end of the inviscid core region as the starting point of the fully developed region. That is not true. Actually, after the inviscid core terminates, the fluid still needs to travel through a mixing region to achieve the final balance between the differential pressure and the viscous forces. When the velocity profile no longer changes in the streamwise direction, the flow is said to be fully developed (Figure 6.15).

A simple argument can be presented based on two facts: the flow in the inviscid core satisfies Bernoulli's equation, and the pressure in the fully developed region declines linearly in the streamwise direction.

The first fact states that near the end of the inviscid core region, the flow parameters at the center-line satisfy the following relation:

$$\mathrm{d}p = -\rho V \mathrm{d}V.$$

The second states that at the starting point of the fully developed region, the velocity at the center-line no longer changes in the streamwise direction (i.e., $\mathrm{d}V = 0$) and the pressure declines linearly in the streamwise direction (i.e., $\mathrm{d}p < 0$).

Assume that the inviscid core region is immediately followed by the fully developed one. If the pressure has been declining linearly before the junction point, the velocity will increase by 1/2, which cannot guarantee the continuity with the downstream constant velocity. Therefore, it can be concluded that after the inviscid core terminates, the fluid has to travel through a mixing region before it becomes fully developed. In this mixing region, the flow does not satisfy Bernoulli's equation, with pressure decreasing and velocity at the center-line increasing.

As shown in Equation (6.26), for the flow through a pipe of given size, pressure drop has a linear relationship with average velocity. For practical design problems, the known quantity is normally the flow rate. With the volume flow rate denoted by Q, Equation (6.26) can be rewritten as:

$$\frac{\Delta p}{l} = 128\frac{\mu Q}{\pi D^4}.$$

As can be seen, *for pipes transporting a specific flow rate, pressure drop per unit length is inversely proportional to the fourth power of the pipe diameter*. Therefore, a slight increase in the designed pipe diameter will result in much greater benefits.

In engineering, the pressure loss coefficient, commonly used to evaluate the pressure loss along pipes, is defined as:

$$C_\mathrm{p} = \frac{\Delta p}{\rho V^2/2}.$$

Substituting Equation (6.26) into the above equation yields:

$$C_\mathrm{p} = \frac{64}{Re_\mathrm{D}}\frac{l}{D} = f\frac{l}{D}.$$

f is generally known as the *friction factor*. For a laminar flow, we have:

$$f = \frac{64}{Re_\mathrm{D}}. \tag{6.27}$$

The above analysis for a laminar flow is not valid for turbulent flows in which the wall shear stress is not solely determined by the average velocity gradient. Prandtl proposed a semi-empirical formula of friction factor for a turbulent flow in a pipe:

$$\frac{1}{f^{1/2}} = 2.0\log\left(Re_\mathrm{D} f^{1/2}\right) - 0.8. \tag{6.28}$$

Equation (6.28) is relatively accurate for turbulent flows in smooth pipes. But it is not easy to use since it is not a simple formula. Prandtl's student, Heinrich Blasius, came up with another useful empirical formula for engineering applications:

$$f = 0.316 Re_\mathrm{D}^{-1/4}, \quad 4{,}000 < Re_\mathrm{D} < 10^5. \tag{6.28a}$$

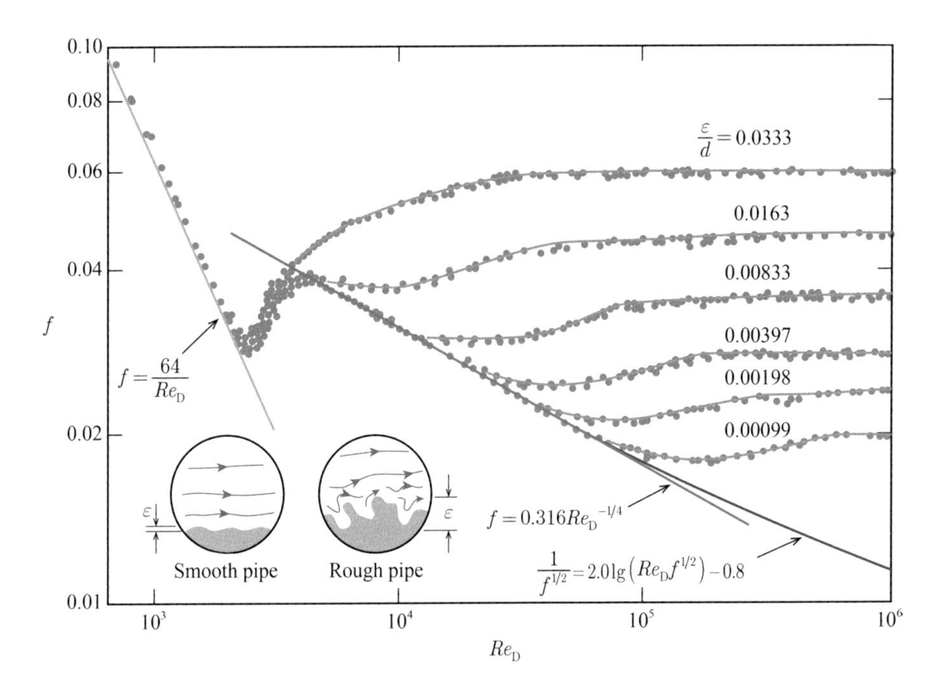

Figure 6.20 Nikuradse's experimental results and relevant theoretical and empirical formulas of the friction factor for pipe flows.

Equation (6.28a) is relatively consistent with Equation (6.28) within a specific range of Reynolds numbers. This equation states that the pressure drop in a turbulent pipe flow is approximately proportional to the 1.75th power of velocity (i.e., $\Delta p \sim V^{1.75}$), which agrees well with experimental results. A previous analysis concluded that the pressure drop in a laminar pipe flow is linear to velocity. Apparently, increasing pipe diameter will be more beneficial for turbulent flows than for laminar flows.

In engineering, the inner surface of pipe walls is usually not completely smooth, causing the friction factor to be inconsistent with Equation (6.28). Another of Prandtl's students, Nikuradse, performed an experiment to investigate this. He glued sand grains of different sizes to the inner wall of a pipe to simulate degrees of roughness. Based on his experiment, the friction factor, f, is measured for various flow rates of water through the pipe. The experimental results are shown in Figure 6.20.

The results of experiments on laminar flow along pipes agree well with Equation (6.27), and those on turbulent flow along smooth pipes are also consistent with Equations (6.28) and (6.28a). However, there is no theoretical formula for turbulent flow in rough pipes. When surface roughness is less than a specific value, the friction factor depends only on the Reynolds number. A pipe whose surface roughness is equal to or less than this critical value is considered to exhibit "smooth pipe" characteristics, commonly known as *hydraulic smooth pipe*. That is to say, a pipe

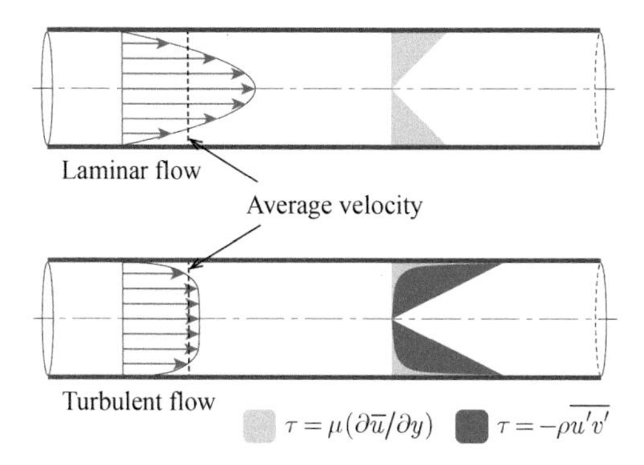

Figure 6.21 Velocity and shear stress distribution for pipe flow.

is considered hydraulically smooth for a flowing fluid that cannot "feel" the small surface roughness.

Following Nikuradse's experimental results, several empirical formulas were devised to calculate the friction factor for a turbulent flow in rough pipes. One of the most famous empirical formulas and charts is the chart proposed by Moody. In engineering, it is very convenient to directly query the friction factor from the *Moody chart* to calculate the head losses along pipes.

According to the control volume analysis in Figure 6.19, we can further analyze the shear stress distribution inside the pipe. A control volume in static equilibrium satisfies the following relation:

$$\sum F = F_{\text{left}} + F_{\text{right}} + F_{\text{side}} = p_1 \cdot \pi r^2 - (p_1 - \Delta p) \cdot \pi r^2 - \tau \cdot 2\pi r l = 0,$$

which can be simplified as:

$$\tau = \frac{\Delta p r}{2l}.$$

If the length of the control volume, l, is replaced by dx, we obtain

$$\tau = \frac{r}{2}\frac{\mathrm{d}p}{\mathrm{d}x}. \tag{6.29}$$

It follows that the shear stress in pipe flow depends only on both pressure gradient and radius. This is true for both laminar and turbulent flows.

A direct conclusion can be drawn from this analysis: *The shear stress in a pipe flow varies linearly with radius, reaching its maximum value at the walls and zero at the center*. For a laminar flow, the radial linear distribution of shear stress directly yields a parabolic velocity profile across the cross-section. For a turbulent flow, the shear stress depends on both average velocity gradient and Reynolds stress generated by fluctuating velocities. So, the average velocity profile for a turbulent flow is not directly related to the shear stress. Similar to the flat plate boundary layer velocity profile, the velocity profile for a turbulent pipe flow is much fuller than that for a laminar pipe flow. Figure 6.21 shows the shear stress distribution with radial distance, implying that the shear stress for a turbulent flow is for the most part contributed by the Reynolds stress.

6.5 Jets and Wakes

Jets and wakes are two common flow phenomena. Different from boundary layer and pipe flows that are caused by wall surfaces, jets and wakes are generated by the difference in velocity between adjacent fluid layers, and known as *free shear flows*.

6.5.1 Jets

A jet is characterized by a high-speed core region and a lower-speed outer region. Figure 6.22 shows a typical jet in which a stream of one fluid is injected from a nozzle into a surrounding still fluid. Once the jet flow leaves the nozzle, there will be a strong shear action at the boundary between the jet and the surrounding still fluid. The viscous effects of this shear layer not only decelerate the jet, but also entrain the surrounding fluid to the jet. If the Reynolds number is low enough, the jet can remain laminar over a long distance. At usual Reynolds numbers, the laminar-to-turbulent transition occurs. Therefore, common jets are characterized as turbulent. The time-averaged velocity profiles in Figure 6.22 do not represent any instantaneous velocity.

The inviscid core of a jet experiences no shear effect and will stay laminar. The fluid travels downstream and grows continually to form a transitional region where viscous effects are dominant. A certain distance further downstream, the normalized average velocity profile no longer changes in the streamwise direction, and the jet is said to be fully developed.

Although there are inviscid cores in the initial region of both jet and pipe flows, the corresponding boundary conditions are different. For flow in a constant cross-section pipe, the average velocity remains unchanged in the streamwise direction, accompanied by the decelerated flow near the pipe walls and the accelerated flow in the inviscid core. On the other hand, the fluid flowing out of a nozzle naturally entrains the

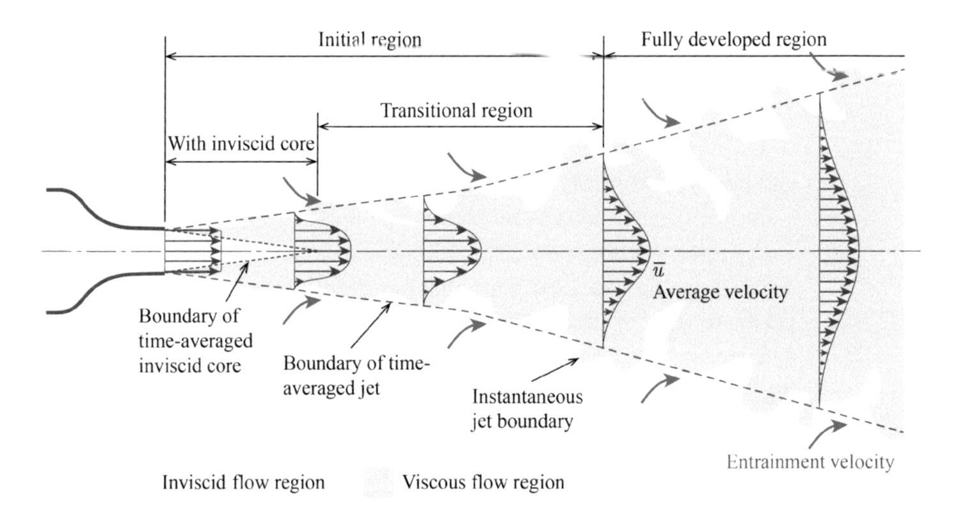

Figure 6.22 Schematic of velocity profiles in an axisymmetric free jet.

surrounding ambient fluid, so neither average velocity nor flow rate remains constant in a jet flow. The boundary condition for a jet is that its surrounding pressure stays constant. Therefore, the fluid in the inviscid core experiences neither viscous forces nor differential pressure forces, and maintains uniform translational motion. In other words, *for a jet, the velocity and pressure in the inviscid core remains unchanged in the streamwise direction.*

There is no suitable theory to describe the transitional region from the end of the inviscid core to the beginning of the fully developed region. Based on the similarity of average velocity profiles in the fully developed region, a method similar to the Blasius solution for laminar boundary layer flow can be used to draw some useful conclusions.

If the flow is laminar, the jet boundary expands slowly. If it is turbulent, the jet boundary expands rapidly due to the intense effect of shear and entrainment. An important conclusion for turbulent flow is that *an axisymmetric turbulent jet expands linearly in both initial and fully developed regions, with a higher expansion rate (divergence angle) in the latter.*

It is generally difficult for a jet to remain laminar over a long distance. This is because the shear layer at the jet boundary has a weak ability to resist external disturbances, and the fluid cannot get any help from the wall to reduce disturbances, as it does in the boundary layer. It has been theoretically proved that a jet is unconditionally unstable. That is to say, *a jet will become turbulent even at very low Reynolds numbers.*

There exists a type of laminar flow fountain, which can be laminar over a long distance, with large jet scale and high jet speed. A water jet in air is different from jets in the same fluid (water jet in water, air jet in air, etc.). The shear force exerted by the ambient air on the water jet is negligible as long as the jet speed is not too large. So, the laminar water jet can maintain its original flow state over a long distance. The method used to make the flow laminar at the nozzle is also simple. First, the water should be relatively pure, without disturbance sources such as air bubbles. Subsequently, to generate laminar flow, a honeycomb can be installed upstream of the nozzle. The small cross-sections of the honeycomb reduce the Reynolds number of the flow, just like the flow in many small-diameter tubes. Then the water can be a laminar flow before it leaves the nozzle. Figure 6.1 shows this kind of fountain, which keeps the water flow laminar. With the addition of colorful lights, the water becomes crystal clear.

6.5.2 Wake

Jet and wake are like two sides of a mirror. The characteristic of a free jet is that the innermost parts flow faster than their surroundings, while that of a wake is just the opposite, the innermost flow is slower than the surrounding parts. A wake always forms downstream from an object immersed in a stream. Figure 6.23 shows the two-dimensional wake behind a circular cylinder at a Reynolds number of around 4,000. The boundary layer flow over the windward side of the cylinder is basically laminar,

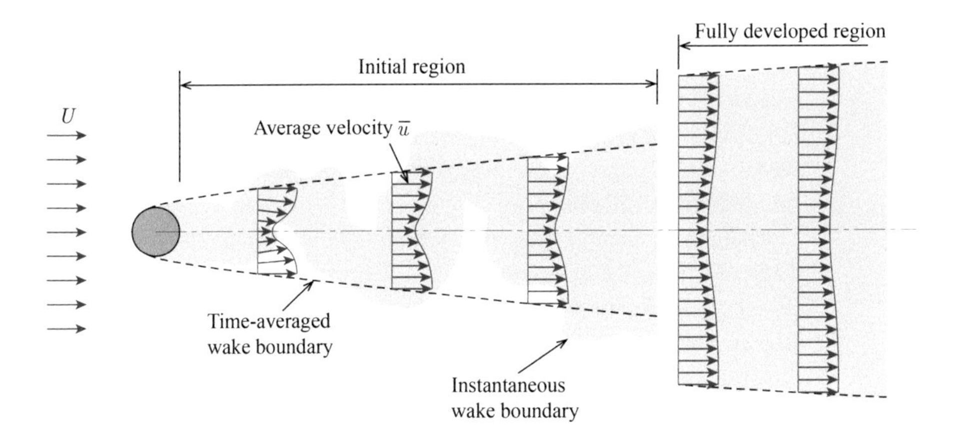

Figure 6.23 Schematic of velocity profiles in the two-dimensional wake behind a circular cylinder.

and the turbulent wake starts as the flow turns onto the leeward side of the cylinder. A highly unsteady Karman vortex street is observed in the wake of the cylinder. However, the velocity profiles averaged over a long enough time period are still regular.

The velocity deficit in the wake originates from two low-velocity regions: the boundary layer over the surface of the object, and the flow separation region at the rear portion of the object. Just downstream from the object, the flow separation region is situated in the middle of the wake, and the boundary layer region is located in both sides of it. These two regions merge to form a unified velocity deficit region downstream.

For the flow over a blunt body shown in Figure 6.23, the pressure in the wake region is generally lower than the ambient pressure, so the fluid immediately behind the body tends to converge toward the wake center-line. Far enough downstream, the pressure in the fluid becomes uniform and the fluid moves in parallel layers, but the velocity deficit still exists

Just as in the case of a free jet, a wake can be divided into initial and fully developed regions. If the wake originates from the trailing edge of an object with zero thickness, and no separation occurs on the side surface of the object, the velocity profiles in the initial wake region are determined solely by the boundary layer. Far enough downstream (approximately 100 times the cylinder diameter), the shape of the wake is completely determined by the free shear layer. The velocity profiles in the wake are similar to those in the free jet, suggesting that these two flow mechanisms are more or less the same. In contrast to the free jet, the wake originates from the boundary layer or separation point, so most of the time it is already turbulent from the beginning.

An analogous solution can be found for the fully developed wake region. The width of the two-dimensional wake grows more slowly than linearly in the streamwise direction:

$$\delta \propto x^{1/2},$$

where δ represents the width of the two-dimensional wake and x represents the streamwise coordinate. This law is valid for both laminar and turbulent wakes, but the width of the turbulent wake grows faster.

6.6 Boundary Layer Separation

For boundary layer flow, the wall shear force is equivalent to the frictional drag exerted on the fluid by the wall. We know that when an object slides on a plane and experiences a frictional force, it will slow down and eventually stop, unless a driving force is exerted to keep it moving. As the object gradually comes to a stop, the frictional force tends to zero. For the boundary layer flow over a flat plate, the fluid particles near the wall are constantly decelerated. As more and more fluid particles are slowed down by wall friction, the thickness of the boundary layer grows. Over a sufficiently long distance, wall friction will decelerate the nearby fluids to near-zero speeds. Then, as the near-wall velocity gradient normal to the wall tends to zero, so does the wall friction. Consequently, the wall will no longer hinder the movement of the fluid. Therefore, the fluid velocity is always positive and never slows down to zero.

The other case is when the mainstream velocity decreases and pressure increases along the wall surface with a given divergence angle or curvature. The fluid particles are therefore subjected to two forces acting opposite to the streamwise direction, namely the wall frictional and differential pressure forces. When the velocity of near-wall fluid particles decreases to nearly zero, the wall frictional force also tends to zero, but the adverse pressure force still exists. Thus, the fluid particles may begin to move in the reverse direction. When the reverse flow mixes with the mainstream, the fluid particles will move away from the wall surface so that the boundary layer may lift off or "separate" from the wall. *Boundary layer separation is the detachment of a boundary layer from the wall surface due to the adverse pressure gradient.*

We have learned from the above analysis that only the adverse pressure gradient may lead to boundary layer separation. Therefore, *one necessary condition for boundary layer separation is adverse pressure gradient.* On the other hand, if the flow is inviscid, the maximum pressure is reached when the flow speed is reduced to zero. No additional pressure can exist pushing the flow to turn around. Therefore, *the other necessary condition for boundary layer separation is the viscous force, which retards the flow together with the adverse pressure gradient.*

The velocity profiles can also be used to prove that a necessary condition for boundary layer separation is the adverse pressure gradient. As shown in Figure 6.24, the upstream fluid moves forward, while the downstream fluid moves backward. Therefore, there must be a position where the streamwise velocity component becomes zero, known as the separation point. At the separation point, we have

$$\left.\frac{\partial u}{\partial y}\right|_{\text{wall}} = 0 .$$

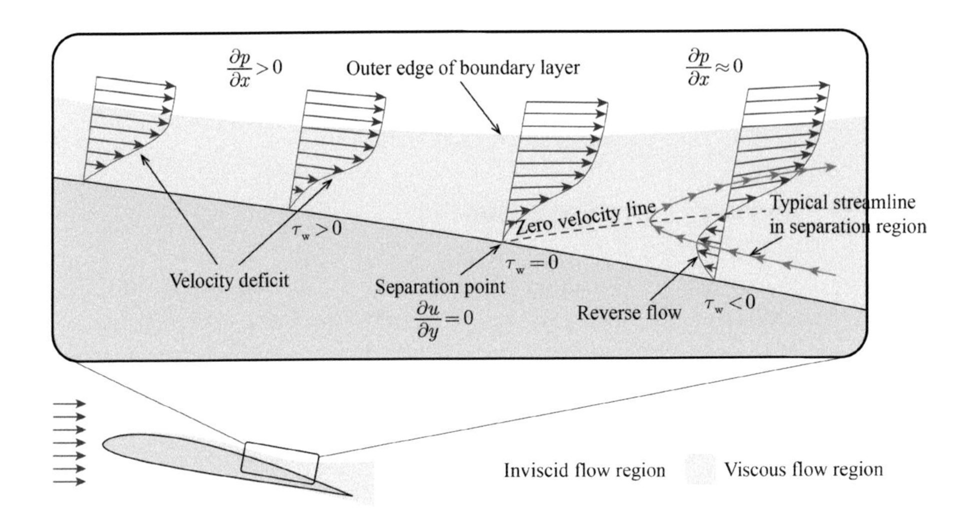

Figure 6.24 Boundary layer separation near the trailing edge of the wing.

As long as no reverse flow occurs, we have $\partial u/\partial y > 0$ throughout the boundary layer. Therefore, the near-wall velocity gradient must increase from zero to some positive value, and the rate of change of the velocity gradient is positive. In other words, the second derivative of the velocity with respect to the coordinate normal to the flow is positive:

$$\frac{\partial^2 u}{\partial y^2} > 0 .$$

For a clearer understanding, Figure 6.25 shows the variations of velocity, velocity gradient (the derivative of the velocity), and the second derivative of the velocity along the direction normal to the wall at the separation point.

According to the boundary layer equation:

$$u\frac{\partial u}{\partial x} + v\frac{\partial u}{\partial y} = -\frac{1}{\rho}\frac{\mathrm{d}p}{\mathrm{d}x} + \frac{\mu}{\rho}\frac{\partial^2 u}{\partial y^2} .$$

At the streamline adjacent to the wall, the velocity components u and v are both zero, and the boundary layer equation reduces to

$$\frac{\mathrm{d}p}{\mathrm{d}x} = \mu\frac{\partial^2 u}{\partial y^2} .$$

If the second derivative of the streamwise velocity component in the normal direction is positive, we have:

$$\frac{\mathrm{d}p}{\mathrm{d}x} > 0 .$$

We have therefore rigorously proved that there must exist an adverse pressure gradient in some neighborhood of the separation point. It should be noted that, since the reduced boundary layer equation only applies to laminar flow, the above proof is not strictly valid for turbulent flow.

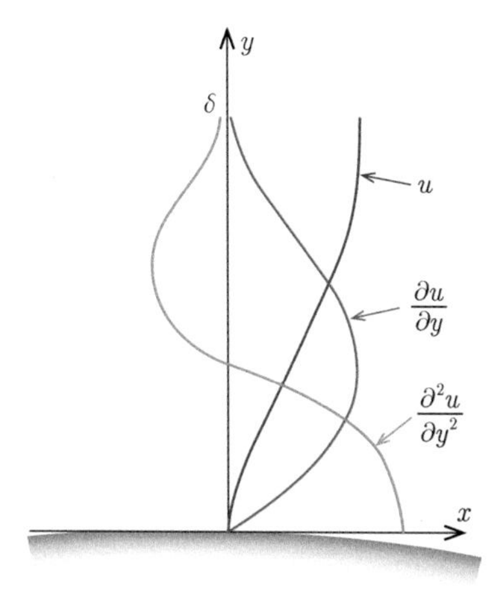

Figure 6.25 Variations of velocity and its derivatives along the direction normal to the wall at the separation point.

As opposed to the clear necessary conditions, the sufficient conditions for boundary layer separation are still unclear. In other words, it is well known that there are situations in which boundary layer separation will certainly not occur, but we do not know for sure exactly when boundary layer separation will occur. For a laminar flow with adverse pressure gradient (such as the two-dimensional flow through a channel with a small divergence angle), the dimensionless pressure gradient parameters are defined as follows:

$$m = \frac{x}{U}\frac{\mathrm{d}U}{\mathrm{d}x} = -\frac{x}{\rho U^2}\frac{\mathrm{d}p}{\mathrm{d}x},$$

where x is the distance from the leading edge and U is the mainstream velocity.

As in the case of the Blasius solution for the laminar boundary layer over a flat plate, a similarity solution for laminar flow with adverse pressure gradient can be obtained to determine the sufficient conditions for boundary layer separation. It is found that the boundary layer separation happens at $m = -0.0904$. This conclusion might not be so accurate because the boundary layers with adverse pressure gradient are nonsimilar ones. According to the similarity solution, the shape factor of the boundary layer prior to the separation point is $H = 4.0$, while the actual one based on a sufficiently large number of measurements is about $H = 3.5$.

Next, we evaluate the sufficient condition for actual flow separation according to the criterion $m = -0.0904$ given by the laminar similarity solution.

Figure 6.26 shows a two-dimensional diverging channel with an inlet width of W_1 and a half-divergence angle of α. According to the continuity equation for an incompressible flow, the dimensionless pressure gradient parameter at a distance x from the channel inlet is:

$$m = \frac{x}{U}\frac{\mathrm{d}U}{\mathrm{d}x} = -\frac{2x}{W_1}\tan\alpha.$$

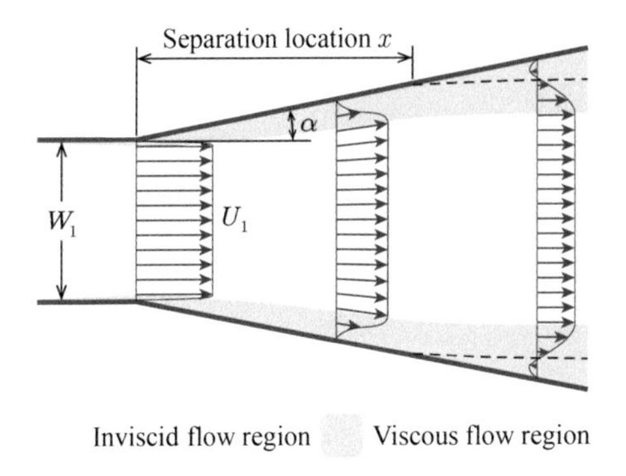

Figure 6.26 Flow separation in a two-dimensional diverging channel.

So, the flow separation condition on this two-dimensional diverging channel is

$$\frac{2x}{W_1}\tan\alpha = 0.0904 .$$

Assuming that the inlet width is 100 mm, the distances of the separation point from the turning point are 52 mm and 260 mm at half-divergence angle of 5° and 1°, respectively.

As can be seen, the laminar boundary layer separation will take place a short distance downstream of the turning point even at very small divergence angles. However, for practical high Reynolds number two-dimensional flows with a small divergence angle of 1°, the boundary layer can remain attached over a long distance. This is because the boundary layers at high Reynolds numbers may transition to turbulent before separation, and turbulent flows are not as easy to separate as laminar flows.

For turbulent flows, there is no good theory to predict flow separation, so predictions rely largely on empirical models obtained from experimental data. It is generally accepted that turbulent boundary layer separation takes places when H reaches or exceeds 2.4. In practice, flow conditions can vary widely, and there is still no good method to determine whether or not a turbulent flow separation will occur. For the two-dimensional diverging channel in Figure 6.26, Reneau et al. published some experimental results in 1967 and described the criteria for flow separation. The following formula is obtained by fitting the experimental data:

$$\frac{x_s}{W_1} = \frac{1}{-0.0031 + 0.0156\alpha^{1.5}},$$

where α is the half-divergence angle and x_s is the position of the separation point (the distance from the turning point to the separation point).

Let us use this formula to calculate the position of the separation point under the same conditions as the aforementioned laminar flow. Assuming that the inlet width is 100 mm, the distances of the separation point from the leading edge are calculated to be 584 mm and 8,000 mm at a half-divergence angle of 5° and 1°, respectively, from which we can see that turbulent flow resists separation much better than laminar flow.

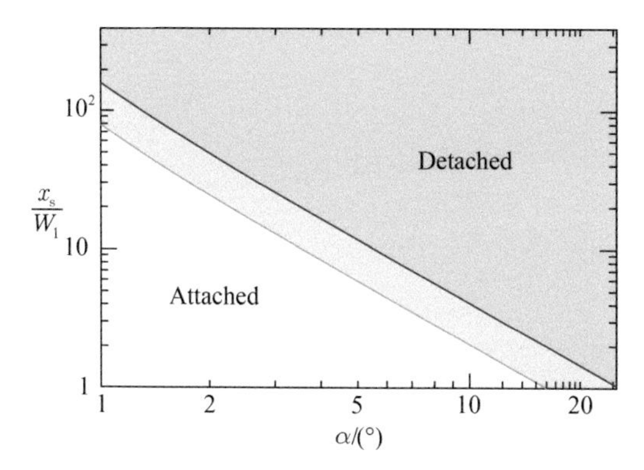

Figure 6.27 The relationship between the position of flow separation and the divergence angle for a turbulent boundary layer through a two-dimensional channel.

The results from this empirical formula may be too conservative. That is to say, it is impossible for boundary layer separation to take place upstream of the separation point calculated by this formula, but it may however occur far downstream from the predicted separation point. The following formula gives the upper limit of the distance between the separation and turning points (that is, the possible longest distance):

$$\frac{x_s}{W_1} = \frac{1}{-0.0016 + 0.0078\alpha^{1.5}}.$$

At a half-divergence angle of 5°, the predicted separation distance is 1,167 mm. Therefore, it is prudent to say that in this case, boundary layer separation may take place somewhere 584–1,167 mm downstream of the turning point.

According to the two above formulas, the relationship between the separation position and divergence angle for a two-dimensional diverging channel flow can be determined. The conclusion is shown in Figure 6.27.

In general, the position of the separation point also depends, either directly or indirectly, on the Reynolds number, the roughness of the wall, the inflow turbulence intensity, the outlet pressure distribution, and other factors. The chart given in Figure 6.27 is for reference only.

Stronger mixing is the main reason why a turbulent flow resists separation much better than a laminar one. For a turbulent boundary layer, the inner fluid element can gain more momentum from the outer layer through the extra shear stress caused by Reynolds stress, allowing it to sustain an adverse pressure gradient over a longer distance without separating. The fuller velocity profile of the turbulent boundary layer also allows the outer layer to exert a larger viscous force on the inner one and drag it forward.

A direct effect of separation is to generate additional drag and losses. Most of the time, flow separation needs to be prevented to reduce drag and losses. For example, all kinds of objects moving at high speeds in fluid are streamlined in shape to reduce the drag, and the local adverse pressure gradient in fluid machinery should be minimized to improve the efficiency. Sometimes, flow separation is deliberately exacerbated to

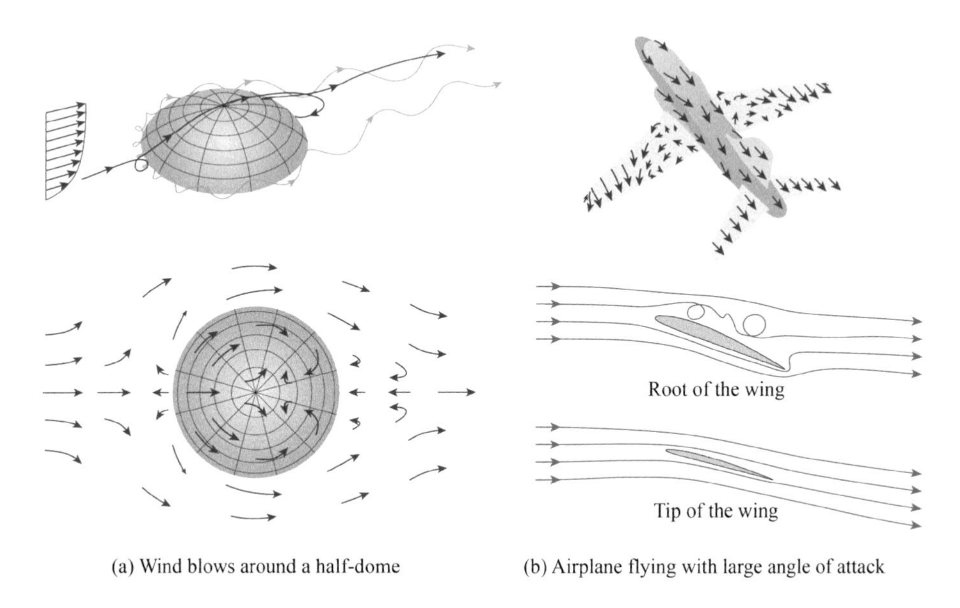

Figure 6.28 Two typical three-dimensional flow separations.

increase drag or losses. For example, the parachute and the break flap opened during landing of an aircraft introduce separation to increase drag; the valves in pipelines produce local separation to increase losses and reduce total pressure, thereby controlling the flow rate.

The separations mentioned above refer to two-dimensional ones, and are characterized by the occurrence of flow reversal. For practical three-dimensional flows, there may not always be reversed flow when flow separation happens; the fluid may move laterally to maintain flow continuity. Figure 6.28 shows two types of flow separation in three-dimensional flow fields. Figure 6.28(a) shows the three-dimensional flow separation on the roof and the ground when wind blows around a half-dome building. Figure 6.28(b) shows the three-dimensional flow separation over the wing of an aircraft flying at high angles of attack. These two types of flow are highly three-dimensional. The fluid does not necessarily leave the surface along the normal direction, but may flow laterally.

In Figure 6.28(a), as the airflow near the ground approaches the half-dome, the adverse pressure gradient causes the separation. The air is lifted upward and flows past both sides of the building, forming a vortex flow (called a *horseshoe vortex*). After passing over the highest point on the roof, the airflow separates somewhere and forms a series of complex vortices below it. The airflow can reattach to the ground behind the building at a certain point, called the *reattachment point* (i.e., the separated boundary layer is reattached to the wall surface). In Figure 6.28(b), if the angle of attack is too large, separation first takes place at the wing root just behind the leading edge. The reversed flow due to flow separation moves outwards along the wingspan and reattaches to the surface at the middle of the wing. At the tips of the wing, there is also a vortex where the high-pressure air below the wing spills onto the low-pressure area above it to form a flow separation within a small scale, called a *wing tip vortex*.

6.7 Drag and Losses

Drag and losses are the eternal problems of fluid mechanics in engineering applications. There is still no general agreement about how to solve the problems related to flow drag and losses. At present, we can only solve specific problems through special analysis, but the accuracy of simulation results cannot be guaranteed. In most cases, we still rely on the data obtained from experiments and empirical models.

We know that an inviscid fluid produces no flow drag or losses. Therefore, it is necessary to study the working mechanism of viscosity in order to calculate the flow drag or losses. This problem can become exceedingly complicated in turbulent flow, which is characterized by the enhanced viscous shear force and dissipation. So far, the extra viscosity in turbulent flow, known as *eddy viscosity*, cannot be properly estimated, which is the bottleneck to a better estimation of drag or losses.

6.7.1 Drag

Although viscosity is the cause of drag, from the perspective of direct effect, the biggest contributor to drag is commonly pressure difference force. As shown in Figure 6.29, there will be a large drag created by a zero-thickness flat plate normal to the flow. Since there is no side surface for the plate, the viscous force has no component in the streamwise direction. Therefore, in this case drag is not caused by viscous force, but entirely due to pressure difference force.

The pressure at the center of the plate on the windward side is equal to the freestream total pressure. After stagnation, the fluid accelerates along the radial direction, accompanied by the decrease in pressure. After flowing over the flat plate, the fluid moves roughly in the streamwise direction. We know that when streamlines are

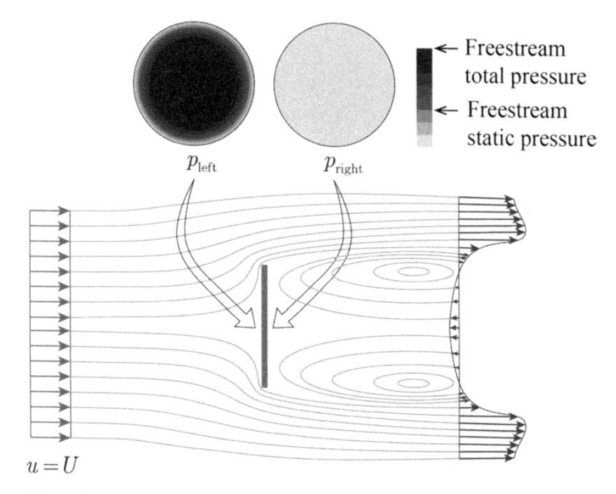

Figure 6.29 An actual flow over a zero-thickness flat plate held normal to a freestream.

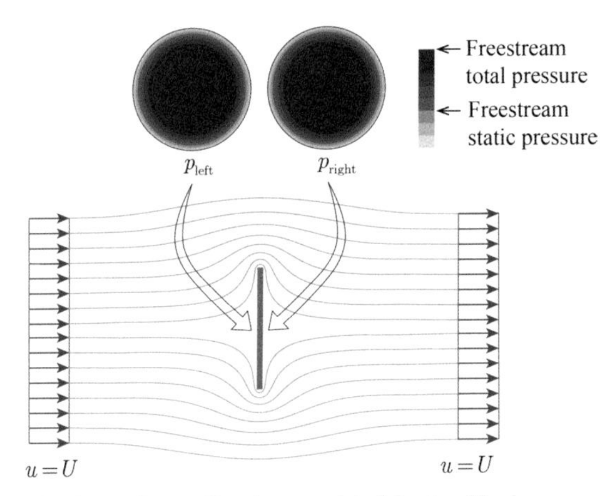

Figure 6.30 Ideal potential flow over a zero-thickness flat plate held normal to a freestream.

parallel, the pressure across them is constant. Therefore, it is fair to speculate that the pressure behind the flat plate is approximately constant and equal to that of the mainstream, which is a little lower than the far ahead freestream static pressure, since speed increases as the fluid flows past the flat plate. The pressure on the windward side of the flat plate is mostly close to the freestream total pressure, and the pressure on its leeward side is lower than the freestream static pressure. Consequently, we can presumably conclude that the drag acting on the flat plate should be slightly greater than the product of the freestream dynamic pressure times the plate area (the measured drag on the flat plate has been proved to be about 1.1 times this value).

In engineering, the flow drag due to the differential pressure between windward side and leeward side surfaces of an object is called pressure drag or shape drag. Although completely created by differential pressure force, the pressure drag is not meant to be independent of viscous force. As shown in Figure 6.30, if the flow is inviscid, according to potential flow theory it is stopped and stagnated by the flat plate, travels around it, converges in the back, and then moves downstream. The flow patterns over the front and rear of the cylinder are perfectly symmetrical. From the integral point of view, the inviscid fluid has no change in momentum, so no external force acts on the flat plate. From the differential point of view, the pressure distributions on both the windward and leeward sides of the flat plate are exactly the same, exerting no external force on the plate. Therefore, the pressure drag is actually closely related to the viscosity, which affects pressure distribution and consequently produces drag.

Next, we analyze the drag for flow over a horizontal flat plate, as shown in Figure 6.31. Since the windward area is zero, the pressure drag must be zero, and the viscous shear stresses on the upper and lower surfaces constitute the total drag exerted on the flat plate. *The drag due to viscous shear stresses acting on the lateral surface of an object is called frictional drag.* Common sense dictates that the total drag for flow over a horizontal flat plate is much smaller than that for flow over a normal flat plate.

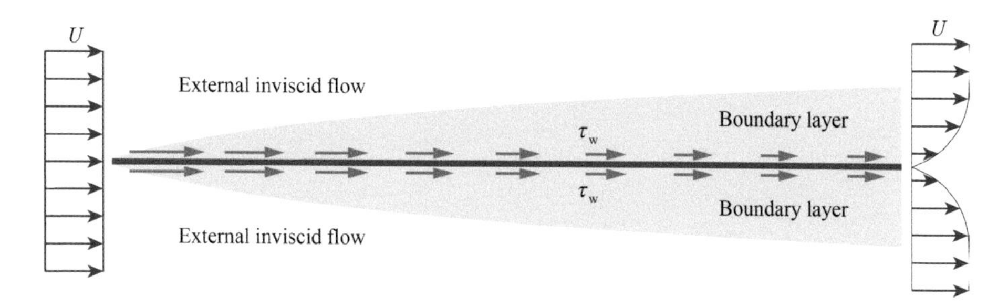

Figure 6.31 Flow over a horizontal flat plate and wall shear stress.

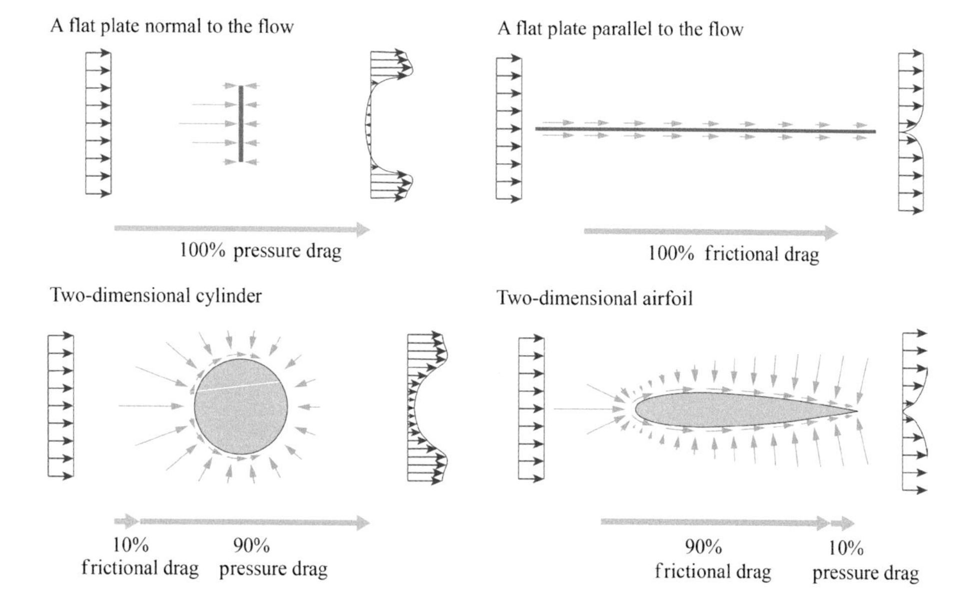

Figure 6.32 The ratios of pressure drag and frictional drag to total drag for several typical objects.

In other words, the drag due to differential pressure force is much greater than that due to viscous force. This common sense is based on high Reynolds number flows in which the differential pressure force is much greater than the viscous force.

For a general object, the windward and lateral areas of which are both not zero, its total drag is the sum of pressure and frictional drag. Which of the two types of drag makes the dominant contribution depends on the distribution of pressure and viscous forces and their corresponding action areas. Figure 6.32 shows the ratios of pressure drag and frictional drag to total drag for several typical objects. It should be noted that these different types of drag are usually related to the Reynolds number, so the solutions of the flows over a cylinder and an airfoil are obtained at a specific value of this number. As we will see later, the ratio of pressure drag to total drag is 90% for flow over a cylinder at Reynolds number $Re = 2{,}025$. For flow over an airfoil, the ratio is related not only to the Reynolds number, but also to the shape of the airfoil (affecting the position of separation). Therefore, the values in Figure 6.32 are not of

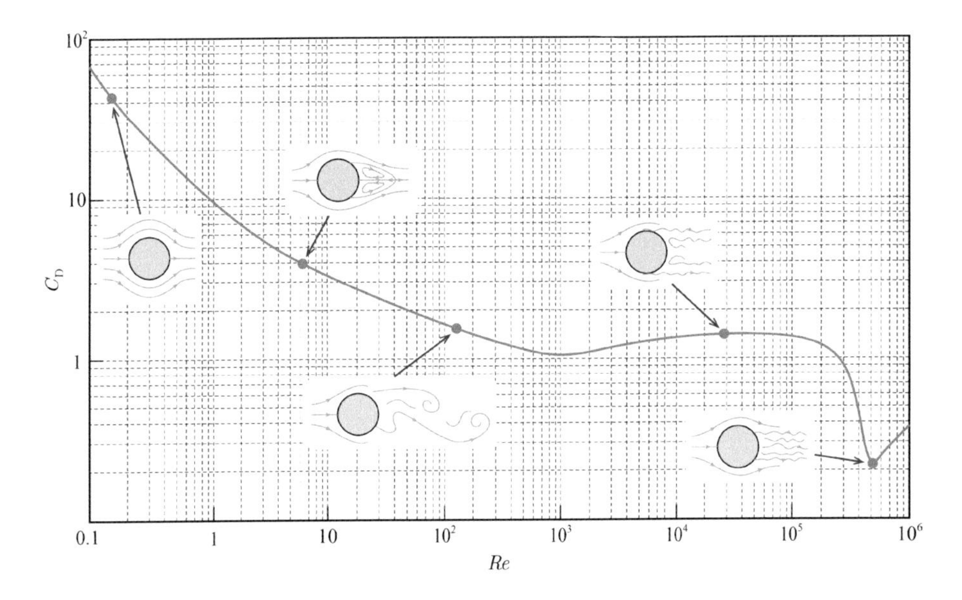

Figure 6.33 Drag coefficient as a function of Reynolds number.

general significance. Nevertheless, they provide insights into the ratio of these two types of drag.

Taking the flow over a two-dimensional cylinder as an example, the frictional drag is mainly determined by the wall shear stress acting on the windward area. On the leeward area, the boundary layer separation occurs and wall stress is negligible. Furthermore, the difference between the high pressure in the front stagnation region and the low pressure in the rear separated region creates a large pressure drag.

The drag coefficient is defined as follows:

$$C_{\mathrm{D}} = \frac{F_{\mathrm{D}}}{\frac{\rho V^2}{2} A}, \tag{6.30}$$

where F_{D} is the drag; V is the freestream velocity; and A is the windward area of the object.

Figure 6.33 depicts the drag coefficient for a cylinder as a function of the Reynolds number, which is determined by fluid velocity, flow scale, and fluid viscosity. Here, the variation of fluid velocity is selected as an example for analysis.

(1) At $Re \ll 1$: At extremely low velocity, the fluid flows over the cylinder and converges at its rear. The seemingly symmetrical flow has an asymmetric pressure distribution with pressure higher on the windward than on the leeward side. For the fluid, viscous force is directly balanced by differential pressure force, while inertia force is almost negligible. This type of flow is also known as creeping flow. Both pressure drag and frictional drag are very large, leading to a significant drag coefficient.
(2) At $Re \sim 1$: As the freestream velocity increases, the near-wall velocity gradient normal to the wall increases in an approximately linear way, so that the frictional

drag is directly proportional to the freestream velocity. Therefore, the drag coefficient is approximately inversely proportional to the freestream velocity (Reynolds number).

(3) At $1 < Re < 10^3$: As the freestream velocity increases, flow separation occurs on the rearward surface of the cylinder, and the pressure drag becomes the dominant contributor to total drag. Due to the forward shift of the separation point, the pressure drag increases. The drag coefficient shows a more gentle decline, approximately inversely proportional to the 0.5th square of freestream velocity.

(4) At $10^3 < Re < 3\times10^5$: The pressure drag that is mostly related to the position of the separation point becomes significantly greater than the frictional drag. In this case, the laminar boundary layer flow on the windward area of the cylinder separates at a fixed point of about 82° from the stagnation point. Therefore, the drag coefficient within this range of *Re* is largely independent of the Reynolds number.

(5) At $Re \approx 3\times10^5$: As the Reynolds number increases, the drag will drop sharply. This is because the boundary layer on the windward area of the cylinder has already changed from laminar to turbulent at some point upstream of separation. Due to the stronger resistance of turbulent boundary layer flow against separation, the flow now separates further downstream (at about 125°). Therefore, the rear separation region with low pressure becomes smaller, and the pressure drag drops sharply.

(6) At $Re > 3\times10^5$: As the freestream velocity increases further, the position of the separation point remains unchanged. The pressure drag has little influence on the drag coefficient. However, an increase in the frictional drag would increase the drag coefficient.

Historically, the phenomenon of the drag dropping sharply above a certain Reynolds number has been known as *drag crisis*. The so-called crisis is not a flow phenomenon with any bad effects, but a puzzle that challenged the traditional theories in fluid mechanics, leading to a crisis in the academic community. It is quite clear to us now that this phenomenon is due to the laminar boundary layer transition to turbulence upstream of the separation point.

We can define two drag coefficients, namely the pressure drag coefficient C_{Dp} and the frictional drag coefficient C_{Df}, to study the relationship between the two types of drag. For flow over a cylinder at $10^3 < Re < 3\times10^5$, the boundary layer flow upstream of the separation point is laminar, so there is an approximate theoretical solution which can be verified by experimental data. The relationship between frictional drag and the Reynolds number is expressed as

$$C_{\mathrm{Df}} = \frac{5.9}{\sqrt{Re}}.$$

Within this range of *Re*, the pressure drag is independent of the Reynolds number, largely considered to be $C_{\mathrm{Dp}} = 1.2$. The ratio of frictional drag to total drag is

$$\frac{F_{\mathrm{Df}}}{F_{\mathrm{D}}} = \frac{C_{\mathrm{Df}}}{C_{\mathrm{Dp}} + C_{\mathrm{Df}}} = \frac{5.9/\sqrt{Re}}{1.2 + 5.9/\sqrt{Re}} \approx \frac{1}{1 + 0.2\sqrt{Re}}.$$

$$\text{At } Re = 10^5,\ F_{\mathrm{Df}}/F_{\mathrm{D}} \approx 14\%.$$

$$\text{At } Re = 10^3,\ F_{\mathrm{Df}}/F_{\mathrm{D}} \approx 1.6\%.$$

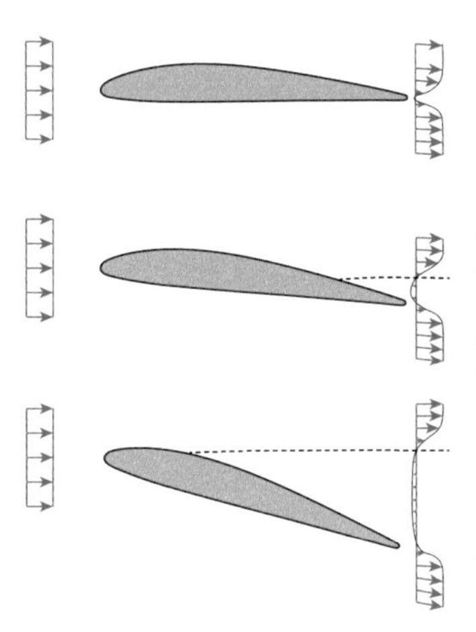

Figure 6.34 Analysis on the drag of a two-dimensional airfoil through an integral approach.

We know that, in most actual flows, Reynolds numbers are larger than 10^5. The above analysis shows that for blunt objects similar to a cylinder, the frictional drag is less than 2% of the total drag. An effective method used to decrease the drag acting on an object is to reduce the pressure drag – that is, to delay or prevent flow separation on part or all of its surface. One of the effective methods to control the flow separation is to "trip" a laminar boundary layer into turbulence in advance. Other common methods include increasing turbulence intensity of the freestream, making the wall surface rougher, and adding local small disturbances on the wall. For example, a golf ball with dimples can travel 4–5 times farther than a smooth ball.

Since there is no large separation region on a streamlined object such as an airfoil, the dominant source of drag can be the frictional drag. We know that the frictional drag in a turbulent boundary layer is much greater than that in a laminar one, hence it is not a good choice to let the transition of the boundary layer occur too early. One way to reduce the frictional drag is to delay the laminar-to-turbulent transition as long as possible, until the boundary layer separation is about to occur. The airfoils based on this principle are called *controlled diffusion airfoils* (CDA) or *laminar airfoils*. These concepts are widely used in wings and compressor vanes design.

If we analyze the drag based on the momentum integral equation, there is no need to distinguish frictional drag from pressure drag, and the total drag can be obtained directly through the magnitude of velocity deficit of the wake. A thick boundary layer indicates a large frictional drag, while a broad separation region indicates a large pressure drag. The velocity deficit of the wake actually reflects the combined effects of the above two factors. Figure 6.34 shows the relative contribution of the boundary layer and separated region to the momentum loss in the wake of an airfoil at different angles of attack. Under normal working conditions, drag is dominated by the frictional drag, and only when a significant flow separation occurs will the pressure drag become the dominant source of drag.

It is said that the pressure drag experienced by an object depends for the most part on the shape of its rear end rather than that of its front end. There is some truth in this. Figure 6.35 shows the shape effects on the amount of drag produced, applicable only to subsonic flows.

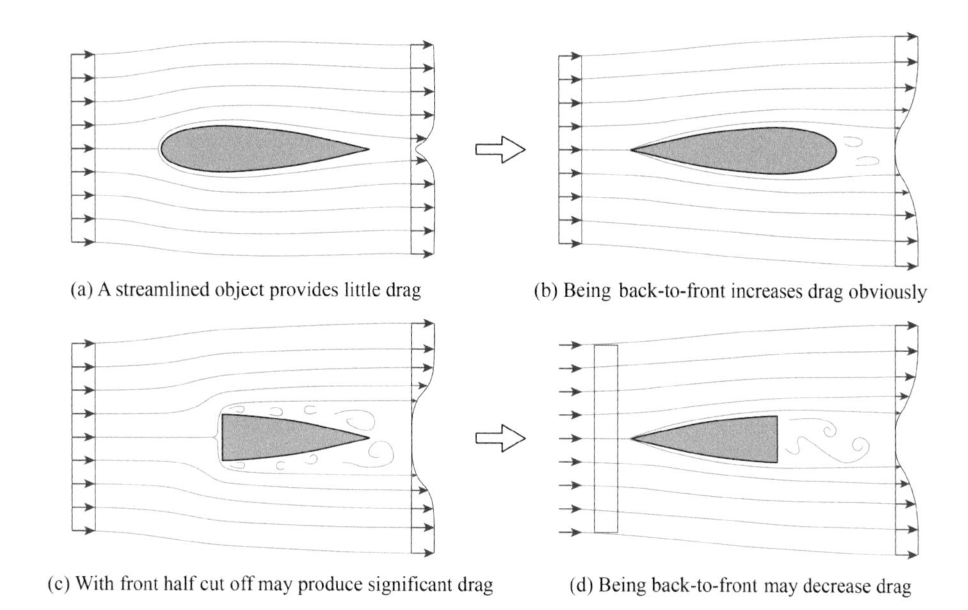

Figure 6.35 The effects of forebody and afterbody shape on drag.

Streamlined designs for objects, round at the front and tapering toward the rear, are implemented in practice to minimize drag. Many animals in nature, such as birds and fish, have streamlined bodies to reduce drag as they move through air or water. Figure 6.35(a) shows the streamlines around a streamlined object. Since the adverse pressure gradient along the surface of the airfoil with a tapering rear end is relatively small, the boundary layer does not separate except for a tiny separated region due to the sudden expansion at the trailing edge. To minimize the drag, the streamlined object should gradually become narrower at the rear. Being back-to-front, however, as shown in Figure 6.35(b), it intensely changes the flow pattern at the rear of the streamlined object. The resulting adverse pressure gradient can become so large that the boundary layer separates just shortly downstream of the location of the maximum airfoil thickness. A broader low-pressure area forms in the wake of the back-to-front streamlined object, which provides a substantially large pressure drag.

However, the shape of the object's front end can also have a great effect on the total drag. As shown in Figure 6.35(a), for the shape effect on the drag it might not be important whether the front end of the streamlined object is tapering or round. However, flow inevitably separates for an object with a square front end, as shown in Figure 6.35(c). If the separated flow over the sharp corners cannot reattach itself to the surface until reaching the trailing edge, a separated region is seen to extend significantly at the rear of the object, resulting in a quite large drag. For this case, being back-to-front might decrease drag, as shown in Figure 6.35(d). This is because the separated region at the rear of the inverted object is of the same order of size as the windward area of the object, smaller than that in Figure 6.35(c).

A change in the shape of the front end in Figure 6.35(a), from round to tapering, may even cause the drag to increase. As long as there is no separation, the shape of the front end negligibly influences the pressure drag. However, the additional frictional drag due to the increased surface area by tapering the front end causes the total drag to increase.

In summary, in subsonic flow, the shape of the object's rear end plays a decisive role in drag, but the shape of its front end is also important. In some cases, an object with a round front and tapering rear end experiences a smaller drag, whereas in other cases the opposite is true. The key to reducing drag is to control boundary layer separation. For general objects, pressure drag is the main contributor to total drag, followed by frictional drag. Both types of drag can be measured by the wake size.

Attempts have been made for years to reduce the aerodynamic drag on cars. It's pretty clear that a streamlined car experiences the least drag, but that type of shape is certainly not the most practical choice. Aided by a deeper insight into flow phenomena, today's cars can be made more aerodynamically optimal without sacrificing practicality. A conventional method is to optimize the shape of the front and rear surfaces, roofs, chassis, etc. For a large container truck with a square head, an air deflector is often mounted on top of its cab, which reduces drag by changing its front shape; there is presently not much change regarding the shape of the rear end of trucks, but it can be expected that with the worsening of the energy crisis increasing attention will be paid to optimizing the aerodynamics here also.

Experiments were performed to obtain the drag coefficient of several typical objects at usual Reynolds numbers, as shown in Figure 6.36. The drag coefficient of the semi-spherical shell that experiences the highest drag is 1.42, while that of the

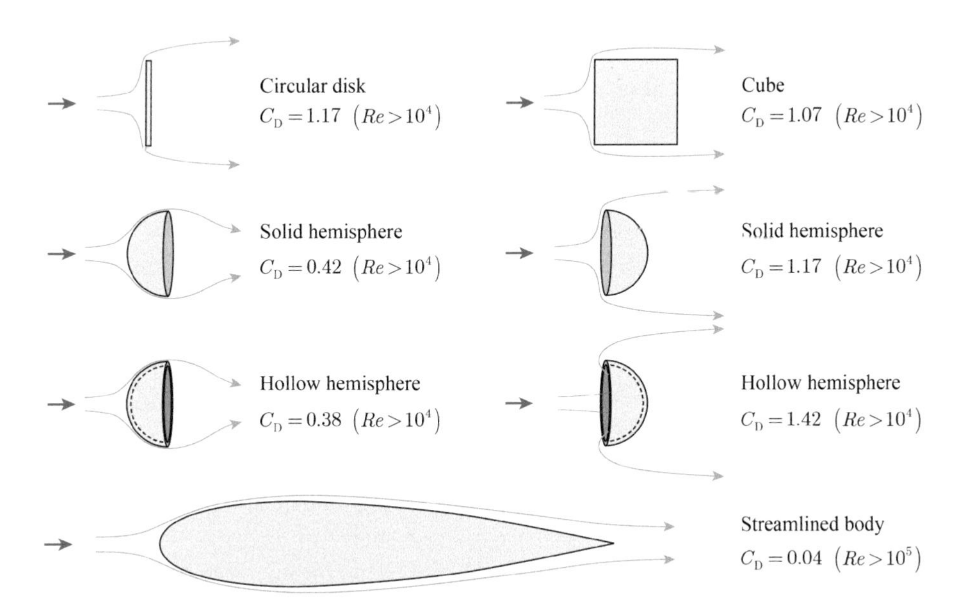

Figure 6.36 Drag coefficients of several typical three-dimensional objects at common Reynolds numbers.

streamlined body experiencing the lowest drag is only 0.04, a difference of a factor of 35 for these two objects. A streamlined aircraft is designed to experience the lowest drag, while a semi-spherical shell-shaped parachute is designed to experience the highest drag.

Tip 6.4: Interesting Air Drag Problems

Let us try to answer the following interesting questions about air drag.

(1) Can a sturdy enough umbrella be used as a parachute?
(2) Can a falling bullet kill a person?
(3) How fast do raindrops fall? Why can fog droplets float in the air?

These questions concern much the same issue, namely the final speed of a falling object when the air drag equals the gravity. For question 1, assume that a 60 kg person holds an umbrella with a diameter of 1.5 m, and the umbrella's shape is a semi-spherical shell with a drag coefficient of 1.42. Then, the following calculations can be performed:

$$D = C_D \times \frac{\pi}{4} d^2 \times \frac{1}{2} \rho V^2 = 1.42 \times \frac{3.14}{4} \times 1.5^2 \times \frac{1}{2} \times 1.2 \times V^2 = 1.5V^2 ,$$

$$D = mg = 60 \times 9.8 = 588\text{N} ,$$

$$V \approx 20 \text{ m/s} .$$

That is to say, a fall from a high place with an umbrella will result in a velocity of roughly 20 m/s, which is equivalent to the final free fall speed of a person jumping from the top of a six-story building. Therefore, an umbrella cannot be used as a parachute because it is not large enough.

The example for question 2 is a pistol bullet with a diameter of 9 mm and a mass of 6 g. Since the falling bullet may roll, its drag coefficient can be estimated from that of a sphere, which is about 0.45 at Reynolds numbers between approximately 10^3 and 10^5. The calculated final speed of the falling bullet is between approximately 50 and 60 m/s, which is much lower than its initial speed. It may cause an injury to a human, but it is usually not life-threatening.

For question 3, assume that raindrops form in a roughly spherical structure due to surface tension. Raindrops vary between 0.5 and 5 mm in diameter, and the Reynolds number lies between 40 and 4,000. They fall at speeds of roughly 2–11 m/s for droplet sizes of 0.5–5 mm in diameter.

Fog droplets vary between 1 and 120 μm in diameter. As estimated by the drag of a sphere, the final speed of the largest droplet is only 0.2 m/s. Since the Reynolds number associated with the smallest droplet is very small, its drag coefficient can be as high as 35,000, while its final speed is about only 0.1 m/h. A drop with an approximate diameter of 1 μm near the ceiling will need a full day until it lands on the floor.

The above analysis, however, is not accurate for very small particles since they do not obey the hypothesis of continuous medium. The final speed of a particle with diameter of 1 μm should be calculated using rarefied gas dynamics.

6.7.2 Flow Losses

For internal flows we are more concerned with flow losses than flow drag. Although in most cases a large loss corresponds to a large drag, the two different concepts essentially correspond to the changes in energy and momentum, respectively. Both momentum and mechanical energy decrease within the boundary layer, so drag and loss have almost the same change trend for boundary layer flow. In the wake region, the momentum deficits basically stay constant downstream of the object, but the corresponding energy deficits increase along with the flow. In other words, *the drag is only generated on the surface in contact with the object. Once the fluid leaves the object, there is no additional drag on the object. However, the velocity deficit downstream in the wake region continually generates losses inside the fluid.*

We start with the simplest examples of ball collisions to understand the distinction between momentum and energy. On a frictionless plane, a ball with a certain speed strikes another identical ball that was initially at rest. If the first ball stops completely, the second ball must have the same speed as the first one. Both kinetic energy and momentum are conserved during this process, which is called a perfectly elastic collision. If the two balls stick together and move at half the initial velocity of the first ball, the conservation of momentum still holds and the kinetic energy is not conserved anymore, but decreases by one-half. This is a perfectly inelastic collision where the kinetic energy is lost or transferred into internal energy. As can be seen, the change trend of mechanical energy is not necessarily consistent with that of momentum. The momentum of a system is conserved if there are no external forces acting on the system. However, much stricter conditions are required for the conservation of mechanical energy.

The inelastic collisions between balls result in the irreversible conversion of mechanical energy into internal energy, with an effect similar to that of viscosity. The energy equation states that the viscous dissipation is primarily caused by the shear motion between fluid particles, so flow losses always occurs wherever there is velocity gradient. For flow around an airfoil with no separation, there are two dominant sources of loss: the frictional loss resulting from the object's surface and the mixing loss in the wake behind the object. In most actual flows, the maximum velocity gradient occurs within the boundary layer, where there is large loss. On the other hand, although the flow experiences the largest local loss within the boundary layer, this region is limited, with a very small amount of fluid involved. For a flow with large separation, the separated region, characterized by multiple, large-scale, unsteady vortices and intense shearing action, becomes the dominant source of loss because it entrains a large amount of fluid from the mainstream.

Besides the frictional loss within the boundary layer, the mixing loss in the wake and that due to flow separation, there is also a considerable amount of mixing loss caused by the secondary flow in fluid machinery, such as fans, pumps, turbines, etc. The secondary flow is characterized in terms of the transverse component of the average velocity, perpendicular to the designed flow direction. The mixing of the mainstreams and the secondary flows would result in a considerable loss.

Flow losses are directly related to entropy rise. When there is no exchange of heat and work between a flow and its surroundings, the change of entropy can be expressed as the change in the total pressure of the fluid as follows:

$$s_2 - s_1 = -R\ln\frac{p_{t2}}{p_{t1}}, \tag{6.31}$$

where p_{t2}/p_{t1} is the *total pressure recovery coefficient.*

Many flows have approximately a negligible exchange of heat and work with their surroundings. Therefore, Equation (6.31) has a wide range of applications. In engineering, the total pressure loss is often used to directly represent the mechanical energy loss. For many flows, the loss is not particularly large and the total pressure recovery coefficient is very close to 1, so that Equation (6.31) reduces to:

$$s_2 - s_1 \approx R\left(\frac{p_{t1} - p_{t2}}{p_{t1}}\right),$$

Notice that the reduction in the total pressure can also represent the magnitude of flow loss. Then, the *total pressure loss coefficient* is a more convenient representation of the energy loss. It is defined as the ratio of thc reduction of the total pressure to the freestream dynamic pressure:

$$\frac{p_{t1} - p_{t2}}{\rho V_1/2}.$$

The total pressure recovery coefficient and the total pressure loss coefficient have different meanings. The former represents the ratio of the residual mechanical energy of a fluid to its original total mechanical energy, while the latter represents the ratio of the mechanical energy loss of a fluid to its original kinetic energy.

For an incompressible flow in a pipe, the total pressure recovery coefficient can be approximately transformed as follows:

$$\frac{p_{t2}}{p_{t1}} = \frac{p_2 + \rho V^2/2}{p_1 + \rho V^2/2} = \frac{p_1 - \Delta p + \rho V^2/2}{p_1 + \rho V^2/2} = 1 - \frac{\Delta p}{p_{t1}}.$$

The pressure drop Δp varies with flow speed. If more of the freestream total pressure is stored in static pressure, there are much smaller flow losses. *The key to reduce flow losses is to store and transport energy in the form of pressure energy rather than kinetic energy*. For the same flow rate, a pipe with greater diameter can reduce loss more effectively because more energy can be stored in the form of pressure energy (with higher static pressure and lower flow speed).

6.7.2.1 Flow Losses in a Pipe with Constant Cross-Sectional Area

For a fully developed laminar pipe flow, the velocity profile remains the same at all sections along the pipe, and all fluid particles move in parallel layers. Since in any streamline the differential pressure is balanced by the viscous shear force, the velocity of each fluid particle remains constant with time, and the static pressure drops in the streamwise direction. Obviously, Bernoulli's equation does not apply, that is to say, the total pressure would not be conserved. Assuming the pipe wall is adiabatic, the decrease in total pressure is directly related to flow losses. Since fluid velocity is constant, the decrease in total pressure is completely reflected by the decrease in static pressure. Therefore, the reduction in static pressure directly represents the flow losses.

In the energy equation, the so-called loss is represented by the dissipation term:

$$\Phi_V = 2\mu\left(\frac{\partial u}{\partial x}\right)^2 + 2\mu\left(\frac{\partial v}{\partial y}\right)^2 + 2\mu\left(\frac{\partial w}{\partial z}\right)^2 + \mu\left(\frac{\partial v}{\partial x} + \frac{\partial u}{\partial y}\right)^2 + \mu\left(\frac{\partial w}{\partial y} + \frac{\partial v}{\partial z}\right)^2 + \mu\left(\frac{\partial u}{\partial z} + \frac{\partial w}{\partial x}\right)^2 .$$

Since the two-dimensional fully developed flow in a pipe is axisymmetric, the above equation reduces to:

$$\Phi_V = \mu\left(\frac{\partial u}{\partial r}\right)^2 .$$

In other words, for incompressible, laminar flows in a straight pipe, the loss is related only to the radial velocity gradient.

Since shear flow exists everywhere in a pipe, the flow losses occur over all regions. Any factor enhancing such shearing action always leads to the increase in flow losses. If we raise the average velocity or lower the pipe diameter, the radial velocity gradient will grow, leading to the increase in flow losses.

In theory, the flow losses mean that the kinetic energy is irreversibly converted into internal energy. However, the velocity stays the same throughout the length of the pipe, while static pressure decreases along it. How can this be explained?

It can be understood in this way: The viscous forces in a fluid have been transforming kinetic energy into internal energy through shearing action. Simultaneously, the pressure potential energy of the fluid continuously transforms into kinetic energy, implying a dynamic equilibrium among the three kinds of energy. The pipe flow can be regarded as a continuous conversion of pressure potential energy into internal energy, the process being realized through viscous stress.

6.7.2.2 Flow Losses in a Pipe with Sudden Expansion

As the fluid goes from a smaller to a larger pipe, the separated flow which resulted from the sudden expansion geometry will generate great energy loss. Different from the flow losses in a constant cross-section pipe, the flow losses in a pipe with sudden expansion is mainly caused by separation-induced mixing rather than wall friction.

Figure 6.37 schematically shows the flow through a horizontal pipe with sudden expansion. In this case, the mixing loss is much larger than the pipe friction loss, so it is assumed that the wall shear stress acting against the entire flow is negligible. For the control volume shown, the continuity equation is:

$$A_1 u_1 = A_2 u_2 .$$

For the control volume, only the differential pressure force is exerted, and its magnitude is expressed as:

$$\sum F = p_1 A_2 - p_2 A_2 = (p_1 - p_2) A_2 .$$

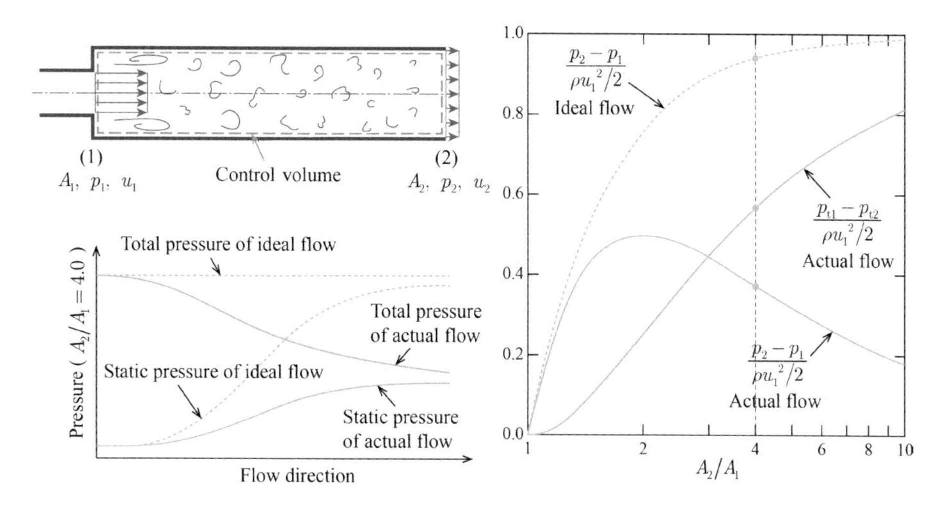

Figure 6.37 Flow losses in an axisymmetric pipe with sudden expansion.

The net momentum flow rate out of the control volume is:

$$\dot{m}(u_2 - u_1) = \rho A_1 u_1 (u_2 - u_1).$$

The momentum equation is:

$$(p_1 - p_2) A_2 = \rho A_1 u_1 (u_2 - u_1).$$

Combining the momentum and continuity equations gives the total pressure loss coefficient for a sudden expansion:

$$\frac{p_{t1} - p_{t2}}{\rho u_1^2/2} = \left(1 - \frac{1}{AR}\right)^2.$$

where $AR = A_2/A_1$ represents the expansion area ratio.

The above formula computes the magnitude of head loss due to sudden expansion. The theoretical analysis that neglects the wall friction agrees well with experimental results. As can be verified, if the area ratio is 4, the total pressure loss coefficient is 0.563. That is to say, the head loss due to sudden expansion is more than half the incoming flow dynamic pressure. If the area ratio approaches infinity, the total pressure loss coefficient is 1 – that is, 100% of the dynamic pressure is lost; this is equivalent to the situation of a jet entering infinite space.

For incompressible flows, the outlet velocity is determined by the continuity equation, or the inlet-to-outlet area ratio. Therefore, the dynamic pressure also depends on this ratio. The outlet dynamic pressure should stay the same for both ideal and viscous flows. The loss causes the total pressure to decrease, and the corresponding static pressure rise cannot reach the ideal level of the diffuser. According to the above analysis, the static pressure rise coefficient for sudden expansion can be calculated:

$$\frac{p_1 - p_2}{\rho u_1^2/2} = \frac{2}{AR}\left(1 - \frac{1}{AR}\right),$$

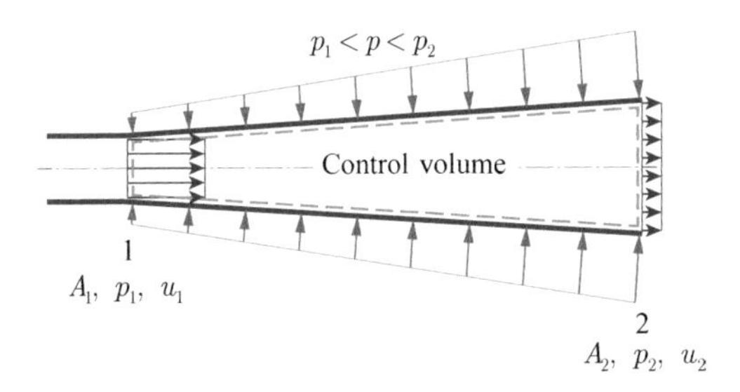

Figure 6.38 Flow and pressure distribution in a diverging pipe.

Figure 6.37 depicts the static pressure rise as a function of area ratio, obtained by using this formula. The static pressure rise coefficient reaches its maximum value of 0.5 at an inlet-to-outlet area ratio of 2. In other words, the maximum dynamic pressure recovery for sudden expansion can only be 50%. A further increase in area ratio will increase the losses and reduce the capability of the diffuser.

Just as in the case of the flow in a constant cross-section pipe, the above analysis derived from the momentum equation cannot reflect the essence of loss. Next, we shall examine in detail the mechanism of losses in this case.

First, let us look at how a simple control volume analysis for inviscid flow can be used to compute the magnitude of losses. The crucial point for force field analysis is that the pressure on the left side of the control surface refers to the pressure inside the smaller diameter pipe, p_1, while the action area refers to the cross-sectional area of the larger diameter pipe, A_2. This assumption is based on the fact that any fluid flowing in the smaller diameter will enter the larger diameter pipe in parallel streamlines. Therefore, the actual flow through the sudden expansion does not conform to Bernoulli's equation. In other words, here the effect of viscosity is taken into account, which causes a loss in the separated region and irreversibly converts part of the mechanical energy into internal energy.

Of course, the actual losses do not occur immediately at the sudden expansion, but are continuously produced by downstream mixing. Consider the pipe flow through a sudden expansion having an area ratio of 1:4, as shown in the lower left of Figure 6.37. The mixing region starts just downstream of the sudden expansion and extends a distance of about 4–5 times the pipe diameter. Within this distance, the static pressure rises, while the total pressure drops.

If the pipe is a gradually diverging pipe rather than a sudden expansion, the above-mentioned losses will not take place. Consider the control volume chosen from a diverging pipe, as shown in Figure 6.38. Besides the inlet and outlet, there is also pressure on the circular wall; which gradually increases from inlet to outlet. It is this pressure that guarantees the conversion from kinetic energy into pressure potential energy, making the flow obey Bernoulli's equation. Actually, this model is often used in the derivation of that equation.

We can analyze the flow at the differential level. As fluid flows from a smaller into a larger pipe through sudden expansion, at the beginning, both its static pressure and

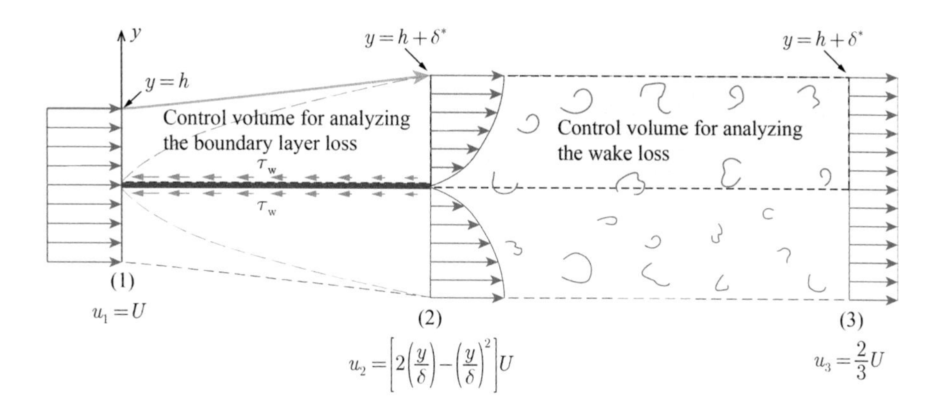

Figure 6.39 Friction loss in the boundary layer over a plate and mixing loss in the wake behind it.

velocity stay the same, thereby generating no losses. As it flows farther downstream, the fluid slows down due to two factors: one is the shearing action of viscous forces exerted on the jet by the low-speed fluid in the surrounding separated region, which irreversibly converts kinetic energy into internal energy. In this case, a decrease in dynamic pressure does not cause the static pressure to increase, so there is a loss of total pressure. The other is the obstruction of the downstream pressure to the jet. For an incompressible flow, this work done by the pressure difference force conforms to Bernoulli's equation. For a compressible flow, the compression work causes the internal energy of the fluid to increase. This completely reversible conversion results in no loss of total pressure. As an extreme case, we consider the jet flow into an infinite space. The expansion ratio is infinite, and the outlet pressure is equal to the pressure further downstream. Then, there is no lossless deceleration caused by pressure difference force. The jet is completely decelerated by viscous shear force at the jet's side surface, therefore 100% of dynamic pressure is lost.

6.7.2.3 Losses in the Wake

A wake is the flow pattern that develops behind an object. The viscous force not only causes energy dissipation over the object's surface, but also produces a certain amount of velocity deficit behind the object after the fluid leaves it. As the flow proceeds downstream, the wake gradually mixes with the mainstream. This mixing process is almost completely dominated by viscous forces, resulting in an obvious loss of energy.

The loss due to viscous mixing in the wake is analyzed for the flow along a zero-thickness plate parallel to the freestream, as shown in Figure 6.39. First, consider the loss generated by the boundary layer over the plate. The inlet of the control volume is taken at the leading edge of the plate, the outlet of the control volume is taken at the trailing edge, the lower control surface is taken at the wall surface, and the upper control surface is taken at the streamline that intersects with the outer edge of the boundary layer at the outlet.

Let the total and static pressure at the inlet be p_t and p_s, respectively. The outlet static pressure is still p_s, but there is a certain loss of total pressure. The magnitude of this loss depends entirely on the velocity deficit. At the wall surface, the dynamic pressure is completely lost, that is, $p_{t,wall} = p_s$. At the outer edge of the boundary layer, there is no loss of total pressure, that is, $p_{t,edge} = p_t$. The outlet total pressure is reasonably estimated as a mass weighted average:

$$\overline{p_{t2}} = \frac{1}{\dot{m}} \int_0^\delta p_{t2} \cdot \rho u_2 \mathrm{d}y\ ,$$

where the outlet and inlet flow rates are equal. The inlet height of the control volume is assumed to be h, which is equal to the outlet boundary layer thickness δ minus the outlet displacement thickness δ^*. Then,

$$\dot{m} = \rho h U = \rho U\left(\delta - \delta^*\right).$$

At the outlet, the total pressure is the sum of the static and dynamic pressures:

$$p_{t2} = p_s + \frac{1}{2}\rho {u_2}^2\ .$$

Combining the above three formulas, the outlet averaged total pressure is:

$$\overline{p_{t2}} = \frac{1}{\rho U\left(\delta - \delta^*\right)} \int_0^\delta \left(p_s + \frac{1}{2}\rho {u_2}^2\right) \cdot \rho u_2 \mathrm{d}y\ .$$

Typical velocity profiles for laminar boundary layers are close to a quadratic curve, expressed as:

$$\frac{u}{U} = 2\left(\frac{y}{\delta}\right) - \left(\frac{y^2}{\delta}\right), \quad 0 \le y \le \delta\ .$$

Substituting the velocity profile into the above formula for averaged total pressure gives:

$$\overline{p_{t2}} \approx p_s + 0.69\left(\frac{1}{2}\rho U^2\right).$$

The total pressure loss coefficient within the boundary layer is:

$$\frac{p_{t1} - \overline{p_{t2}}}{\rho U^2/2} \approx 0.31\ .$$

We shall next analyze the wake loss. The static pressure is still reasonably assumed to remain the same everywhere, so the fluid in the wake flows in parallel layers. In addition, it is assumed that the fluid in the wake no longer experiences the shearing action exerted by the mainstream after leaving the plate. With increasing downstream distance, turbulent mixing gradually decomposes the wake, which becomes uniform through the entrainment with the surrounding mainstream.

For the control volume shown in Figure 6.39, the continuity equation states that the mass flow rates at the inlet and outlet are equal:

$$\int_0^\delta \rho u_2 \mathrm{d}y = \rho u_3 \delta,$$

where the velocity term inside the left integral sign, u_2, takes the fluid velocity at the end of the boundary layer. The uniform velocity at the outlet can be calculated using the assumed quadratic velocity profile and continuity equation:

$$u_3 = \frac{2}{3}U \, .$$

Thus, the total pressure at the outlet can be calculated:

$$p_{t3} = p_s + \frac{1}{2}\rho u_3{}^2 \approx p_s + 0.44\left(\frac{1}{2}\rho U^2\right).$$

According to the undisturbed freestream dynamic pressure upstream of the leading edge of the plate, the total pressure loss coefficient is calculated as follows:

$$\frac{\overline{p_{t2}} - p_{t2}}{\rho U^2/2} \approx 0.24 \, .$$

To sum up, the loss generated by the dissipation within the boundary layer is equal to 31% of the freestream dynamic pressure; the loss generated by dissipation in the wake is equal to 24% of the freestream dynamic pressure, and the total losses are 55%.

The above results are only valid for the fluid within the boundary layer. For specific flow problems, the amount of total pressure loss will vary depending on the percentage of boundary layer mass flow in the total mass flow. What is emphasized here is that the wake loss cannot be ignored compared to the boundary layer loss. In the present example, the contributions of both boundary layer and wake to the total loss are expressed as:

$$\text{Boundary layer loss: } \frac{Loss_{\text{BL}}}{Loss_{\text{Total}}} = \frac{0.31}{0.31 + 0.24} \approx 56\%,$$

$$\text{Wake loss: } \frac{Loss_{\text{Wake}}}{Loss_{\text{Total}}} = \frac{0.24}{0.31 + 0.24} \approx 44\% \, .$$

Expanded Knowledge

The Theory of Homogeneous Isotropic Turbulence

As the complex motion of a fluid, turbulence flow is difficult to describe with a general theory. So far, a theory has been established for homogeneous isotropic turbulence. Andrey Nikolaevich Kolmogorov (1903–1987), a scientist from the former Soviet Union, made the greatest contributions to this theory. On the basis of previous turbulence theories, in 1941 Kolmogorov proposed a model describing a universal structure of small-scale turbulent motion within a homogeneous fluid under high Reynolds numbers, which is sometimes called the *K41 theory*. Here we present a brief introduction to this theory. Interested readers can gain a deeper insight from textbooks on turbulent flow.

The aim of turbulence theory is to investigate the transportation of energy through the formation, interaction, and dissipation of eddies. The so-called energy refers to the *turbulent kinetic energy*, which is related to the fluctuating velocity components. For homogeneous isotropic turbulence, the averaged fluctuating velocity components in all directions remain the same, and the turbulence kinetic energy is expressed as:

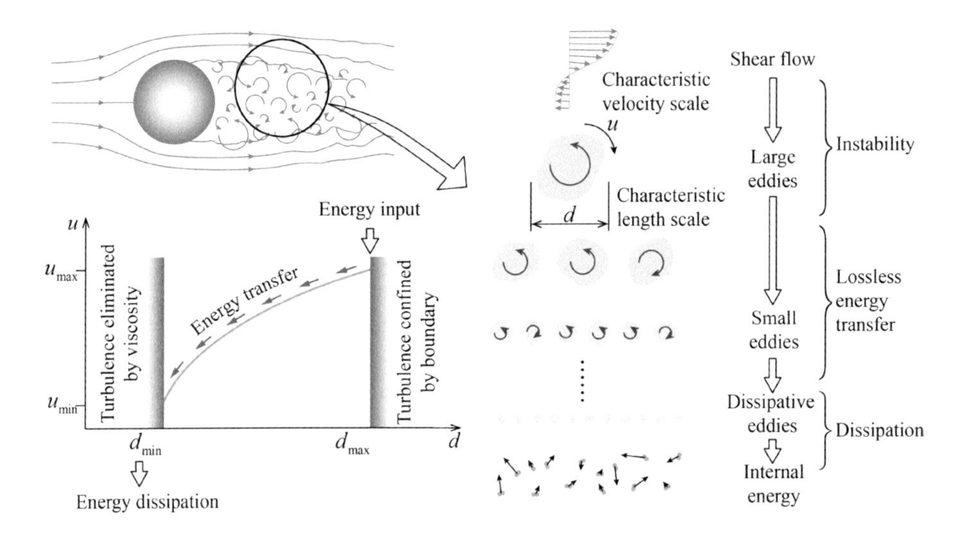

Figure 6.40 Energy transfer and energy spectrum in turbulence.

$$k = \frac{1}{2}\left(\overline{u'^2} + \overline{v'^2} + \overline{w'^2}\right) = \frac{3}{2}\overline{u'^2} .$$

In classical theory, turbulent flow is viewed as a flow made up of eddies of different scales. For a particular flow, the largest-scale eddies represent the shear and rotation of the average flow. For high Reynolds number flows, such large eddies tend to be unstable and break into a number of small ones with strong unsteadiness, which converts the kinetic energy of the average flow into turbulent kinetic energy. These small eddies might still be unstable and further break into smaller ones. In this region, turbulent kinetic energy is not dissipated but continuously transported from large-scale to small-scale eddies, until the scales are small enough (corresponding to very small Reynolds numbers) for the dominant viscous forces to dissipate the kinetic energy into internal energy. This process, which consists of the generation, transportation, and dissipation of turbulent kinetic energy, was first proposed by the British scientist Lewis Fry Richardson (1881–1953) in 1922, as shown in Figure 6.40.

In classical turbulence flow theory, the smallest eddy in a turbulent flow is called a *dissipative eddy*, with a Reynolds number of 1. Its size is called dissipative scale, or Kolmogorov scale η. On this smallest scale, turbulent kinetic energy will be rapidly dissipated by viscosity. The Reynolds number based on the dissipative eddy is expressed as:

$$Re_k = \frac{u_k \eta}{\nu} \sim 1 ,$$

where $\nu = \mu/\rho$ is the kinematic viscosity of the fluid and u_k may be simply understood as the rotational velocity of the dissipative eddy. This formula apparently states that the higher the viscosity, the larger the dissipative eddy size. However, this is not the case, because u_k is also related to the viscosity. The dissipation rate of turbulent kinetic energy, denoted by ε, is the rate at which the turbulent kinetic energy is converted into internal energy. The following relation can be obtained:

$$\eta \sim \left(\frac{\nu^3}{\varepsilon} \right)^{1/4}.$$

The spectral energy transfer states that the dissipation rate of turbulent kinetic energy is equal to its production rate, which is the net conversion rate per unit mass from kinetic energy of the average flow into turbulent kinetic energy. For a fluid, the greater the turbulent kinetic energy taken from the mean flow, the smaller the dissipation eddy size. Suppose you are stirring a glass of water. The faster you stir, the more chaotic the water becomes – that is, the fluid eddies cover a wider range of spatial scales. More intense stirring results in a greater production rate of turbulent kinetic energy, and the wider range of spatial scales imply a smaller size of the smallest eddy (dissipative eddy).

The size of the largest eddies in turbulent flow, denoted by L, is set by that of the system boundary, such as pipe diameter or boundary layer thickness. Its relationship with the size of the dissipative eddies is expressed as:

$$\frac{L}{\eta} \sim Re_{\mathrm{L}}{}^{3/4}.$$

For a flow with a given geometry, this formula states that the higher the Reynolds number, the smaller the size of the dissipative eddies. In other words, the flow at high Reynolds numbers covers a wider range of spatial scales. When the Reynolds number is small, the sizes of the large and small eddies are similar. In this case, the kinetic energy begins to dissipate in the large eddies, and the transportation process of energy cannot form. This type of flow is a laminar flow with eddies, rather than a turbulent flow.

Numerical Computation of Turbulent Flows

Computational fluid dynamics (CFD) has been developed rapidly in recent years to make quantitative predictions of fluid flow phenomena, but turbulence is still an obstacle that engineers must face. From the perspective of physics, the laminar flow at high Reynolds numbers becomes unstable and transitions to turbulence. The mathematical properties of the N-S governing equations are such that these equations are extraordinarily sensitive to small perturbations in initial and boundary values at high Reynolds numbers, producing irregular differences in the long-term behavior of the solutions. In other words, the fluid appears to behave in a chaotic fashion. Even if a steady solution of the N-S equations can be obtained, it represents only one of an infinite number of solutions. In practical problems, the flow pattern corresponding to this solution is ephemeral and of little value. When solving a practical problem, one usually cares more about the time-averaged flow field and the spectral information of the fluctuating terms.

In CFD, direct numerical simulation (DNS) is used to solve instantaneous N-S equations. It can obtain the whole range of spatial and temporal scales of the turbulence. Reynolds averaged Navier–Stokes (RANS) equations are used to estimate the

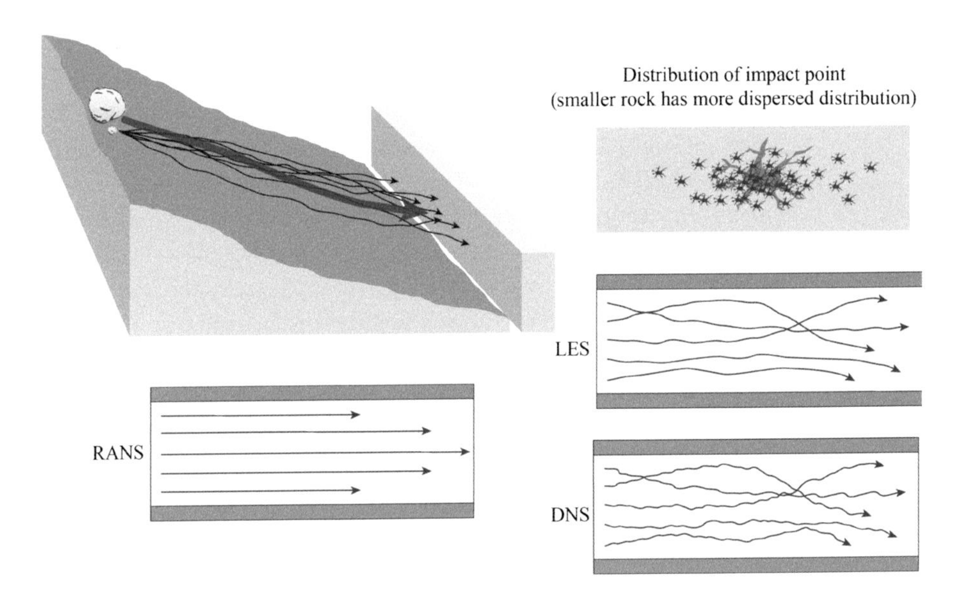

Figure 6.41 Comparison of numerical methods for the calculation of turbulence.

effect of fluctuating terms with mathematic models. As a third option, large eddy simulation (LES) is used to solve N-S equations for the large eddies in turbulence, while small isotropic eddies are estimated with mathematical models.

Due to the very small spatial and temporal scales of small eddies in high Reynolds number flows, an astronomical amount of computer memory and time are needed to perform a computation. Limited by current computing power, most of the time there is no choice but to adopt the RANS method. In this situation, it is crucial to choose the proper turbulence model within CFD. At present, DNS is typically used for the study of turbulence itself. DNS involves the unsteady solution of the N-S equations directly, without any assumptions or simplifications to obtain the exact instantaneous motions. These instantaneous solutions can be averaged to theoretically yield a time-averaged motion of turbulence, provided their results are accurate and sufficient in number. At present, the rapidly developing LES approach can lower the computational cost compared to DNS, and accurately model time-varying turbulent flows of features at smaller scale compared to RANS.

Here is an example to illustrate the comparison of these three calculation methods. As shown in Figure 6.41, stones are rolled down from the top of a hill, and we are to estimate where they will hit the wall. Due to the uncertainty of the shape of the stones and the nonuniform surface roughness, this problem is difficult to solve in a theoretical way. Moreover, the collision position is not truly random, and errors may be introduced by statistical theory. Rolling a small stone gives drastically different results every time. Since the rolling of a large stone is not sensitive to small disturbances on the ground, the result of a single measurement is close to the average of several measurements of the rolling of smaller stones. Since DNS takes all details of a flow field into account, the solutions are greatly affected by small perturbations, resulting

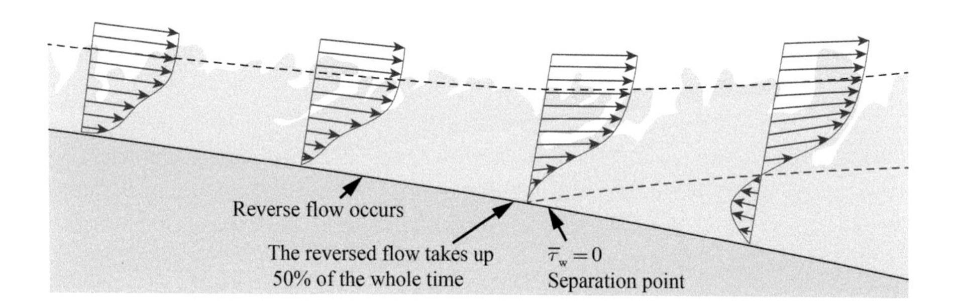

Figure 6.42 Turbulent boundary layer separation.

in a large dispersion of the solutions. One needs to average the time-dependent values to obtain the averaged solution. Since RANS ignores the details and impacts of small turbulent fluctuations, the solution might be closer to the average result. So, the RANS approach appears to be sufficient if one is interested only in the averaged flow. However, there is no perfect turbulence model to simulate the effect of the small-scale turbulence on the average flow. Therefore, there is no guarantee that the RANS results will reproduce the averaged DNS results. In general, the time-averaged flow field obtained using DNS or LES should be more accurate.

In conclusion, RANS yields the averaged flow field just in theory, while DNS and LES are able to provide more details of the flow field. For the majority of engineering problems, the most important result should be an accurate time-averaged flow field. RANS will be the major method for a rather long period of time.

Turbulent Boundary Layer Separation

Separation of turbulent boundary layers is not as strictly defined as that of two-dimensional laminar boundary layers. In general, the magnitude of the fluctuating velocity within the boundary layer is distinctly lower than the average velocity, so no reverse flow occurs for an attached turbulent boundary layer. Near the separation point, the average velocity decreases considerably, while the fluctuating velocity does not change that much, leading to the occurrence of instantaneous reverse flow. In general, the separation point is the location of zero average wall shear stress. The reverse flow already appears at a certain distance upstream of the separation point, as shown in Figure 6.42. In other words, the occurrence of instantaneous reverse flow cannot be used as the separation criterion for turbulent boundary layers.

Questions

6.1 In general, the critical Reynolds number for pipe flow is about 2,300, while its value for the boundary layer over a flat plate is about 10^5. Why is the difference so great?

6.2 The velocity profile in Newton's shear stress experiment (Figure 1.7) is linear, while the velocity profile in the boundary layer over a flat plate is close to a quadratic curve. Why?

6.3 Laminar flow tends to occur through pipes with small diameter. For the boundary layer over a flat plate at high Reynolds numbers, the turbulence intensity is largest near the wall, while the mainstream is laminar. Will the turbulence be enhanced or weakened due to the presence of the wall?

6.4 It is difficult to define a turbulent flow, but we can usually distinguish laminar from turbulent flow at a glance. Summarize a few characteristics which determine if a flow is laminar or turbulent.

7 Fundamentals of Compressible Flow

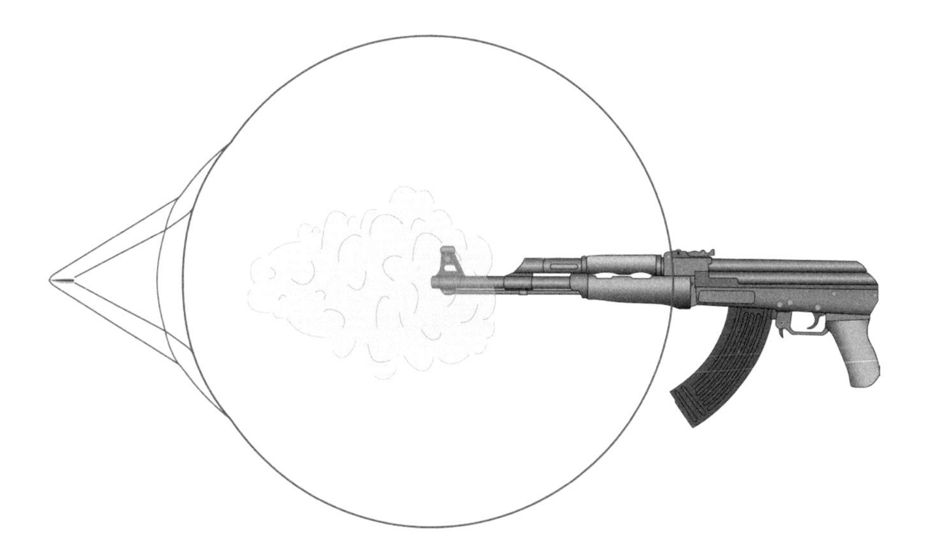

A rifle's bullets will arrive before one can hear the sound.

7.1 Sound Speed and Mach Number

The only criterion for determining the compressibility of a flow is whether the volume of a fluid element changes as it moves. During compression of a fluid, the surroundings do compression work on the fluid, while during expansion of the fluid, the system does expansion work on the surroundings. Consequently, in these processes there is a conversion between mechanical and internal energy. The flow can still be isentropic, but it does not obey the conservation of mechanical energy described by Bernoulli's equation anymore.

The volume of a fluid element changes with the external forces acting on it. A change in temperature causes the fluid to have a corresponding change in pressure. If the fluid element is not trapped and is able to change volume, its pressure tends to be consistent with the ambient pressure, resulting in thermal expansion or contraction. This type of compressible flow is due to temperature change. As air rises from the Earth's surface to a height of several thousand meters, it expands because of lower atmospheric pressure. This is a type of compressible flow due to body forces (gravity). For flow through a rotating pipe or duct, the fluids moving from areas of small diameter to those of large diameter will be decelerated and consequently compressed. This is a type of compressible flow due to inertial forces.

In practical problems, many compressible flows are caused by inertial forces. As the fluid moves at high speeds, the inertial forces often play a crucial role in the compression and expansion of the fluid elements. In particular, as the fluid velocity approaches or exceeds the speed of sound, the impact of compression and expansion due to inertial forces on the flow will be determinant. Therefore, *the so-called compressible flows often refer to transonic and supersonic flows*. Mach number is used to determine if a flow can be treated as incompressible or compressible.

7.1.1 Speed of Sound

An infinitesimal pressure disturbance transmits through fluids as longitudinal waves that consist of a repeating pattern of compressions and rarefactions. The rate of propagation of the pressure disturbance depends only on the thermodynamic property of the fluid rather than the type of disturbance. Sound waves, whether those audible to human ears, infrasound, or ultrasound, travel at the speed of sound.

Longitudinal wave propagation can be analyzed by the deformation transfer in a spring. Clearly, the greater the stiffness of the spring, the greater the force causing the deformation transfer; the greater the mass of the spring, the greater the inertia dragging the deformation transfer. According to Newton's second law, an object subjected to a greater force will experience a larger acceleration, while an object with greater mass will experience a smaller acceleration. If E is the modulus of elasticity and ρ is the mass per unit length of the string, the propagation speed of an elastic deformation should be positively correlated with E/ρ. For springs of the same shape and size, longitudinal waves always propagate faster in steel springs than in copper springs because steel has a higher modulus of elasticity but a lower density than copper. Longitudinal waves propagate at significantly different speeds through springs of the same material and mass per unit length but with different springs and wire dimeters, as shown in Figure 7.1. The longitudinal waves travel more quickly through a thinner spring made of thicker wire due to its larger modulus of elasticity.

In Hooke's law, the modulus of elasticity is defined as the ratio of the stress to the strain (i.e., $E = \sigma/\varepsilon$). For fluids, the stress refers to the pressure difference, denoted by $\mathrm{d}p$, and the strain refers to the fractional change in volume, denoted by $-\mathrm{d}B/B$. The specific volume (B) of any fluid is the reciprocal of its density (ρ), so the fractional change in volume is expressed as

$$-\mathrm{d}(1/\rho)/(1/\rho),$$

and the modulus of elasticity of the fluid is

$$E = \frac{\mathrm{d}p}{-\mathrm{d}(1/\rho)/(1/\rho)} = \rho\frac{\mathrm{d}p}{\mathrm{d}\rho}.$$

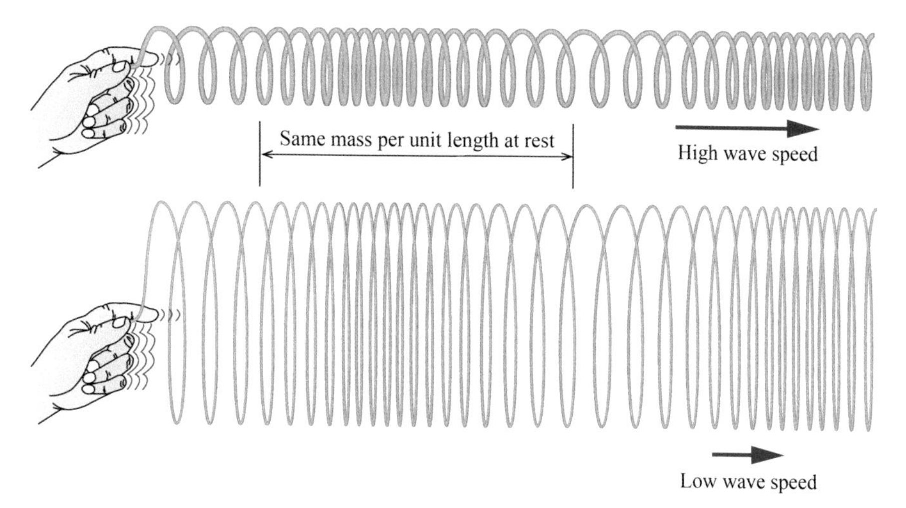

Figure 7.1 Wave speed in two springs of the same material and mass but different modulus of elasticity.

As in the case of the spring, the velocity of propagation of a disturbance depends upon the E/ρ in the medium. That is to say, the speed of sound in a fluid can be qualitatively expressed as:

$$a \sim \mathrm{d}p/\mathrm{d}\rho.$$

The control volume integral technique can be used to derive the formula for the speed of sound from the continuum and momentum equations for one-dimensional flows. This derivation can be found in textbooks on aerodynamics, and the general formula is:

$$a = \sqrt{\mathrm{d}p/\mathrm{d}\rho}. \tag{7.1}$$

As can be seen, the speed of sound is positively correlated with $\mathrm{d}p/\mathrm{d}\rho$, which is consistent with the qualitative analysis.

Equation (7.1) represents the general formula for the speed of sound. Given the expression for pressure as a function of density, a more practical formula for the speed of sound through specific gases and liquids can be obtained. Isaac Newton considered the propagation of sound waves through gases as an isothermal process, and obtained the following formula for the speed of sound:

$$a_{\text{isothermal}} = \sqrt{RT}.$$

The isothermal assumption seems reasonable because we cannot feel the change in temperature due to the propagation of sound waves. However, the speed of sound calculated by this formula is much smaller than the measured one. For example, the actual speed of sound in air at 15°C is 340 m/s, while that calculated by the formula is only 288 m/s.

After Newton, there were many attempts to calculate or directly measure the speed of sound, among which the greatest contribution was made by Pierre-Simon Marquis de Laplace (1749–1827), and eventually a more accurate formula was derived. Now we know that the pressure and density of an isentropic flow are related as follows: $p/\rho^k = \text{const}$. We can use this relationship to easily obtain the formula for the speed of sound:

$$a = \sqrt{kRT}. \tag{7.2}$$

In a given ideal gas, the speed of sound depends only on its temperature; the higher the temperature, the greater the speed of sound.

We can examine the speed of sound in terms of molecular motion. Gas molecules move faster as a result of a higher temperature. On a microscopic scale, disturbances are transmitted through a gas by the collisions between the molecules in random motion. Consequently, it can also be concluded that *the propagation velocity of a small disturbance through a gas corresponds to the average velocity of gas molecules*. The thermal speed of the gas molecules at room temperature obeys the Maxwell–Boltzmann distribution. At the same temperature, different types of gas molecules travel at different speeds. The air we breathe is a mixture of various gases, so the gas molecules in air move at a wide range of thermal speeds. However, the average speed of gas molecules is found to roughly correspond to the speed of sound.

7.1.2 Mach Number

In previous chapters, the Reynolds number often appears as an important dimensionless number in incompressible flow analysis. Here is another important dimensionless number in compressible flow analysis: the *Mach number*, which is defined as a speed ratio, referenced to the local speed of sound:

$$Ma = \frac{V}{a}.$$

Since the speed of sound is not a constant, the same Mach number does not describe the same speed. For example, a fighter jet cruising at Mach 2 is traveling at twice the local speed of sound in the stratosphere, rather than at twice the speed of sound in air at sea level (i.e., 680 m/s). We know that the average temperature of the stratosphere is about −60°C, and the local speed of sound calculated by Equation (7.2) is about 290 m/s. So, the aircraft travels at 580 m/s.

Since the Mach number does not represent the actual aircraft's speed, why is it used in aerodynamics to represent the magnitude of velocity? Similar to the Reynolds number that describes the ratio of inertial forces to viscous forces, *the dimensionless Mach number describes the ratio of inertial to elastic forces*. At larger Reynolds numbers, the effects of viscous forces are relatively smaller, while at larger Mach numbers, the effects of elastic forces are relatively smaller. At high Mach numbers, gas is similar to a heavy spring with a small elastic modulus, which is significantly easier to compress and stretch. For example, if the air accelerates isentropically from rest to a Mach number of 3.0, its density will be reduced by about 92%, equivalent to being greatly stretched.

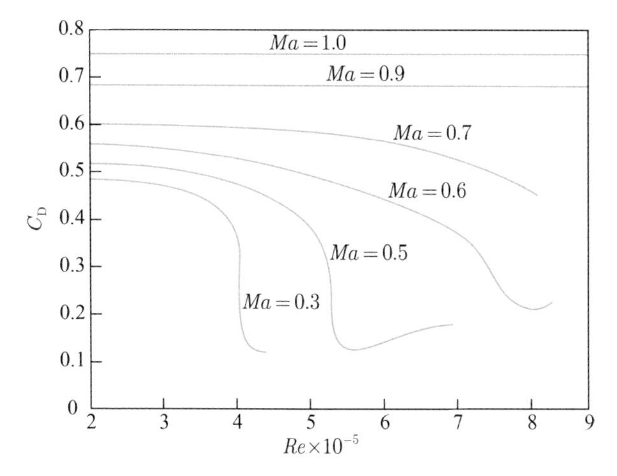

Figure 7.2 The drag coefficient for a ball as a function of Reynolds number and Mach number.

For actual flows at high Mach numbers, compressibility has far greater effects on flows than viscous effects. Figure 7.2 shows the effect of the Mach number on the drag coefficient of a ball in a particular range of Reynolds numbers. As can be seen, the drag on a ball described in Figure 6.33 is only applicable to low Mach number flows, or incompressible flows. At higher Mach numbers, the effect of the Reynolds number can be ignored. In other words, when compressibility plays a significant role in high-speed flows, the viscous effects take a back seat.

The relationship between compressibility and Mach number can be derived from the Eulerian equation for one-dimensional compressible flows:

$$\frac{\mathrm{d}p}{\rho} + V\mathrm{d}V = 0,$$

rearranged as:

$$\frac{\mathrm{d}p}{\mathrm{d}\rho} \cdot \frac{\mathrm{d}\rho}{\rho} = -V\mathrm{d}V.$$

We can substitute the formulas for speed of sound ($\mathrm{d}p/\mathrm{d}\rho = a^2$) and Mach number ($Ma = V/a$) into the above equation:

$$\frac{\mathrm{d}\rho}{\rho} = -Ma^2 \frac{\mathrm{d}V}{V}.$$

This formula gives the expression for fluid density as a function of fluid velocity. At $Ma = 0.3$, the change of relative density is 9% of that of relative velocity. That is to say, the density will change by less than 1% when the velocity changes by 10% in the flow field. Therefore, the density can be assumed to be nearly constant. This formula also states that for $Ma > 1$, the decrease in density is much greater than the increase in velocity, leading to a decrease in mass flux, ρV. According to the continuity equation, $\rho VA = \text{const}$, the supersonic flow will accelerate through the diverging duct.

7.2 Steady Isentropic Flow Equations

To focus on the compressibility effects, let us consider a steady, inviscid flow where there is no exchange of heat or work between the fluid and its surroundings. The total energy of the fluid is conserved as a consequence of the absence of heat transfer into or out of the flow and no work done on or by the flow. At the same time, there is no irreversible conversion from mechanical to internal energy due to the lack of viscous forces, so the flow is isentropic.

Bernoulli's equation is only applicable to steady, inviscid, and incompressible flows along a streamline. However, the adiabatic condition is not mentioned, implying that the mechanical energy of the fluid can be conserved even if there is exchange of heat with its surroundings. This conclusion has been obtained earlier, in the chapter discussing the energy equation. Ignoring chemical, electrical, and other energy, the sum of the mechanical and internal energies of the fluid is conserved as total energy. There are two types of energy conversion: one is the reversible conversion between mechanical and internal energy due to compression and expansion of the fluid; the other is the irreversible conversion of mechanical energy into internal energy owing to viscous dissipation. For inviscid and incompressible flows, where the mechanical and internal energies are independent, the heat transferred between the fluid and its surroundings affects only its internal energy rather than its mechanical energy.

Tip 7.1: Entropy and Process Parameters for an Ideal Gas

Work and heat are two different ways of transferring energy from one system to another. In thermodynamics, the p–v diagram is used to represent the transfer of work between two systems, and the T–s diagram is used to represent the transfer of heat between two systems. The volume work and the heat transfer in a reversible process can be expressed as:

$$\mathrm{d}w = p\mathrm{d}v \text{ and } \mathrm{d}q_{\mathrm{rev}} = T\mathrm{d}s \left(v = 1/\rho \text{ is the specific volume}\right).$$

The heat absorbed by an ideal gas partly increases the internal energy of the gas and partly does expansion work on its surroundings. Therefore, the entropy change and internal energy are related as follows:

$$T\mathrm{d}s = \mathrm{d}\hat{u} + p\mathrm{d}v.$$

For an ideal gas, we have:

$$\mathrm{d}\hat{u} = c_{\mathrm{v}}\mathrm{d}T,\ \mathrm{d}h = c_{\mathrm{p}}\mathrm{d}T,\ pv = RT,\ c_{\mathrm{p}} - c_{\mathrm{v}} = R,\ c_{\mathrm{p}}/c_{\mathrm{v}} = k.$$

Substituting these relations into the above equation gives the change of gas parameters in a reversible adiabatic process ($\mathrm{d}s = 0$):

$$\frac{p_2}{p_1} = \left(\frac{\rho_2}{\rho_1}\right)^k,\ \frac{T_2}{T_1} = \left(\frac{\rho_2}{\rho_1}\right)^{k-1},\ \frac{T_2}{T_1} = \left(\frac{p_2}{p_1}\right)^{\frac{k-1}{k}}.$$

The density, pressure, and temperature of an isentropic flow are related by the above equations. For an ideal gas with specific heat ratio $k = 1.4$, if the density increases by a factor of 2, the temperature and pressure increase by factors of 1.32 and 2.64, respectively. The equation of state of an ideal gas states that an increase in pressure depends on both density (representing the number of gas molecules) and temperature (representing the kinetic energy of a single molecule):

$$p_2/p_1 = (\rho_2/\rho_1)(T_2/T_1) = 2\cdot 1.32 = 2.64.$$

In an irreversible process, there will be a certain amount of losses (the entropy of the system increases). As the pressure increases by the same factor (the compression work done on the system remains the same), an irreversible process is accompanied by a larger increase in temperature and a smaller increase in density than a reversible one. When the gas is allowed to expand back to its original volume, its final temperature is higher than its original temperature, reserved as a part of unusable internal energy.

The dissipation term, ϕ_v, in the energy equation represents the work done by viscous forces. The work done due to shearing motion increases only the temperature but not the pressure of the gas. In other words, there is always a certain amount of loss in viscous flows (the entropy of the flow increases).

(In an ideal gas, there is a change in temperature upon compression or expansion. See Tip 4.3 in Chapter 4).

In compressible flows, the mechanical and internal energies can be converted from one form to another through compression/expansion processes. For the sake of simplicity, if the flow is also assumed to be adiabatic, inviscid, and with no work exchange, then the change in internal energy in these processes is reversible.

7.2.1 Static and Total Parameters

From either Bernoulli's equation for compressible flows or thermodynamic relations, we can derive the one-dimensional, steady, isoenergetic flow energy equation for a gas:

$$c_p T + \frac{V^2}{2} = \text{const} \quad \text{or} \quad h + \frac{V^2}{2} = \text{const}.$$

This formula represents the fact that the sum of the enthalpy and kinetic energy remains constant when there is no exchange of energy between the fluid and its surroundings. In a way similar to the definition of total pressure in incompressible flows, we define the total enthalpy as:

$$h_t = h + \frac{V^2}{2}. \tag{7.3}$$

The total enthalpy represents the total energy of a flowing fluid. Using this definition, the total temperature can be defined as:

$$T_t = T + \frac{V^2}{2c_p}. \tag{7.4}$$

Consequently, the total temperature of a particular gas represents its total energy. It should be noted that there is no concept of total temperature for incompressible flow since the temperature is independent of its velocity.

We know that the total pressure represents the total mechanical energy of a flowing fluid, so a necessary flow condition under which the total pressure of a fluid element remains constant over time should be inviscid, or frictionless. Viscosity does not change the total energy of a flowing fluid, so the total temperature remains the same as long as there is no energy exchange between the fluid and its surroundings.

Substituting the definitions of specific heat capacity at constant pressure and Mach number into Equation (7.4) yields the equations relating Mach number and total/static temperatures (see Maths 7.1). (There are many derivations of equations in gasdynamics. In order to make the discussion concise, in this chapter all mathematical derivations are expressed as graphical representations.) The relationship between total and static temperatures is:

$$\frac{T_t}{T} = 1 + \frac{k-1}{2} Ma^2. \tag{7.5}$$

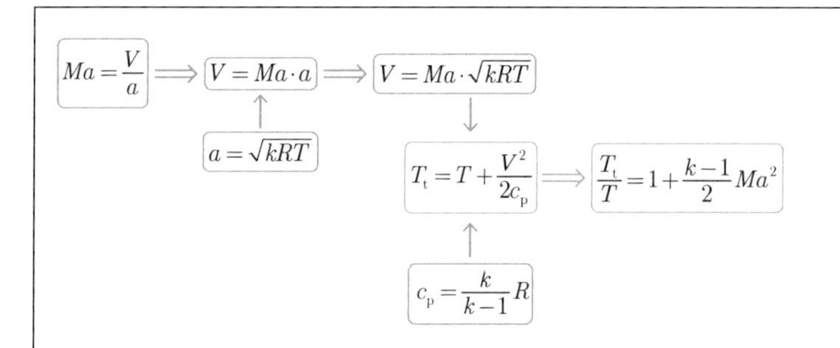

Maths 7.1 Derive the relationship between the total and static temperatures for a compressible flow.

The Mach number represents the degree of the compressibility of a gas, which is also reflected by the relationship between total and static temperatures. The stronger the compressibility of the gas, the greater the difference between total and static temperatures. For example, if a car travels at a speed of 120 km/h, equivalent to $Ma = 0.1$, then the total temperature is higher than the static temperature by 0.2%. If you put your hand out of the window while the car is in motion, the temperature on the windward side will be about 0.6°C warmer than the surrounding air. Compared with the heat exchange from the airflow, this type of temperature rise is imperceptible, so we cannot feel the temperature rise on the windward side. At high Mach numbers, the temperature rise of gas is considerable. For example, when a fighter plane is cruising at $Ma = 2.0$, we obtain from Equation (7.5) that the ratio is $T_t/T = 1.8$. Even if the average temperature of the stratosphere is −60°C, the total temperature can be 110°C at the leading edge of the plane. The tip of a missile that flies even faster will have a much higher temperature rise. The re-entry spacecraft capsule needs thermal protection systems, and meteors mostly burn up in the Earth's atmosphere. This kind of phenomenon is named "aerodynamic heating."

Substituting the equation of state of an ideal gas, $p = \rho RT$, and the isentropic condition, $p/\rho^k = \text{const}$, into Equation (7.5) gives the pressure and density as a function of Mach number in compressible flows as:

$$\frac{p_t}{p} = \left(1 + \frac{k-1}{2} Ma^2\right)^{\frac{k}{k-1}}, \tag{7.6}$$

$$\frac{\rho_t}{\rho} = \left(1 + \frac{k-1}{2} Ma^2\right)^{\frac{1}{k-1}}. \tag{7.7}$$

Note that the value of total temperature and total pressure varies with the choice of a reference coordinate system. The thermodynamic properties of a flowing fluid (static temperature and static pressure) remain constant regardless of the selected coordinate system.

For incompressible flows, the total pressure is written as the sum of the static and the dynamic pressure:

$$p_t = p + \frac{1}{2}\rho V^2. \tag{7.8}$$

What is the relationship between Equations (7.6) and (7.8)? We must understand that the incompressible flow theory is not completely accurate for gas. Therefore, Equation (7.8) is only approximately valid, while Equation (7.6) is accurate.

The relationship between the two formulas can be proved (see Maths 7.2) as

$$p_t = p + \frac{1}{2}\rho V^2 \left(1 + \frac{1}{4} Ma^2 + \frac{2-k}{24} Ma^4 + \cdots\right). \tag{7.9}$$

This is the exact relationship between the total pressure, p_t, and the dynamic pressure, $\rho V^2/2$.

When the Mach number is much smaller than 1, Equation (7.9) can be simplified to Equation (7.8). The incompressible flow assumption brings larger errors at higher Mach numbers. In engineering, flow velocity is commonly calculated by measuring the difference between the total and static pressures and using Equation (7.8). The velocity calculated by Equation (7.8) generates a 1% error at $Ma = 0.3$, which is acceptable in general problems. This is why $Ma = 0.3$ is set as the threshold between compressible and incompressible flows.

From Equation (7.9) we can draw two useful conclusions: (1) knowing only the total and static pressures of a compressible flow is not enough to calculate its velocity – its temperature (or density) is also needed; and (2) the aforementioned dynamic pressure of air, $\rho V^2/2$, is only approximate – a more accurate formula is:

$$p_d = \frac{1}{2}\rho V^2\left(1+\frac{1}{4}Ma^2+\frac{2-k}{24}Ma^4+\cdots\right). \tag{7.10}$$

The actual dynamic pressure is slightly higher than its generally defined value if the compressibility effect is taken into account. The extra pressure produced due to the compressibility of a fluid can be interpreted as an increase in inertia force attributed to the increase in density during a stagnation process.

It should also be noted that the binomial series used in Maths 7.2 converges only when $x^2 < 1$, so Equations (7.9) and (7.10) can only be used when $Ma < 2.24$. For compressible flows, many authors actually prefer to use the concept of total pressure to static pressure ratio, defined by Equation (7.6), instead of dynamic pressure.

Binomial theorem: $(1+x)^n = 1+nx+\frac{n(n-1)x^2}{2!}+\frac{n(n-1)(n-2)x^3}{3!}+\cdots$

Let: $\frac{k-1}{2}Ma^2 = x$, $\frac{k}{k-1} = n$

↓

$\frac{p_t}{p} = \left(1+\frac{k-1}{2}Ma^2\right)^{\frac{k}{k-1}}$ ⟹ $\frac{p_t}{p} = 1+\frac{k}{2}Ma^2+\frac{k}{8}Ma^4+\frac{2-k}{48}kMa^6+\cdots$

⇓

$Ma^2 = \frac{V^2}{a^2} = \frac{V^2}{kp/\rho}$ ⟶ $p_t = p+\frac{k}{2}Ma^2 p\left(1+\frac{1}{4}Ma^2+\frac{2-k}{24}Ma^4+\cdots\right)$

⇓

$p_t = p+\frac{1}{2}\rho V^2\left(1+\frac{1}{4}Ma^2+\frac{2-k}{24}Ma^4+\cdots\right)$

Maths 7.2 The relationship between total and static pressure for an incompressible flow.

Temperature influences the speed of sound, which varies from one location to the next in the flow field. Although the concept of Mach number is clear, it is just a convenient way to represent the compressibility of the flow. Another dimensionless number that is often used in gasdynamics is the *coefficient of velocity, λ*. This coefficient is directly related to velocity. Next, let us examine its definition and the aerodynamic theory equations expressed in terms of this coefficient.

7.2.2 Critical State and Coefficient of Velocity

The total temperature of an isentropic flow remains constant. The acceleration of the flow causes the static temperature to continuously decrease. The limiting case is when all the internal energy of the flowing fluid changes to kinetic energy. In other words, when temperature drops to absolute zero, the resultant maximum velocity is called the *limiting velocity*. According to the expression for total temperature:

$$T_t = T + \frac{V^2}{2c_p} = T + \frac{V^2}{\frac{2k}{k-1}R}.$$

We obtain the limiting velocity as:

$$V_{max} = \sqrt{\frac{2k}{k-1}RT_t}. \tag{7.11}$$

Of course, this theoretical value is impossible to achieve. Technically speaking, even if the temperature of a gas could be lowered to absolute zero by accelerating it, Equation (7.11) can no longer be true, since the gas may be liquefied or too thin to behave as a continuum fluid. The significance of Equation (7.11) is to tell us that *there is an upper limit to the velocity in an isentropic expansion flow*. For example, when air at 15°C at rest expands adiabatically, the limiting velocity is 761 m/s. However, the Mach number can be as high as infinity due to the infinitesimal speed of sound at absolute zero.

The temperature in the equation of sound speed can be substituted by total temperature to define another reference velocity, which is an invariant in an isentropic flow:

$$a_t = \sqrt{kRT_t}.$$

It can be called the *stagnant speed of sound*, and it is equal to the speed of sound in a gas at rest. When the gas begins to flow, its temperature decreases, and so does the actual speed of sound in it.

A dimensionless number similar to the Mach number can be defined in term of stagnant speed of sound:

$$Ma_t = \frac{V}{a_t} = \frac{V}{\sqrt{kRT_t}}.$$

Since the total temperature remains constant, this dimensionless number depends only on the velocity and seems to be more convenient than the Mach number. However, there is a problem with using this definition. As we have seen, the Mach number has a clear physical meaning. A flow is called sonic when $Ma = 1$, subsonic when $Ma < 1$, and supersonic when $Ma > 1$. On the other hand, we have $Ma_t = 0.913$ at the sonic condition, which is not very intuitive. Consequently, another reference quantity has been suggested, namely the *critical speed of sound*. The speed of sound in a gas at rest is maximum – that is $a_t = \sqrt{kRT_t}$; the speed of sound in a gas at a limiting speed of V_{max} is minimum, which is 0. In the process of accelerating airflow starting from rest, the velocity increases from 0 to V_{max}, while the speed of sound decreases from $a_t = \sqrt{kRT_t}$ to 0. There exists a state in which the velocity is exactly equal to the speed of sound, called the critical state. The sound speed at this state is called the critical speed of sound.

By appropriate derivation (see Maths 7.2), we can obtain the critical speed of sound as:

$$a_{cr} = \sqrt{\frac{2k}{k+1}RT_t}. \tag{7.12}$$

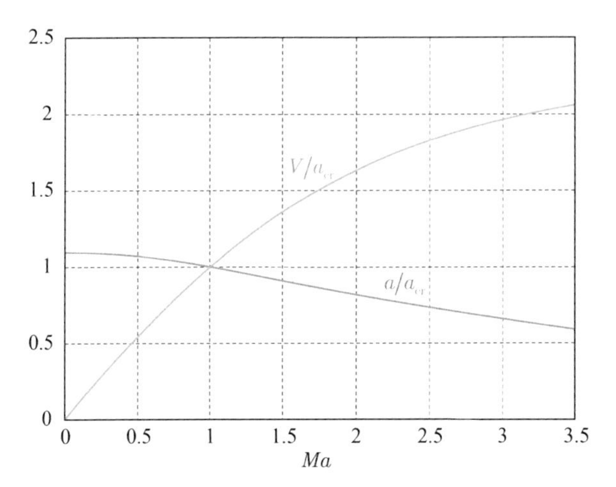

Figure 7.3 The velocity and speed of sound of an isentropic flow as a function of Mach number.

The subscript cr represents the critical state. Notice that the critical speed of sound depends only on the total temperature of the gas.

Using the critical speed of sound as a reference, a *coefficient of velocity*, λ, may be defined as:

$$\lambda = \frac{V}{a_{cr}}.$$

Now, when a flow is at sound speed, both λ and Ma are equal to unity:

$$\lambda = \frac{V}{a_{cr}} = \frac{V}{\sqrt{\frac{2k}{k+1}RT_t}} = \frac{V}{\sqrt{\frac{2k}{k+1}R\left(1+\frac{k-1}{2}\cdot 1^2\right)T}} = \frac{V}{\sqrt{kRT}} = Ma = 1.$$

This conclusion is obvious because the critical sound speed is just the actual speed of sound at $Ma = 1$. Figure 7.3 shows the relationship between flow speed and sound speed for the isentropic flow of an ideal gas. As the flow speed increases, the speed of sound decreases. At critical state, the velocity and sound speed are the same. As the flow accelerates further, the flow speed approaches the limiting speed of V_{max}, and the sound speed approaches zero.

Equations (7.5)–(7.7) describe the temperature, pressure, and density of an isentropic flow as a function of the Mach number. From these formulas, the isentropic flow relations between stagnation and static parameters at critical state can be calculated:

$$\frac{T_{cr}}{T_t} = \frac{2}{k+1} \approx 0.833,$$

$$\frac{p_{cr}}{p_t} = \left(\frac{2}{k+1}\right)^{\frac{k}{k-1}} \approx 0.528,$$

$$\frac{\rho_{cr}}{\rho_t} = \left(\frac{2}{k+1}\right)^{\frac{1}{k-1}} \approx 0.634.$$

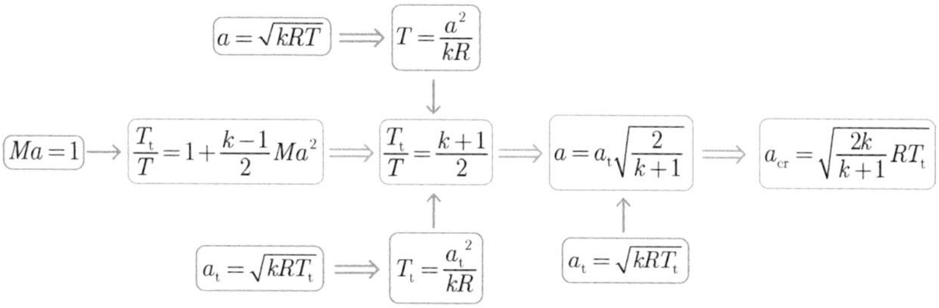

$$a = \sqrt{kRT} \Longrightarrow T = \frac{a^2}{kR}$$

$$Ma = 1 \longrightarrow \frac{T_t}{T} = 1 + \frac{k-1}{2} Ma^2 \Longrightarrow \frac{T_t}{T} = \frac{k+1}{2} \Longrightarrow a = a_t \sqrt{\frac{2}{k+1}} \Longrightarrow a_{cr} = \sqrt{\frac{2k}{k+1} RT_t}$$

$$a_t = \sqrt{kRT_t} \Longrightarrow T_t = \frac{a_t^2}{kR}$$

$$a_t = \sqrt{kRT_t}$$

Maths 7.3 Derivation of the critical sound speed.

When compressed air flows frictionlessly out of a large vessel, the outlet velocity can reach the sound speed as long as the ratio of the pressure in the vessel to the atmospheric pressure is equal to or greater than 1/0.528 = 1.89. When you pump air into a tire, the pressure inside the pump can easily reach 2 atm or more. Since the tire pressure is relatively low at the beginning, the air may enter the tire with sound speed. The pressure inside a natural-gas or oxygen tank can be much higher, and the velocity at the nozzle outlet can also reach sound speed.

In an isentropic expansion, an increase in the Mach number is due to two factors: increased velocity and decreased speed of sound. At hypersonic speeds, the decrease in speed of sound becomes the leading factor causing the Mach number to increase. Although the limiting velocity is a finite speed, the Mach number approaches infinity because the corresponding speed of sound is zero. If the flow is described in terms of the coefficient of velocity, the maximum attainable value of this coefficient is finite, since the critical speed of sound is constant:

$$\lambda_{max} = \frac{V_{max}}{a_{cr}} = \frac{\sqrt{\frac{2k}{k-1} RT_t}}{\sqrt{\frac{2k}{k+1} RT_t}} = \sqrt{\frac{k+1}{k-1}}.$$

If the specific heat ratio k is roughly considered to have a constant value of 1.4, then the maximum coefficient of velocity is about $\lambda_{max} = 2.45$.

There is a fixed relationship between the coefficient of velocity and the Mach number (see Maths 7.4). The final result can be calculated:

$$Ma^2 = \frac{2\lambda^2}{(k+1)-(k-1)\lambda^2}, \tag{7.13}$$

$$\lambda^2 = \frac{(k+1)Ma^2}{2+(k-1)Ma^2}. \tag{7.13a}$$

Figure 7.4 depicts Ma and λ as a function of velocity as the airflow accelerates. The specific heat ratio is assumed to be constant and equal to 1.4. In subsonic flow, both Ma and λ are less than 1. The difference between these two dimensionless parameters is insignificant. In supersonic flows, both Ma and λ are greater than 1. As the velocity

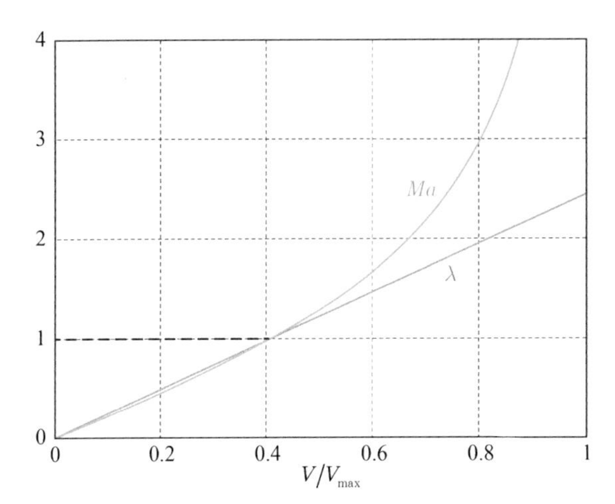

Figure 7.4 Ma and λ as a function of velocity.

increases, the Mach number rapidly increases, while the coefficient of velocity and the velocity are linearly related. As can be seen, subsonic and supersonic flows can also be conveniently characterized by the coefficient of velocity.

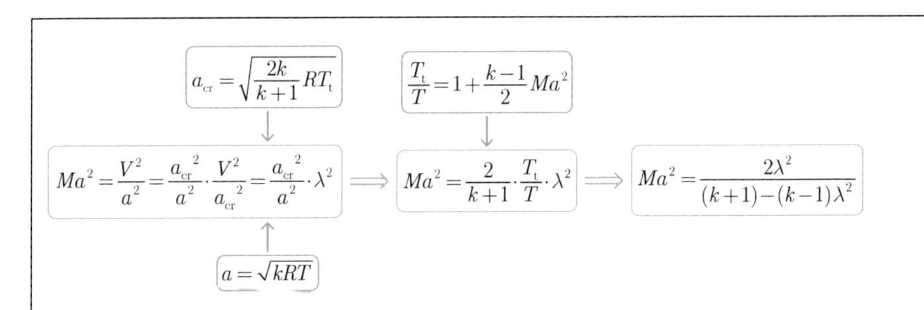

Maths 7.4 The relationship between the coefficient of velocity and the Mach number.

7.2.3 Gasdynamic Functions

Gasdynamic formulas as a function of λ can considerably simplify calculations. The most fundamental gasdynamic functions describe the ratio between static and total parameters, including the temperature, pressure, and density ratios:

$$\tau(\lambda)=\frac{T}{T_t}=1-\frac{k-1}{k+1}\lambda^2, \tag{7.14}$$

$$\pi(\lambda)=\frac{p}{p_t}=\left(1-\frac{k-1}{k+1}\lambda^2\right)^{\frac{k}{k-1}}, \tag{7.15}$$

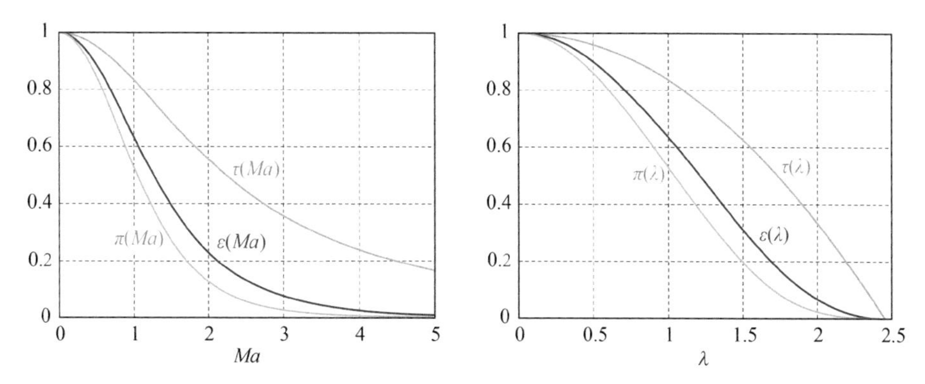

Figure 7.5 Three gasdynamic functions versus *Ma* and λ.

$$\varepsilon(\lambda)=\frac{\rho}{\rho_{\mathrm{t}}}=\left(1-\frac{k-1}{k+1}\lambda^{2}\right)^{\frac{1}{k-1}}. \tag{7.16}$$

In fact, these relations have been given in Equations (7.5)–(7.7) with the Mach number as the argument. They are plotted in Figure 7.5 versus *Ma* and λ.

Another important gasdynamic function is the *mass flow function*. For compressible flows the density is not constant, and it is inconvenient to calculate the flow rate of a fluid using the flow rate formula: $\dot{m}=\rho AV$. So, a dimensionless form of mass flow rate using *Ma* or λ as a parameter is defined as (see Maths 7.5):

$$q(\lambda)=\frac{\rho V}{\rho_{\mathrm{cr}}V_{\mathrm{cr}}}=\left(\frac{k+1}{2}\right)^{\frac{1}{k-1}}\lambda\left(1+\frac{k-1}{k+1}\lambda^{2}\right)^{\frac{1}{k-1}}. \tag{7.17}$$

$$\rho/\rho_{\mathrm{t}}=\left(1-\frac{k-1}{k+1}\lambda^{2}\right)^{\frac{1}{k-1}}$$

Define $q(\lambda)=\dfrac{\rho V}{\rho_{\mathrm{cr}}V_{\mathrm{cr}}}$

$$\frac{\rho V}{\rho_{\mathrm{cr}}V_{\mathrm{cr}}}=\frac{\rho}{\rho_{\mathrm{cr}}}\cdot\lambda=\frac{\rho/\rho_{\mathrm{t}}}{\rho_{\mathrm{cr}}/\rho_{\mathrm{t}}}\cdot\lambda \Longrightarrow \frac{\rho V}{\rho_{\mathrm{cr}}V_{\mathrm{cr}}}=\left(\frac{k+1}{2}\right)^{\frac{1}{k-1}}\lambda\left(1-\frac{k-1}{k+1}\lambda^{2}\right)^{\frac{1}{k-1}}$$

$$\rho_{\mathrm{cr}}/\rho_{\mathrm{t}}=\left(\frac{2}{k+1}\right)^{\frac{1}{k-1}}$$

$$q(\lambda)=\left(\frac{k+1}{2}\right)^{\frac{1}{k-1}}\lambda\left(1-\frac{k-1}{k+1}\lambda^{2}\right)^{\frac{1}{k-1}}$$

Maths 7.5 Definition of the mass flow function.

The formula for the mass flow rate is expressed as: $\dot{m}=\rho VA=\dfrac{\rho V}{\rho_{\mathrm{cr}}V_{\mathrm{cr}}}\rho_{\mathrm{cr}}V_{\mathrm{cr}}A$, where $\dfrac{\rho V}{\rho_{\mathrm{cr}}V_{\mathrm{cr}}}$ depends only on λ, and $\rho_{\mathrm{cr}}V_{\mathrm{cr}}$ depends only on total parameters.

Consequently, the mass flow rate can be expressed in terms of total parameters (see Maths 7.6):

$$\dot{m} = K\frac{p_t}{\sqrt{T_t}}Aq(\lambda), \tag{7.18}$$

where, $K = \sqrt{\frac{k}{R}\left(\frac{2}{k+1}\right)^{\frac{k+1}{k-1}}}$. When $k = 1.4$, $K = 0.0404$.

$$V_{cr} = a_{cr} = \sqrt{\frac{2k}{k+1}RT_t}$$

$$\rho_t = \frac{p_t}{RT_t} \quad \dot{m} = \frac{\rho V}{\rho_{cr}V_{cr}}\rho_{cr}V_{cr}A = \rho_{cr}V_{cr}Aq(\lambda) \Longrightarrow \dot{m} = \sqrt{\frac{k}{R}\left(\frac{2}{k+1}\right)^{\frac{k+1}{k-1}}}\cdot\frac{p_t}{\sqrt{T_t}}Aq(\lambda)$$

$$\rho_{cr}/\rho = \left(\frac{2}{k+1}\right)^{\frac{1}{k-1}} \Longrightarrow \rho_{cr} = \frac{p_t}{RT_t}\left(\frac{2}{k+1}\right)^{\frac{1}{k-1}}$$

Maths 7.6 The mass flow rate expressed in the mass flow function.

Equation (7.18) is very useful in gasdynamics because a great number of flow problems approximately obey the isentropic relations. The total parameters of a flowing fluid remain constant during an isentropic process. The continuity equation in fluid dynamics states that at any section we have:

$$Aq(\lambda) = \text{const.}$$

Figure 7.6 shows the mass flow function $q(\lambda)$ as a function of the coefficient of velocity. It can be seen that in subsonic flow, $q(\lambda)$ increases when the velocity increases; in supersonic flow, $q(\lambda)$ decreases with an increase in velocity. Its maximum value is obtained at the sonic condition, where $q(\lambda) = 1.0$. In other words, the density flow, ρV, attains its maximum at the sonic condition, or there is a maximum airflow ability through the same area when the Mach number is equal to 1. For a one-dimensional, steady airflow through a pipe, a decrease in area accelerates a subsonic flow, while an increase in area accelerates a supersonic flow.

Only a converging–diverging duct can smoothly accelerate a subsonic flow through sonic to supersonic for one-dimensional flow. Somewhere in between the inlet and outlet of the duct there is a minimum area called *throat* where the velocity equals sound speed. This device, called the *Laval nozzle*, was first proposed by Carl Gustaf Patrik de Laval (1845–1913), a Swedish engineer and inventor who pioneered the development of high-speed turbines. Figure 7.7 shows the steam turbine designed by Laval. In order to increase the power of a steam turbine, it is necessary to make the velocity of the steam impinge upon the turbine blade as high as possible. A converging–diverging nozzle designed by Laval allows the flow to be

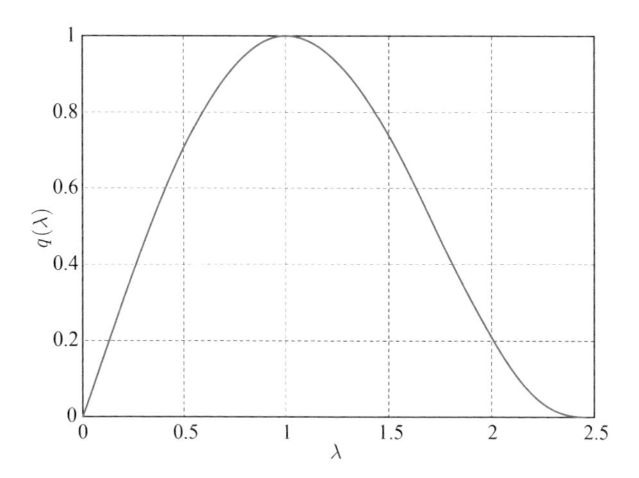

Figure 7.6 Mass flow function as a function of the coefficient of velocity.

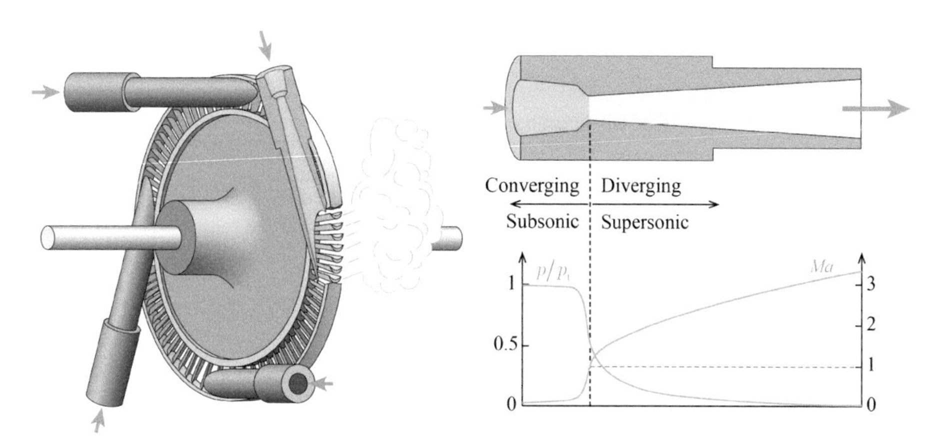

Figure 7.7 The steam turbine designed by Laval and its converging–diverging nozzle.

accelerated to supersonic. As a typical Laval nozzle, the nozzle of a rocket engine ejects high-temperature and pressure gas to produce thrust.

7.3 Expansion Wave, Compression Wave, and Shock Wave

Sound waves propagate through an incompressible fluid at infinite speed as if it were a rigid body. No matter how large the fluid volume, any small displacement at one end must be immediately reflected at the other end. However, any disturbance propagates through a fluid at a finite speed as real fluids are compressible. *If the disturbance is weak and propagates relying on the compressibility of the fluid itself, the speed of propagation is equal to the speed of sound. If the disturbance is strong, the propagating disturbance can travel at a supersonic speed.*

7.3.1 Pressure Waves in Fluids

A pressure rise or fall in a flow field will propagate toward all directions in the form of a pressure wave. A pressure wave, also known as a compression wave or expansion wave, increases or decreases the pressure of fluid as the wave passes through it. Figure 7.8 shows two ways to make fluid flow through a one-dimensional pipe. In both cases, the fluid in the middle moves under the differential pressure. This differential pressure is caused by the piston-generated pressure waves. If the piston is pushed from the left, a series of compression waves continually propagate away from the piston and into the gas to the right. If the piston is pulled from the right, a series of expansion waves continually propagate to the left. The compression and expansion waves propagate through the fluid, thereby generating corresponding pressure distribution throughout the flow field. For clarity, several discrete waves are shown in Figure 7.8. In fact, a continuous pressure field develops in the flow field.

As can be seen, these compression and expansion waves travel at different speeds. A change in internal energy due to compression and expansion will cause the fluid to undergo a corresponding change in the local speed of sound, which is the speed at which the weak disturbance waves propagate away from the source. When the air passes through an expansion wave, the local speed of sound will decrease. Consequently, the expansion waves generated later cannot catch up with the ones generated earlier. When the air passes through a compression wave, the local speed of sound will increase. In a sufficiently long pipe, the compression waves generated later will eventually catch up with the first compression wave. If there are enough weak compression waves they will coalesce to form a strong compression wave, namely a *shock wave*.

Let us consider a piston being pushed intermittently into a sufficiently long pipe, as shown in Figure 7.9. The piston-generated compression waves travel to the right at local sound speed. Even if the piston is not being moved at supersonic velocity, the faster compression waves generated later will constantly catch up with the slower ones generated earlier, merging into a shock wave in the far front. As the piston moves, the shock wave becomes stronger and travels faster.

All the compression waves can only catch up with the shock wave and then move along with it but not surpass it, because before the shock wave in the undisturbed region the fluid temperature is lower and the sound speed is smaller, and after the shock wave the temperature and sound speed increase. Given enough time and distance, the shock wave can catch up with any compression wave in front of it and be caught up by any compression wave behind it.

The shock wave is a strongly compressed layer in which pressure increases suddenly and markedly, while its thickness is merely of the order of the molecular free path. Therefore, the internal structure of the shock wave can no longer be described based on conventional fluid mechanics, because such small-scale changes do not satisfy the assumption of continuity, let alone the equation of state for an ideal gas.

A shock wave travels at supersonic speed with respect to the fluid ahead of it. Due to the continuous coalescence of successive compression waves, the shock wave

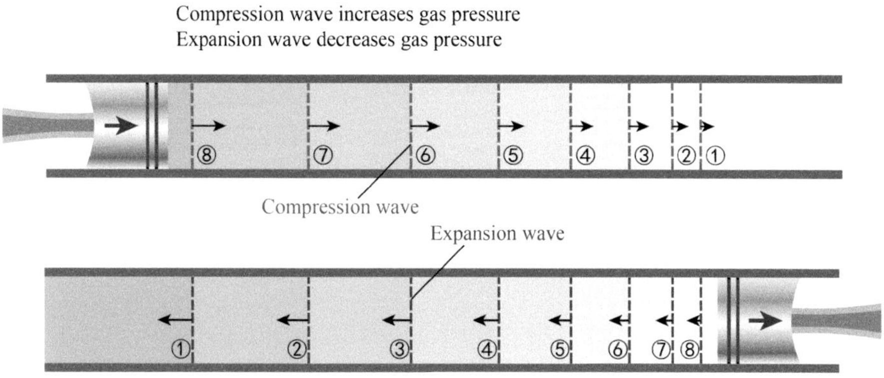

Compression wave increases gas pressure
Expansion wave decreases gas pressure

Compression wave
Expansion wave

Figure 7.8 Piston-generated compression and expansion waves in a one-dimensional pipe.

The compression waves generated later will catch up with the earlier waves and merge into a shock wave.
A shock wave is supersonic relative to the undisturbed gas yet subsonic relative to the compressed gas.

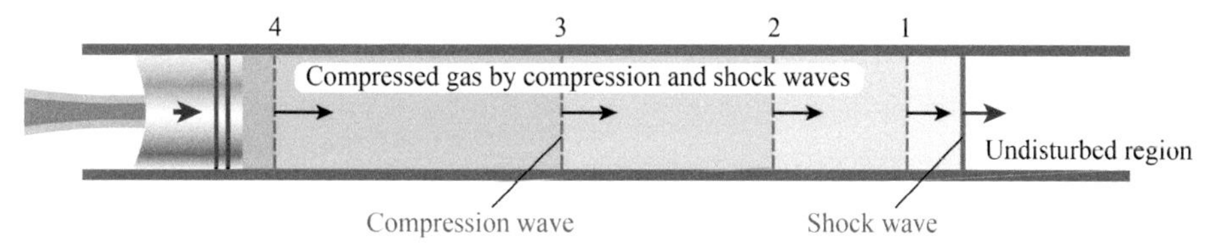

Figure 7.9 Piston-generated weak compression waves merge into a shock wave in a sufficiently long pipe.

travels faster and faster, and the Mach number becomes ever higher with respect to the fluid ahead of the shock wave.

In fact, a weak compression wave also leads to a slight increase in temperature, so this wave travels faster than the speed of sound with respect to the wavefront. On the other hand, the speed of the weak expansion wave is slightly slower than that of sound. Generally speaking, a sound wave consists of an alternating pattern of compression and expansion waves traveling through the medium. On the wavelength scale, the sound wave alternately speeds up and slows down, and the sound speed is an average value. Of course, this is only a theoretical analysis. In practice, the temperature change caused by the weak disturbance waves is so small that its effect on sound speed can be ignored.

The assumption that the above one-dimensional pipe is long enough is required, otherwise the early compression waves will run out of pipe before they are caught up and form into a shock wave. Furthermore, if the flow is not one-dimensional, the spherical compression waves rapidly decrease in strength with distance from the source, resulting in no shock wave.

As an inflated balloon expands, it compresses the surrounding air. Weak compression waves are constantly generated from the outer surface of the balloon and propagate outwards in all directions. Every compression wave originally travels slightly faster than the speed of sound and is later reduced in strength. Therefore, the compression

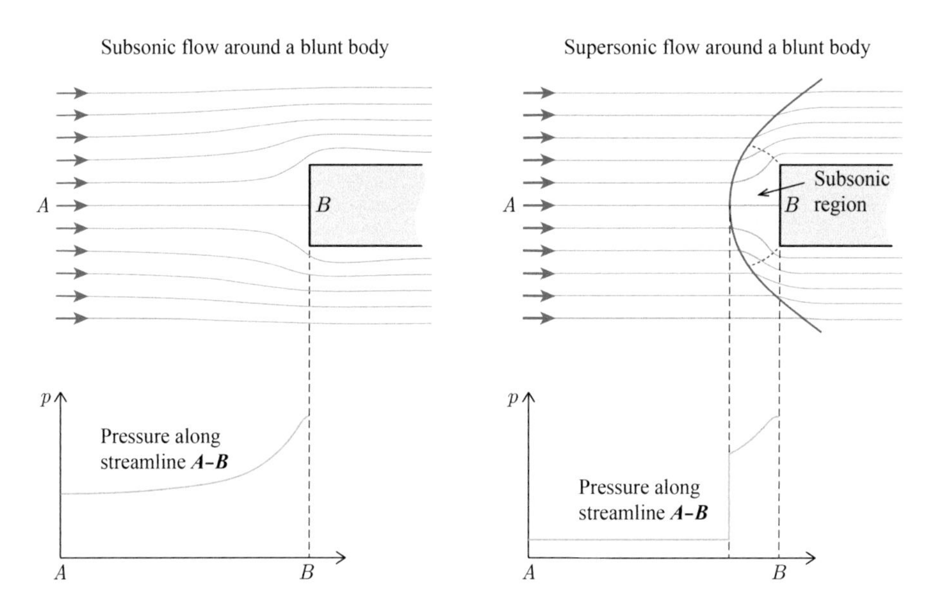

Figure 7.10 A supersonic flow encountering an obstacle downstream forms a shock wave. The flow ahead of the shock wave is not affected by the obstacle.

waves are unable to merge into a shock wave. Furthermore, the continuous inflation generates only compression waves. For the ambient pressure to continuously increase without forming any audible sound waves requires pressure fluctuation.

When the balloon explodes, it produces a loud bang. This is because, as the skin of the balloon disappears, there is a discontinuity interface of pressure between the inside and outside air, which forms a strong compression wave – or a shock wave. The whole wave, named an explosive wave, is composed of a shock wave followed by expansion waves. However, we do not always hear the sound of the shock wave. Since the pressure inside the balloon is not very high, the explosive wave rapidly reduces to a weak disturbance wave with spreading and dissipative processes before reaching our ears. Sometimes we do hear the shock wave itself, such as the explosion of a bomb; this kind of sudden jump in pressure can be quite damaging to the ears.

Let us take a look at another case, that of a flow past a stationary blunt object. For subsonic flow, within a specific distance upstream of the object, the velocity decreases and the pressure increases. The fluid is slowed down before it hits the object, by a series of compression waves transmitted upstream. If the flow is supersonic, since weak compression waves travel at the speed of sound, the pressure perturbations from the object cannot propagate upstream. Seemingly, the incoming flow should directly hit the object without slowing down. However, a subsonic region must exist near the blunt nose where the perturbations can propagate upstream. There exists an interface between the subsonic region and the incoming supersonic flow. After the supersonic flow directly hits this interface, a large number of molecules are squeezed together to form a shock wave, as shown in Figure 7.10.

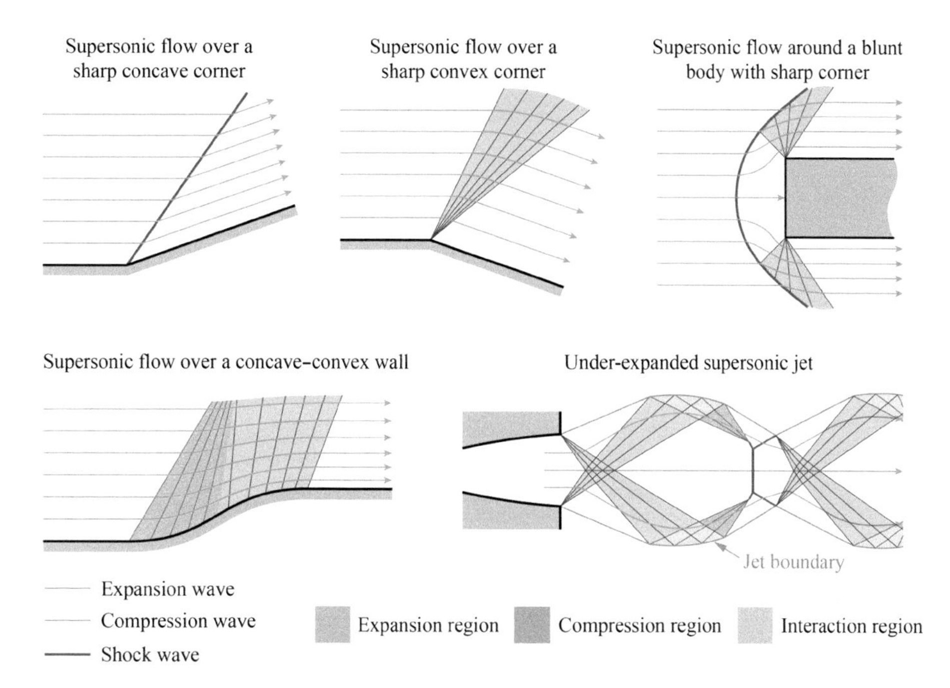

Figure 7.11 Compression and expansion waves occurring in supersonic flows.

Across a compression wave or a shock wave, a supersonic flow decelerates; across an expansion wave, it accelerates. When a supersonic flow accelerates, decelerates, or swerves, compression or expansion waves appear. Figure 7.11 shows some compression and expansion waves occurring in supersonic flows. Sometimes there are only compression or expansion waves in the flow field, yet multiple expansion–compression waves occur simultaneously in most actual flows. For regions of supersonic flow undergoing neither compression nor expansion, the pressure of the gas remains constant. Ignoring the viscous forces and gravity, the fluid element moves in uniform linear motion in these regions.

Tip 7.2: Shock Tube

As shown in Figure T7.2, a shock tube is a device to generate shock waves by the nearly instantaneous rupture of a thin diaphragm that separates a high-pressure and a low-pressure section. Simultaneously, a shock wave is launched into the low-pressure section, and a series of expansion waves are launched backward into the high-pressure section.

The expansion waves travel at the local speed of sound in the high-pressure section. The speeds of both the contact surface and the shock wave depend on the initial differential pressure between the high- and low-pressure sections, and the shock wave travels faster than the contact surface.

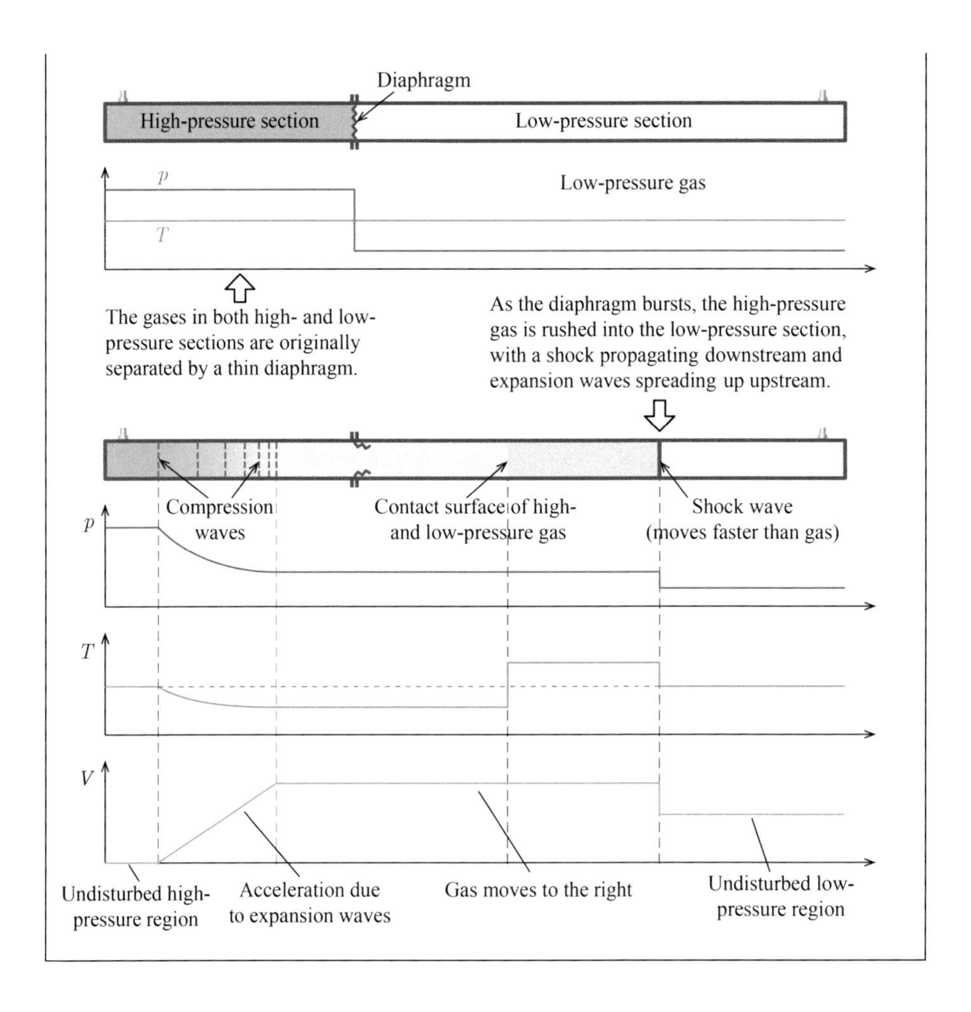

7.3.2 Normal Shock Wave

A typical effect of a normal shock wave is that it slows a flow from supersonic to subsonic. Next, we perform a control volume analysis to quantitatively investigate the variation of the flow parameters for a fluid crossing a normal shock wave.

As shown in Figure 7.12, we define the control volume to encompass a normal shock wave. Since the flow area remains constant across the normal shock wave, we have:

$$\text{continuity equation: } \rho_1 V_1 = \rho_2 V_2,$$
$$\text{momentum equation: } p_2 - p_1 = \rho_2 V_2^2 - \rho_1 V_1^2,$$
$$\text{energy equation: } c_p T_1 + \frac{1}{2} V_1^2 = c_p T_2 + \frac{1}{2} V_2^2.$$

Combining these equations with the gasdynamic functions mentioned before, we can obtain the relationships for flow properties before and after a normal shock wave:

$$\text{coefficient of velocity, } \lambda: \lambda_1 \lambda_2 = 1, \tag{7.19}$$

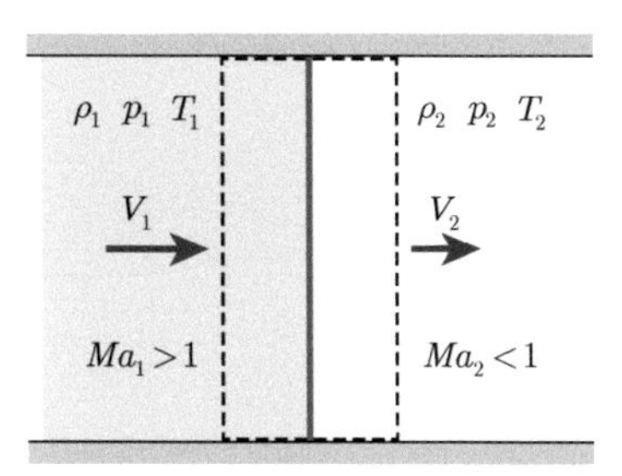

Figure 7.12 A control volume encompassing a normal shock wave.

Mach number, Ma: $Ma_2^2 = \dfrac{Ma_1^2 + \dfrac{2}{k-1}}{\dfrac{2k}{k-1}Ma_1^2 - 1}$,

pressure, p: $\dfrac{p_2}{p_1} = \dfrac{2k}{k+1}\left(Ma_1^2 - 1\right) + 1$,

temperature, T: $\dfrac{T_2}{T_1} = \dfrac{\left(1 + \dfrac{k-1}{2}Ma_1^2\right)\left(\dfrac{2k}{k-1}Ma_1^2 - 1\right)}{\dfrac{(k+1)^2}{2(k-1)}Ma_1^2}$,

density, ρ: $\dfrac{\rho_2}{\rho_1} = \dfrac{(k+1)Ma_1^2}{2 + (k-1)Ma_1^2}$,

total pressure, p_t: $\dfrac{p_{t2}}{p_{t1}} = \dfrac{\left[\dfrac{(k+1)Ma_1^2}{2 + (k-1)Ma_1^2}\right]^{\frac{k}{k-1}}}{\left[\dfrac{2k}{k+1}Ma_1^2 - \dfrac{k-1}{k+1}\right]^{\frac{1}{k-1}}}$,

total temperature, T_t: $T_{t2} = T_{t1}$.

Across a shock wave, the gas is compressed suddenly, accompanied by intense viscous effects and heat transfer. Some of its mechanical energy is irreversibly converted into internal energy, leading to a loss in total pressure. The flow across the shock wave can be considered adiabatic to the surroundings, as the process is done within a very limited time and space, and the total temperature remains constant. Figure 7.13 shows the relationships for flow properties before and after a normal shock wave at different incoming Mach numbers.

7.3.3 Oblique Shock Wave

For a two-dimensional supersonic flow around a blunt object (as shown in Figure 7.10), a normal shock wave is created directly in front of the object, and the flow immediately

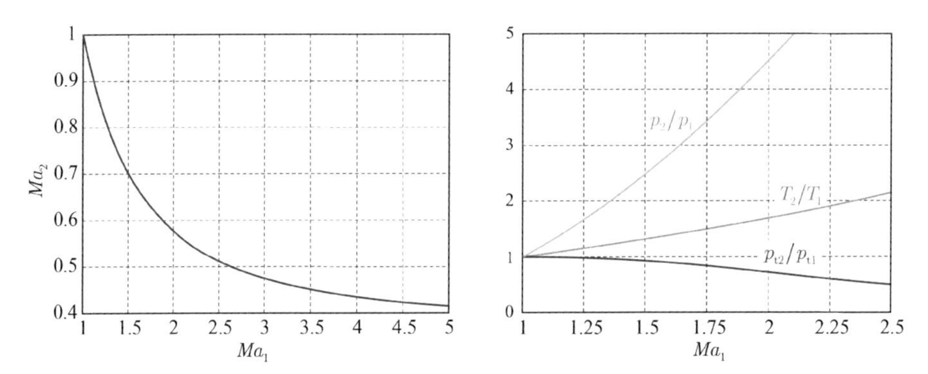

Figure 7.13 Relationships for flow properties across a normal shock wave.

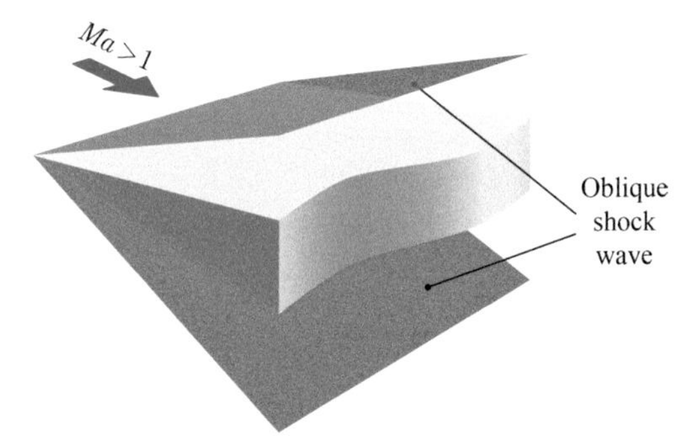

Figure 7.14 Two attached oblique shock waves generated by a two-dimensional wedge.

behind the wave is subsonic. The normal shock wave extends around the body as a curved shock, known as a *bow shock wave*. Bow shock waves are common in supersonic flows over blunt bodies. It is a type of detached shock wave due to the subsonic region ahead of the object.

For a two-dimensional supersonic flow around a sharp front end, the shock wave can attach to the leading edge since there is no stagnation point ahead of it. Unlike a normal shock, an oblique shock wave is inclined with respect to the flow direction. A sharp body, such as a two-dimensional wedge, generates two planar oblique shock waves, as shown in Figure 7.14.

A normal shock wave is generated when a supersonic flow slows down; an oblique shock wave is created when a supersonic flow is deflected. As shown in Figure 7.15, since the decelerated flow across an oblique shock wave is deflected by an angle so as to become parallel to the downstream surface, the deflection angle is determined by the wedge angle. The oblique shock angle depends in a complex way on the deflection angle and the incoming Mach number.

Since the shock wave is very thin, the tangential velocity component should remain constant across the shock wave on account of continuity:

$$V_{t2} = V_{t1}.$$

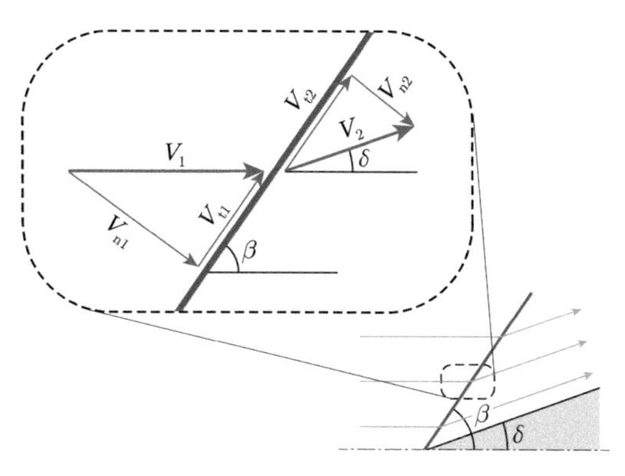

Figure 7.15 The relationship for velocity before and after an oblique shock wave.

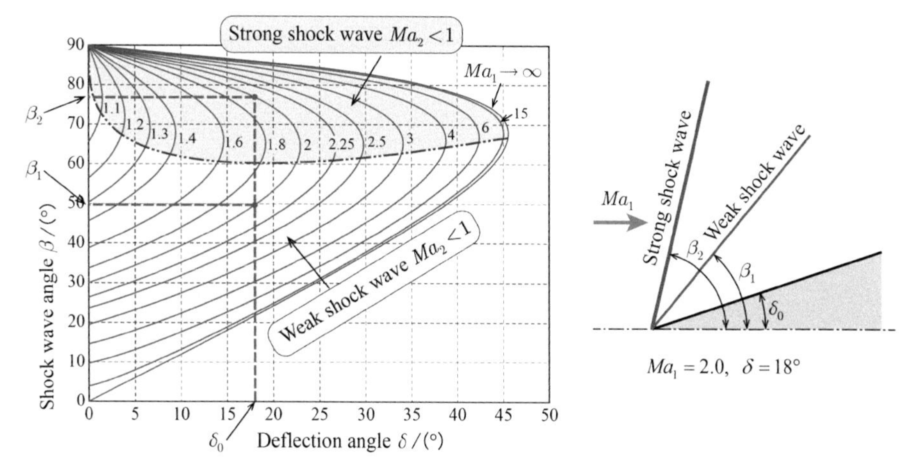

Figure 7.16 The relationship for oblique shock angle in terms of freestream Mach number and deflection angle.

For an oblique shock wave, the incoming Mach number Ma_1, deflection angle δ, and oblique shock angle β are related through the continuity equation:

$$\tan\delta = \frac{Ma_1^2 \sin^2\beta - 1}{\left[Ma_1^2\left(\dfrac{k+1}{2} - \sin^2\beta\right) + 1\right]\tan\beta}.$$

Figure 7.16 shows the relationship for oblique shock angle in terms of incoming Mach number and deflection angle. Each incoming Mach number has two possible oblique shock angles with the same deflection angle. The larger shock angle case is called a *strong shock wave*, and the other one a *weak shock wave*. In actual flows, these oblique shock angles are determined by the relationship for pressure before and after the shock wave. Obviously, the higher the back pressure, the stronger the shock wave. Therefore, there is a large/small pressure rise across the strong/weak shock wave. For the strong shock, the downstream velocity is subsonic; for the weak one, the downstream velocity can be supersonic.

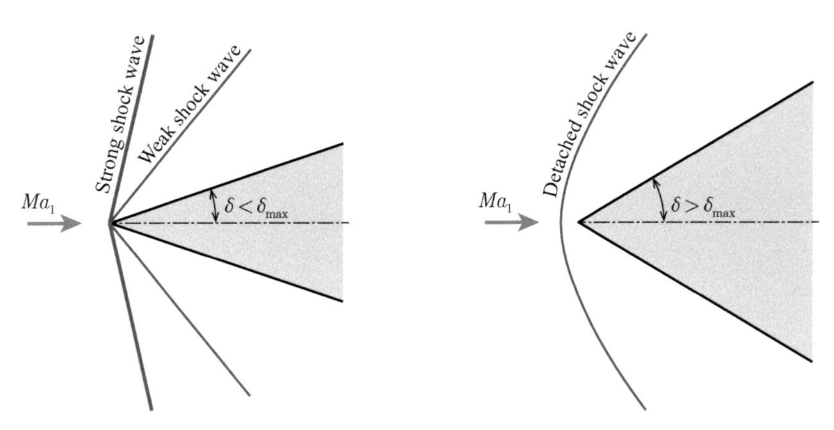

Figure 7.17 A "detached" shock wave forms when the deflection angle exceeds the maximum.

For a given incoming Mach number, there is a maximum deflection angle, δ_{max}, below which the shock remains attached to the leading edge of a wedge. As the deflection angle increases, the weak shock angle also increases. For deflection angles exceeding the maximum value, the oblique shock detaches and stands some distance upstream of the leading edge of the wedge, forming a curved shock, as shown in Figure 7.17. After the shock, the airflow can be further deflected by the continuous change in pressure in the subsonic region between the detached shock wave and the leading edge.

From the equations of continuity, momentum, and energy, we can obtain the relationships for flow properties before and after an oblique shock wave:

$$\text{coefficient of velocity, } \lambda\text{: } \lambda_{n1}\lambda_{n2} = 1, \quad \lambda_{t1} = \lambda_{t2}, \tag{7.20}$$

$$\text{Mach number, Ma: } Ma_2^2 = \frac{Ma_1^2 + \dfrac{2}{k-1}}{\dfrac{2k}{k-1} Ma_1^2 \sin^2\beta - 1} + \frac{Ma_1^2 \cos^2\beta}{\dfrac{k-1}{2} Ma_1^2 \sin^2\beta + 1},$$

$$\text{pressure, } p\text{: } \frac{p_2}{p_1} = \frac{2k}{k+1} Ma_1^2 \sin^2\beta - \frac{k-1}{k+1},$$

$$\text{temperature, } T\text{: } \frac{T_2}{T_1} = \frac{\left(1 + \dfrac{k-1}{2} Ma_1^2 \sin^2\beta\right)\left(\dfrac{2k}{k-1} Ma_1^2 \sin^2\beta - 1\right)}{\dfrac{(k+1)^2}{2(k-1)} Ma_1^2 \sin^2\beta},$$

$$\text{total pressure, } p_t\text{: } \frac{p_{t2}}{p_{t1}} = \frac{\left[\dfrac{(k+1) Ma_1^2 \sin^2\beta}{2 + (k-1) Ma_1^2 \sin^2\beta}\right]^{\frac{k}{k-1}}}{\left[\dfrac{2k}{k+1} Ma_1^2 \sin^2\beta - \dfrac{k-1}{k+1}\right]^{\frac{1}{k-1}}},$$

$$\text{total temperature, } T_t\text{: } T_{t1} = T_{t2}.$$

When $\beta = 90°$, we obtain the change in the flow variables across a normal shock wave.

Using these relations, various curves can be drawn to conveniently analyze and calculate the change in the flow variables before and after all types of shock waves.

7.4 Isentropic Flow in a Variable Cross-Section Pipe

The one-dimensional, inviscid, incompressible flow is the simplest flow in fluid mechanics. This flow has been fully discussed in Chapter 4. In general, compressible flows are much more complicated to calculate than incompressible ones. This section investigates the variation of parameters in a one-dimensional, compressible flow through converging and converging–diverging nozzles.

7.4.1 Converging Nozzle

The mass flow function (7.18) states that a one-dimensional, steady, isentropic flow satisfies the following relation:

$$Aq(\lambda) = \text{const.}$$

As the velocity increases, the mass flow function $q(\lambda)$ increases in subsonic flow but decreases in supersonic flow. That is, the velocity increases in subsonic converging flow but decreases in supersonic converging flow. In extreme circumstances, a subsonic flow will be accelerated in a converging nozzle until it becomes sonic, while a supersonic flow will be decelerated to sonic. We will discuss this phenomenon in subsonic and supersonic flows separately.

7.4.1.1 Subsonic Incoming Flow

Figure 7.18 shows the gas flowing from a tank into the ambient air through a converging nozzle. The most important quantity in this typical accelerated subsonic flow is the mass flow rate, expressed as:

$$\dot{m} = K \frac{p_t}{\sqrt{T_t}} Aq(\lambda).$$

As long as the tank is large enough, its internal pressure and temperature can be used as total pressure and total temperature. Given the nozzle area at outlet A, only the coefficient of velocity λ (or Mach number) at the nozzle outlet is needed to calculate the mass flow rate. The coefficient of velocity can be calculated by the ratio of total pressure to static pressure at the nozzle outlet. Given the total pressure, only the outlet static pressure is required.

As the tank's pressure exceeds a critical value, the outlet velocity reaches the speed of sound. For a converging nozzle, the flow is choked at the nozzle outlet and the flow velocity remains sonic even if the tank's pressure is further raised. As the tank's pressure increases, so does the mass flow rate, which, however, is caused only by the increase in gas density.

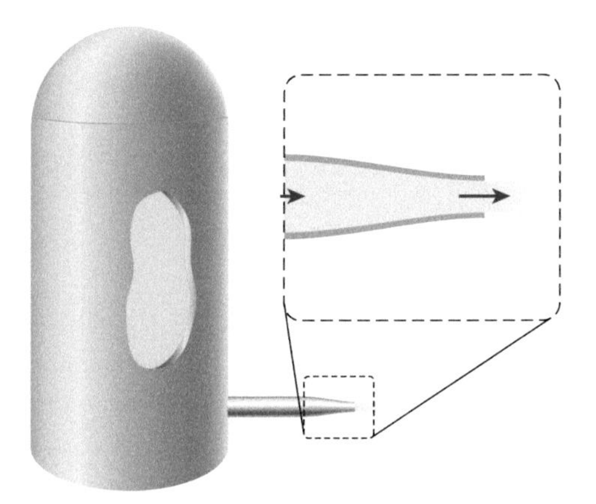

Figure 7.18 The gas in a pressure tank is discharged into the atmosphere through a converging nozzle.

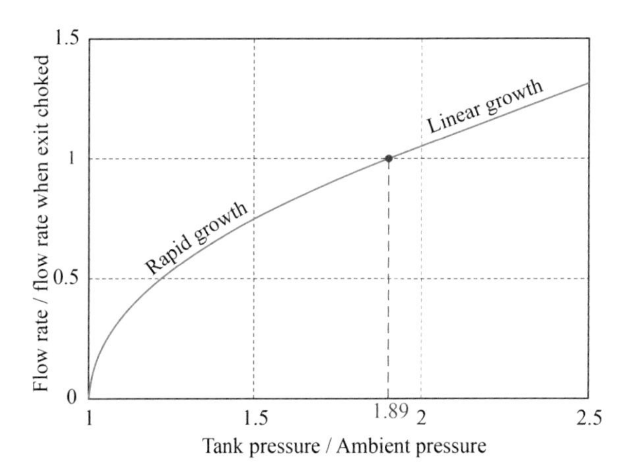

Figure 7.19 Flow rate as a function of freestream total pressure in a converging nozzle.

Figure 7.19 depicts the mass flow rate as a function of the tank's pressure at a given temperature. For a subsonic flow, an increase in the tank's pressure rapidly increases the mass flow rate by increasing the outlet velocity and outlet density together. After the outlet velocity reaches sound speed, the mass flow rate increases linearly because only the density increases with increasing pressure. If the tank's temperature and pressure remain constant, the outlet velocity increases as we lower the ambient pressure. Once the outlet velocity reaches the sound speed, regardless of how much lower we set the ambient pressure, the mass flow out of the tank will no longer increase.

When the fluid velocity reaches sound speed, the ratio of total pressure to static pressure is:

$$\frac{p_t}{p} = \left(1 + \frac{k-1}{2} Ma^2\right)^{\frac{k}{k-1}} = \left(1 + \frac{k-1}{2}\right)^{\frac{k}{k-1}}.$$

At $k = 1.4$, this critical value is approximately equal to 1.89.

When the ratio of tank pressure to ambient pressure is greater than the critical value, the outlet velocity remains sonic, indicating that the outlet static pressure is not equal to the ambient pressure. This is different from the one when the outlet velocity is subsonic, for which the outlet pressure is always equal to the ambient pressure. Downstream of the nozzle outlet, the airflow continues to expand to a lower static pressure, and finally mixes with the ambient air with the same pressure.

7.4.1.2 Supersonic Incoming Flow

A supersonic flow decelerates as pressure increases along a converging nozzle. The flow at the nozzle outlet may be supersonic or sonic, but cannot be subsonic. Given the incoming Mach number, the outlet Mach number depends only on the inlet-to-outlet area ratio:

$$q(Ma_2) = \frac{A_1}{A_2} q(Ma_1).$$

Figure 7.20 depicts the outlet Mach number as a function of inlet-to-outlet area ratio, A_1/A_2, at some fixed inlet Mach number. There is a fixed contraction ratio, known as the *critical contraction ratio*, for each inlet Mach number, at which the outlet Mach number equals to 1. As the outlet area decreases further, one-dimensional, steady, isentropic flow relations cannot be satisfied. What happens then to the flow?

Figure 7.21 shows the changes of flow pattern with decreasing outlet area, as well as the evolution of the compression waves inside the nozzle. When the contraction ratio, A_1/A_2, is smaller than the critical contraction ratio, the outlet velocity is still supersonic, and the exhaust pressure of the nozzle is lower than the ambient pressure. Downstream of the nozzle, there are a series of alternating expansion and compression waves. Among them, normal shock waves may also periodically occur. As the outlet area decreases, the static pressure increases and the expansion and compression waves gradually get close to the nozzle outlet. When the critical contraction ratio is reached, a normal shock wave forms at the nozzle outlet. As we lower the outlet area further, the normal shock wave propagates upstream into the nozzle. However, a shock wave cannot be stably located in the converging portion of the nozzle and it will rapidly move upstream, making the flow through the converging portion become entirely subsonic.

Where does the shock go when it moves upstream and out of the converging portion? To answer this question, we need to consider the actual configuration of the nozzle. There are two possibilities that can make a steady flow at the inlet of a converging nozzle supersonic. One is that the nozzle itself is moving at supersonic speeds, such as those installed on a supersonic airplane. The other is that a Laval nozzle installed upstream of the converging nozzle accelerates flow from subsonic to supersonic speed. If the nozzle is in supersonic motion, the normal shock wave is pushed upstream until it is expelled from the inlet, forming a detached shock wave to make

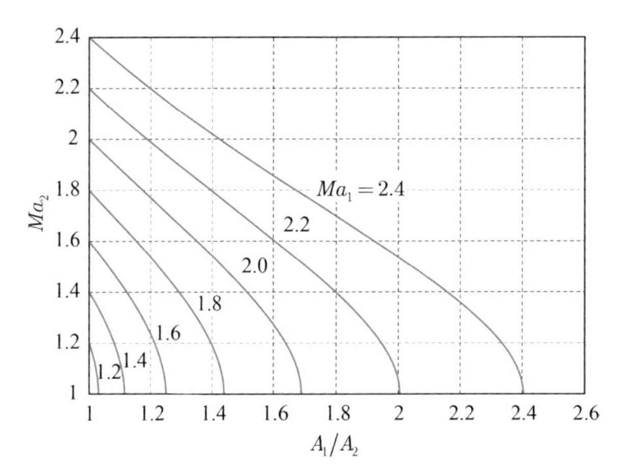

Figure 7.20 Outlet Mach number as a function of inlet-to-outlet area ratio.

At a smaller contraction ratio, a series of alternating expansion and compression waves form outside the nozzle. normal shock waves may exist among them.

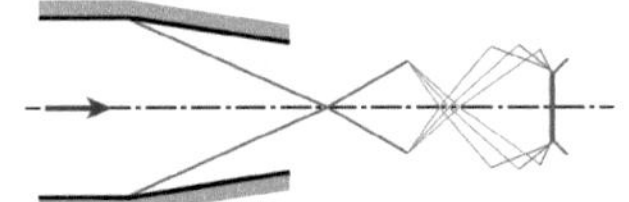

At the critical contraction ratio, a normal shock wave forms at the nozzle outlet. The flow downstream of the normal shock wave is subsonic.

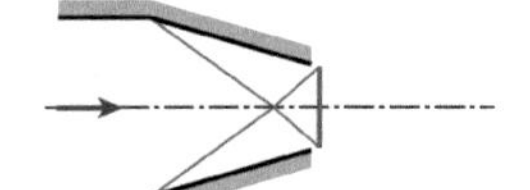

Lower the outlet area further, the normal shock wave moves upstream into the nozzle. The entire flow becomes subsonic.

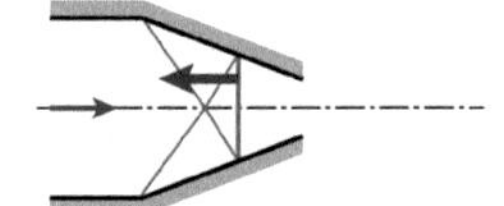

Figure 7.21 The variation of flow pattern with decreasing outlet area.

the flow through the nozzle become entirely subsonic, as shown in Figure 7.22(a). If a Laval nozzle is installed upstream of the converging nozzle, the normal shock wave is pushed upstream and located in the diverging portion of the Laval nozzle, or simply expelled from the Laval throat to make the flow through the Laval nozzle become entirely subsonic, as shown in Figure 7.22(b).

Now let us discuss why it is impossible to establish a standing normal shock wave in a converging nozzle.

The strength of a shock wave is determined by the pressure difference across the shock. A large differential pressure generates a strong shock wave, while a small one generates a weak shock wave. Furthermore, the propagation velocity of the shock wave is directly related to its strength, which can be used to analyze the stability of the shock wave.

If a normal shock wave stands at some location in the converging nozzle, the supersonic flow upstream of the normal shock wave is decelerated, while the subsonic flow downstream of it is accelerated, as shown in Figure 7.23(a). As the outlet pressure decreases slightly, the pressure behind the shock wave will be correspondingly reduced, resulting in a weaker shock wave. Since its propagation velocity relative to the flow decreases, the weaker shock wave moves downstream to a location with

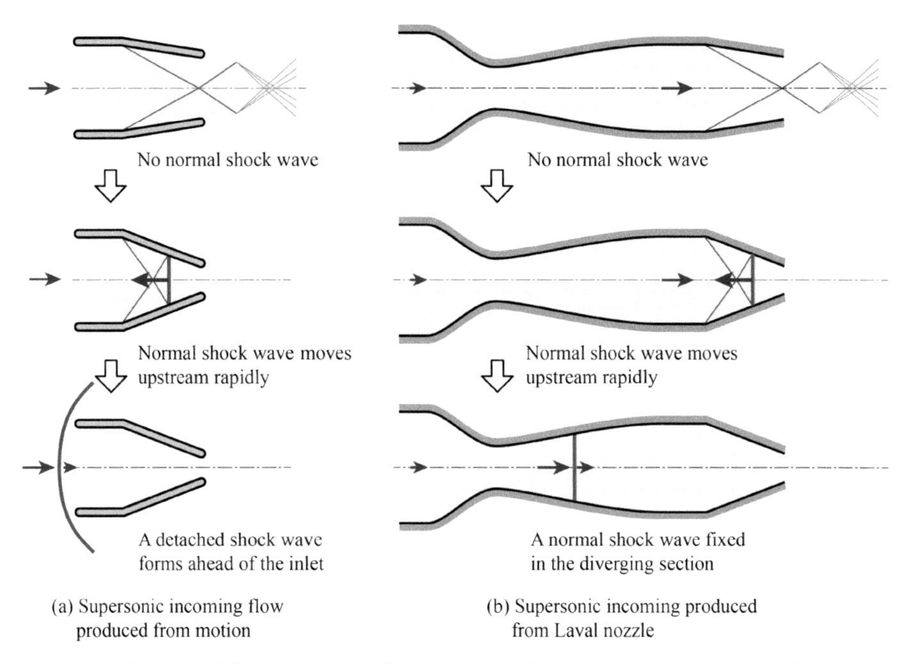

Figure 7.22 Two possibilities in which the normal shock wave in a converging nozzle is pushed upstream.

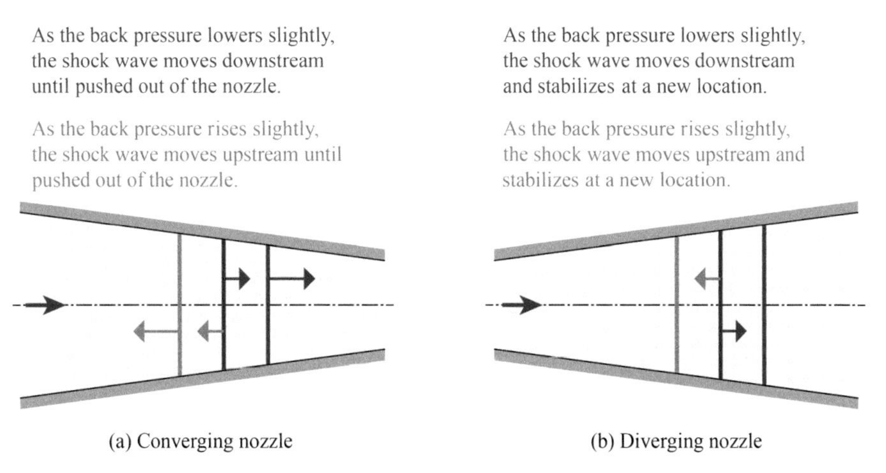

Figure 7.23 A normal shock wave can stand in a diverging nozzle rather than a converging nozzle.

a smaller cross-sectional area. Subsequently, the Mach number ahead of the shock wave decreases, and the shock wave gets weaker and propagates further downstream. Therefore, when the outlet pressure is slightly reduced, the normal shock wave cannot be stabilized in the contraction section until it is pushed out of the nozzle outlet. Similarly, when the outlet pressure is increased slightly, the shock wave will continue to propagate upstream until it is pushed out of the nozzle inlet.

By contrast, for the diverging nozzle shown in Figure 7.23(b), as the outlet pressure is slightly reduced, the shock wave moves downstream to a location with a larger cross-sectional area. Subsequently, the flow Mach number ahead of the shock wave increases, and the pressure rise across the shock wave becomes higher to match the static pressure behind the wave. So, the shock wave will be stabilized at a short distance downstream from the original position. Similarly, as the outlet pressure is slightly increased, the shock wave will be stabilized at a short distance upstream from the original position.

7.4.2 Laval Nozzle

Now let us consider subsonic and supersonic flows through a Laval nozzle.

7.4.2.1 Subsonic Incoming Flow

If the flow through a Laval nozzle begins as a subsonic flow, it accelerates through the converging section. If the flow speed at the throat does not reach the speed of sound, the flow downstream of the throat will decelerate and stay subsonic. If the flow speed at the throat reaches the speed of sound, there can be three flow patterns in the diverging section of the nozzle: subsonic flow; supersonic flow; and supersonic flow + shock wave + subsonic flow, as shown in Figure 7.24.

With viscous effect neglected, the flow through the entire Laval nozzle is close to isentropic and can be calculated using the fundamental gasdynamics equations. For this flow, the total temperature and total pressure are constant at the inlet value. When the Laval nozzle is not choked, the flow through it is entirely subsonic. The inlet, throat, and outlet are denoted by 1, 2, and 3, respectively, and we obtain the isentropic flow relations as:

$$A_1 q(\lambda_1) = A_2 q(\lambda_2) = A_3 q(\lambda_3),$$

$$p_1/\pi(\lambda_1) = p_2/\pi(\lambda_2) = p_3/\pi(\lambda_3).$$

With total pressure remaining unchanged, the outlet coefficient of velocity, λ_3, is determined by the outlet pressure, p_3, and the throat coefficient of velocity, λ_2, is determined by λ_3 and A_3/A_2 together. Therefore, the Mach number at the throat is only related to p_3. Assume the flow is at rest initially, and the outlet ambient pressure is decreased continually; when the ambient pressure drops to a specific value, the throat velocity reaches the speed of sound but the flow is decelerated again to subsonic speed in the diverging section. Except for the throat, the flow through the whole nozzle is still subsonic.

As the ambient pressure continues to decrease, the pressure at a short distance downstream of the throat becomes smaller than the throat pressure, so that the corresponding velocity exceeds the throat velocity – that is, supersonic flow occurs. A normal shock wave forms in the diverging section of the nozzle. Across the normal shock wave, the supersonic flow slows down to subsonic speed. The subsonic flow then continues to decelerate further in the remaining part of the diverging section.

As the ambient pressure decreases, the static pressure behind the shock wave decreases, causing the shock wave to propagate downstream. When the ambient

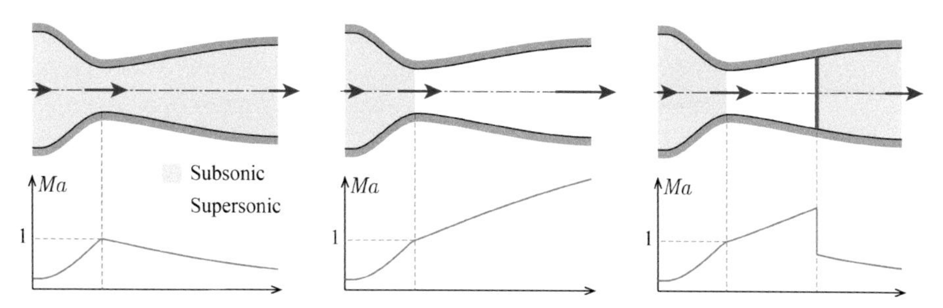

Figure 7.24 Three possible flow patterns if the flow through a Laval nozzle begins as a subsonic flow, and the flow speed at the throat reaches the speed of sound.

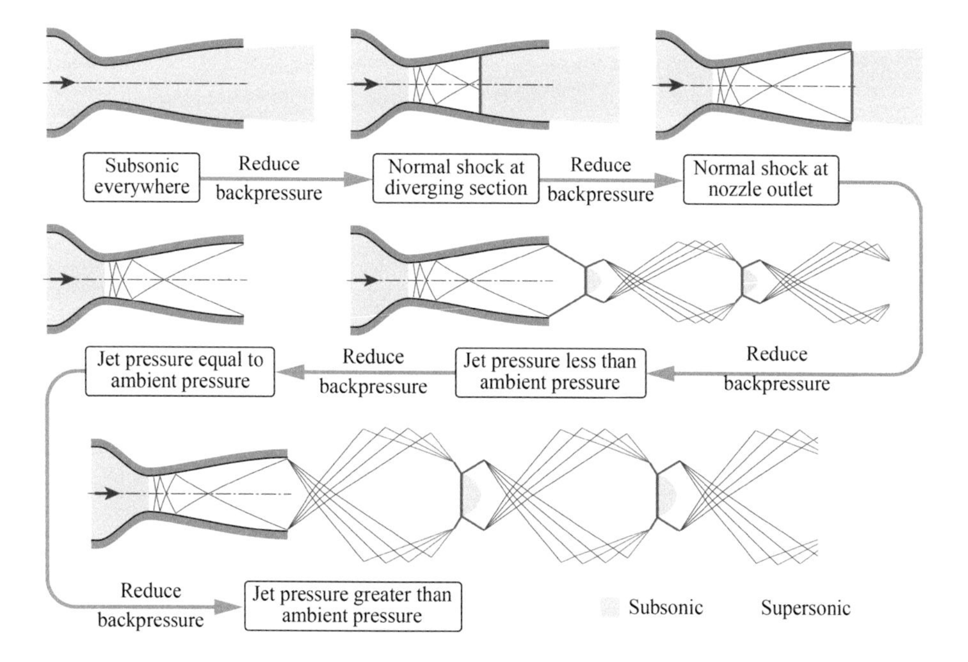

Figure 7.25 The flow pattern downstream of the Laval nozzle (in the diverging portion and jet).

pressure drops to a certain value, the shock wave moves to the outlet, and the flow through the entire diverging section of the nozzle becomes supersonic. The exhaust pressure of the nozzle is still lower than the ambient pressure, and oblique shock waves form at the nozzle outlet to match the exhaust and ambient pressure.

Further lowering the ambient pressure cannot change the flow pattern in the diverging section. As the ambient pressure drops to a value equal to the outlet pressure, there is neither shock wave nor expansion wave at the outlet. The flow forms a parallel jet pattern. Further lowering the ambient pressure, when it is smaller than the outlet flow pressure, expansion waves form outside the nozzle.

Figure 7.25 shows the flow pattern downstream of the throat of the Laval nozzle as the ambient pressure is lowered.

We refer to the case in which the exhaust pressure is lower than the ambient pressure as over-expanded flow, while the case in which the exhaust pressure is higher than the ambient pressure is known as under-expanded flow. A flow structure consisting of compression and expansion waves, generally known as *Mach disks* or *shock diamonds*, forms outside the nozzle of rocket and supersonic aircraft engines. Under-expanded flows are more common than over-expanded ones.

7.4.2.2 Supersonic Incoming Flow

If the flow through a Laval nozzle begins as a supersonic flow, it decelerates through the converging section. The flow speed at the throat may not decelerate to the speed of sound. The throat velocity entirely depends on the throat-to-inlet area ratio. Only at one specific area ratio, also known as the critical area ratio, the flow speed at the throat is exactly equal to the speed of sound. When the throat area is not small enough, the flow at the throat is still supersonic; when this area is too small, the given inlet Mach number cannot be maintained.

If the flow at the throat is supersonic, there may be two flow patterns in the diverging section of the Laval nozzle: supersonic flow or supersonic flow + shock wave + subsonic flow. If the flow is sonic at the throat, three flow patterns are possible in the diverging portion of the Laval nozzle: supersonic flow; supersonic flow + shock wave + subsonic flow; or entirely subsonic flow, as shown in Figure 7.26.

Let us assume that the flow throughout a Laval nozzle (including the throat) is supersonic, and the inlet total temperature, inlet total pressure, and inlet Mach number remain constant. As we raise the outlet ambient pressure, a normal shock wave originally forms at the nozzle outlet, and propagates upstream into the nozzle until it reaches the throat. Once the normal shock wave propagates into the converging section, it cannot stand still and rapidly approaches the inlet to change the inlet Mach number. Then, the flow throughout the Laval nozzle becomes subsonic.

For a Laval nozzle with a specific area ratio and inlet boundary condition, if the flow in the throat is sonic, the flow in the entire diverging section can be subsonic with appropriate ambient pressure. In this case, the flow decelerates throughout the Laval nozzle without a shock wave. This is the ideal method to decelerate the supersonic flow.

However, this flow pattern is unstable. Any increase in outlet ambient pressure may cause the flow upstream of the throat to become subsonic. A shock wave then forms in the converging section of the nozzle. The shock wave cannot stand still in the converging portion of the nozzle and moves out of the inlet, resulting in a phenomenon called inlet "unstart" in the air intake of a supersonic aircraft. Therefore, a reliable way to slow down the airflow with small loss is to have the throat flow velocity supersonic, accompanied with a short supersonic region downstream of the throat, and then a weak normal shock wave slowing the flow to subsonic.

Figure 7.27 shows several forms of flow deceleration in the inlet of a supersonic aircraft. It is necessary to design a high-efficiency and robust inlet. High efficiency means fewer losses, while robustness means inlet "unstart" will not occur due to some disturbances.

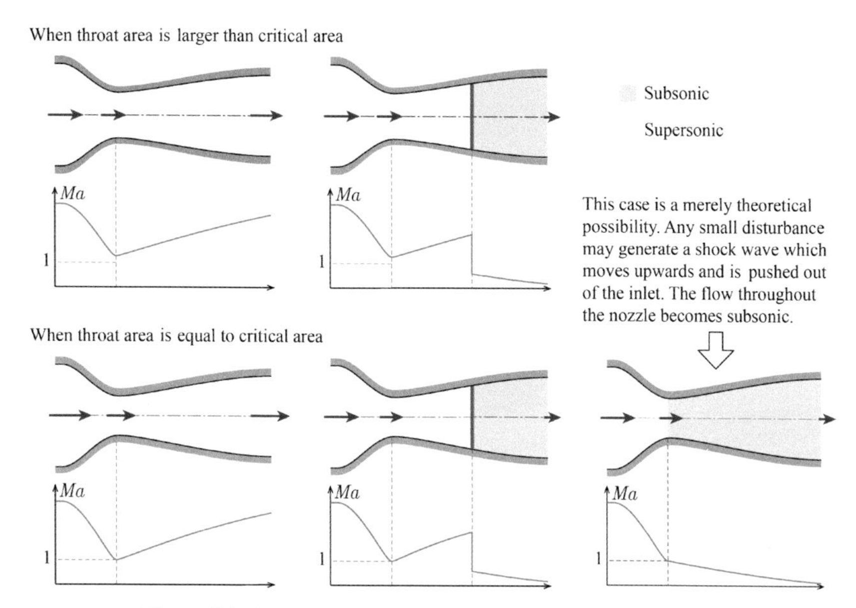

Figure 7.26 All possible flow patterns if the flow through a Laval nozzle begins as a supersonic flow.

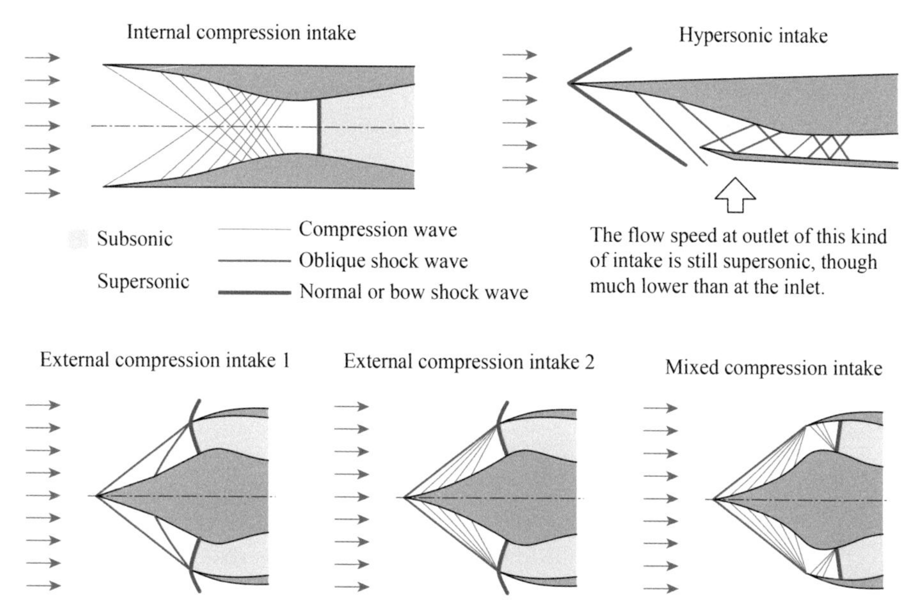

Figure 7.27 Airflow deceleration in several types of supersonic intakes.

We conclude this chapter with an example. As a civil aircraft flying at subsonic speed passes overhead, we can hear whether it is approaching or flying away. However, an aircraft whose velocity is close to the speed of sound cannot be observed in terms of the direction of the sound source, because by the time the noise reaches our ears the aircraft has traveled some distance ahead of the original source of the sound.

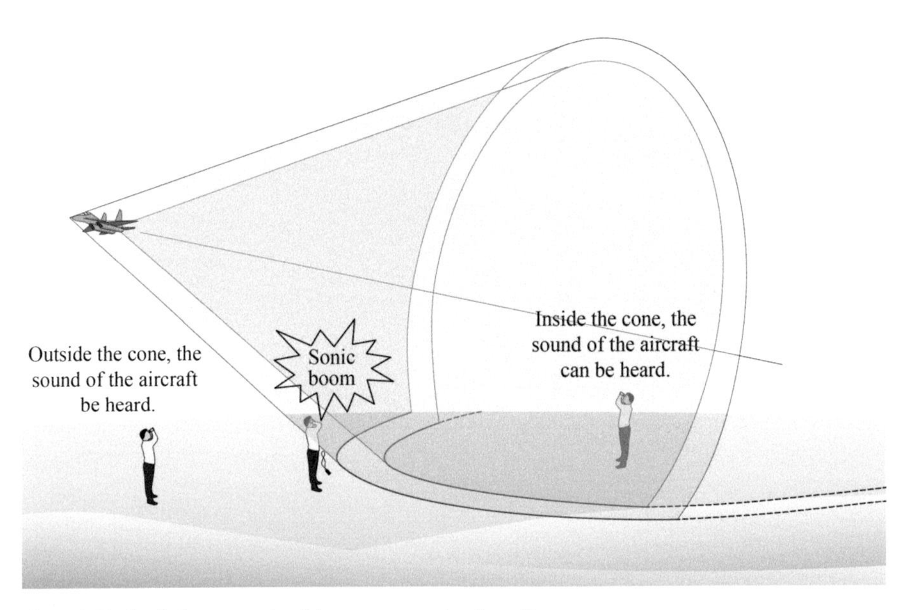

Figure 7.28 Sonic boom emitted by a supersonic aircraft.

As an aircraft flying at supersonic speed passes overhead, no sound is heard as it approaches us because the aircraft is traveling faster than the sound it produces. However, a loud noise will shock our ears after the aircraft has passed by. The sound, like a clap of thunder heard on the ground, also called a *sonic boom*, is tied to the shock waves produced by the aircraft, as shown in Figure 7.28.

Although the aircraft moves at a supersonic speed, the sound of its engines can still travel through still air to our ears after the aircraft flies away. On the other hand, the pilot of a supersonic aircraft cannot hear noises coming from the ground behind him because sound traveling through still air cannot catch up with the aircraft.

Expanded Knowledge

Aerodynamic Heating

Aerodynamic heating is the heating of a solid object produced by the high-speed air passing it, whereby the kinetic energy of the air is converted into heat by compression and skin friction on the surface of the object. Regardless of whether the flow is viscous or not, the maximum temperature that can be reached by air deceleration is the freestream total temperature, and the maximum temperature on the surface of the object is called the recovery temperature, which is generally lower than the freestream total temperature but higher than the local static temperature outside the boundary layer. Figure 7.29 shows the temperature distribution of the gas around a cylinder with a freestream static temperature of 15°C and Mach number of 3.0. As can be seen, the maximum temperature throughout the flow field appears directly in front

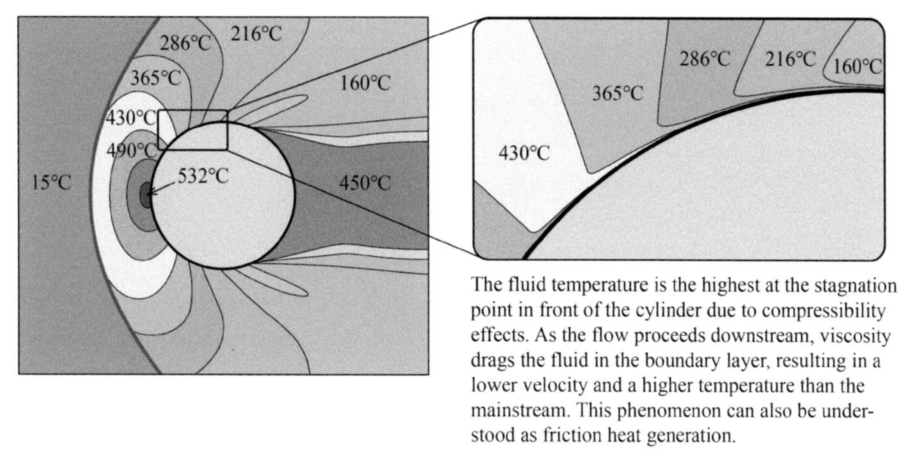

Figure 7.29 Aerodynamic heating occurring in the supersonic flow past a cylinder.

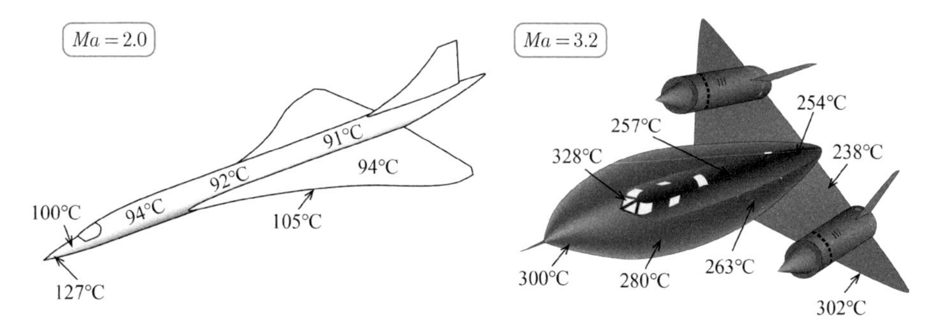

Figure 7.30 Skin temperature contours on the Concorde and SR-71 Blackbird during cruise flight.

of the cylinder, and the aerodynamic heating has the strongest effect near the stagnation point. We therefore conclude that aerodynamic heating is principally caused by compression rather than friction. Therefore, it is not proper to refer to aerodynamic heating as frictional heating.

For inviscid flow, the temperature distribution is similar to the pressure distribution; for viscous flow, the temperature distribution within the boundary layer is strongly affected by viscosity. As can be seen from the zoomed contour on the right of Figure 7.29, the flow temperature within the boundary layer is significantly greater than that of the mainstream at the same streamwise location. The reason for this is that the gas within the boundary layer, which is not accelerated to the speed of the mainstream due to viscosity, is not effectively cooled down. It is also reasonable to interpret this action as frictional heat, since the friction converts part of the mechanical energy into internal energy in the boundary layer.

Figure 7.30 shows the surface temperature distribution of the Concorde and the SR-71 Blackbird with cruising speeds of Mach 2.0 and 3.2, respectively. Aerodynamic heating should be taken into account for supersonic aircraft, and how to effectively dissipate the heat is a problem worthy of research.

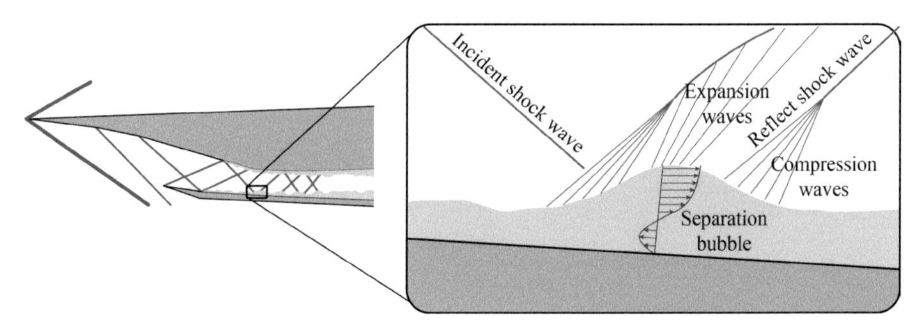

Figure 7.31 The shock wave–boundary layer interaction in the inlet of a supersonic/hypersonic vehicle.

Shock Wave–Boundary Layer Interaction

In Chapter 6 we discussed the effect of viscosity on incompressible flows, and the effect of compressibility on inviscid flows was examined in the present chapter. For practical viscous and compressible flows, the combined effects of viscosity and compressibility must be considered, and the complex interaction between shock waves and boundary layers is a chief concern.

Figure 7.31 shows the flow separation in a hypersonic inlet used in scramjets, caused by the interaction of strong shock waves with the boundary layer. The fluid crossing the shock waves will experience a sudden increase in pressure. The incident shock waves form sharp adverse pressure gradients in local areas, and this may cause the separation of the boundary layer. If properly controlled, only separation bubbles are formed, which will affect the shape of the shock wave in supersonic flows. Shock wave/boundary layer interactions have been broadly investigated in high-speed aircraft and turbomachinery. How to control the boundary layer separation caused by shock waves and how to evaluate the influence of boundary layer on the shock system structure are key design issues.

Questions

7.1 Shock wave drag, also known as wave drag, is the dominant form of drag for a supersonic aircraft. Explain the causes of wave drag using the control volume analysis.

7.2 As seen in this chapter, the total temperature remains unchanged across a shock wave. Explain why the still air is heated significantly after the shock wave generated by an explosion sweeps through it. (Note: Away from the center of the explosion, the shock wave travels faster than the reacted gas, so this temperature rise is not caused by the high-temperature gas.)

7.3 A shock wave is a kind of strong compression wave. In the flow field, a series of compression waves will merge into a shock wave, while a series of expansion waves cannot merge into a strong expansion wave. In fact, a strong expansion wave cannot stably exist. Why?

7.4 In a diverging channel, subsonic flow decelerates, yet supersonic flow accelerates. Explain the role of the diverging walls.

8 Similarity and Dimensional Analysis

Hummingbirds fly more like bees than like albatrosses.

8.1 The Concept of Flow Similarity

Flow similarity theory is mostly developed from experiments. Some experiments cannot be conducted in real-world environments; those that can closely simulate real-world environments require considerable human and financial resources; and some experiments suffer from a certain amount of experimental error due to the inaccurate measurement of the detailed flow information in a real-world situation. Although some model experiments might look completely different from the real thing, if analyzed properly they can still reflect the essence of the problem.

For example, the early attempts to fly made by imitating birds or insects with flapping wings all failed. Later successes in aircraft design would have been virtually impossible without an inexhaustible number of model experiments. Early changes were mainly about perfecting the power and shape of aircraft. If the shapes of two objects are similar, they are supposed to have the same principle of flight. However, we know now that scale also has a great influence. It is impossible to make an aircraft that exactly mimics a bee by magnifying it thousands of times.

Due to the significant change in Reynolds number, the laminar flow over a bee's wing becomes turbulent after amplification, resulting in a different type of flow pattern. In science fiction, a bee exposed to radiation grows much bigger than a person and causes trouble everywhere. This kind of scenario will never happen in reality, because such a giant bee could only crawl along the ground (the other significant reason is the lift-to-weight ratio).

Most fish share perfect streamlining features for ease of movement through water. A streamlined body mimicking the body shape of a fish could move through water or air at low speeds with little drag. However, if the body were to move through air at supersonic speeds, the shape of a fish would not be at all suitable. Subsonic airfoils almost reached their ultimate performance late in World War II. Nowadays, those aircraft are still used in aerobatics shows. However, when an aircraft tries to break the sound barrier, drags produced by these once-excellent subsonic airfoils become very large. A diamond-shaped rather than a streamlined airfoil is a better choice.

When a strong wind blows across a power wire, a whirring sound is caused by the shedding of vortices, known as a Karman vortex street. The Karman vortices are shed at an audible frequency in this case. However, the strong wind blowing across a thicker column, such as a chimney, makes no such sound. Even if the Reynolds numbers are identical, the larger cylinder will still not produce a sound that can be perceived by human ears. Naturally, there are other factors that make the vortex shedding frequency of a large cylinder different from that of a small one.

To sum up, the above examples discuss the similarity of flow in terms of viscosity, compressibility, and flow unsteadiness. In order for a model to represent the original flow, their flows need to be made similar. Essentially, a flowing fluid obeys Newton's laws and the laws of thermodynamics. Although the two flows have different fluid properties, they exhibit similar flow behavior as long as they have similar influencing factors in the governing equations. Most of the time there is only one or two dominant influencing factors, and the influence of other factors is negligible. For example, for a steady, incompressible flow with no body forces, the type of flow pattern is basically determined by its Reynolds number.

8.2 Dimensionless Numbers

Dimensional analysis is a mathematical technique that we shall not discuss in depth in this book. We would rather directly define several dimensionless numbers in fluid mechanics and explore their physical significance. Figure 8.1 shows the commonly used dimensionless numbers, together with their expressions and physical significance.

These dimensionless numbers represent the ratio of two types of force. This is because the flow state is determined by the dominant force exerted on it. Next, we will analyze the physical meanings and applications of these numbers.

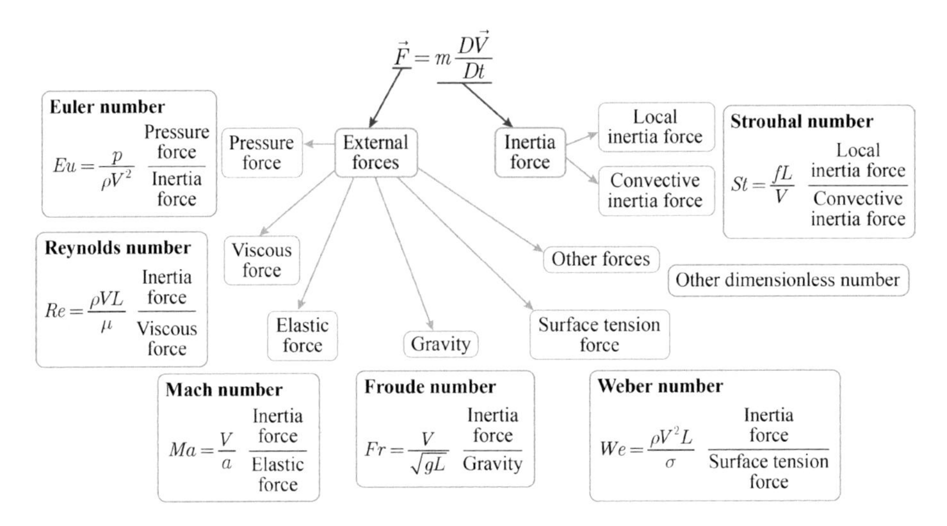

Figure 8.1 Some dimensionless numbers in fluid mechanics.

8.2.1 Reynolds Number

The Reynolds number is

$$Re = \frac{\rho VL}{\mu}, \tag{8.1}$$

where ρ is the fluid density; V is the fluid velocity; L is the characteristic length; and μ is the dynamic viscosity.

The Reynolds number is perhaps the most widely known dimensionless number in fluid mechanics. In the famous experiment conducted by the Manchester professor Osborne Reynolds, the Reynolds number is introduced as a parameter for predicting whether a flow condition will be laminar or turbulent. Therefore, many of those who have studied fluid mechanics end up remembering that the Reynolds number is the parameter used to determine the state of a flow. However, this is not entirely accurate. For example, both honey and water can remain laminar at low speed, but have quite different motion characteristics because they have different Reynolds numbers.

The Reynolds number represents the ratio of inertial to viscous forces. Since the inertial force represents the rate of change of the velocity of a given fluid element (acceleration), the Reynolds number actually indicates the influence of viscous effects on a flow. A gradual increase in Reynolds number leads to a decrease in viscous force. For general flows, the Reynolds number regimes and corresponding flow state can be roughly classified as:

(1) For $Re \ll 1$, the viscous forces are far greater than the inertial forces. This ultra-low Reynolds number flow is named creeping flow or Stokes flow. The movement of bacteria and viruses in liquid is a type of creeping flow. In these

flows, inertia is negligible compared to viscous and pressure forces. Therefore, a fluid moves under the action of the force, and stops when no force acts on it. The same is true for a moving solid where the frictional force is much greater than the inertial force, such as a heavy piece of furniture being dragged over rough ground.

(2) For $1 < Re < 2{,}300$, neither the inertial nor viscous effects in the fluid flow can be completely ignored. The viscous force plays a role in restricting the fluid particles to travel on regular paths – that is, the flow is laminar. (It should be noted that we have $Re_{cr} = 2{,}300$ for a fluid flowing through a constant-diameter pipe. For other flows, the critical Reynolds number may have a significant deviation due to the arbitrariness of the selected characteristic scale and characteristic velocity.)

(3) For $2{,}300 < Re < 10^5$, a laminar flow will become unstable as the viscous force diminishes. Small pressure perturbations may produce long-distance oscillations lasting for a long time, resulting in alternating laminar and turbulent flows. The flow may be either laminar or turbulent, depending upon external factors like the entrance conditions into the pipe and the roughness of the pipe surface.

(4) For $Re > 10^5$, the viscous force decreases dramatically and can be neglected, while the contribution from the inertia force becomes dominant. Under such conditions, the flow transitions into a turbulent state, and the effect of turbulent flow fluctuation on the macroscopic movement of fluids is similar to that of viscosity, so the flow cannot be treated as inviscid.

(5) For $Re \to \infty$, the flow is inviscid in theory. Superfluids probably correspond to this situation, in which it is reasonable to apply inviscid theory. However, studies have shown that superfluids have many new properties, such as the change of electron-spin inside atoms caused by macroscopic shearing, and so forth. A simplified inviscid-flow theory cannot accurately describe the motion of superfluids. It can be said that inviscid flows do not actually exist.

Figure 8.2 shows Reynolds numbers for typical flows. Since the Reynolds number represents the ratio of two types of forces, we may ask: Why does not the inertial force contribution become fully dominant until $Re = 10^5$ and above, rather than near $Re = 1$?

This is mainly due to the selection of characteristic velocity and characteristic length in the expression for the Reynolds number. For engineering convenience, a certain macroscopic size is generally chosen as the characteristic length, such as the pipe diameter for a typical internal pipe flow, the cylinder diameter for a typical external flow around it, and the chord length of a wing for a typical external flow around it. However, it is more reasonable to select the local characteristic length to describe the motion of fluid particles – for example, to select the boundary layer thickness as the characteristic length scale in a boundary layer flow. In a turbulent flow, the Reynolds number calculated by the dissipative vortex diameter is equal to 1, which indicates that both viscous and inertial forces have the same order of magnitude on this length scale.

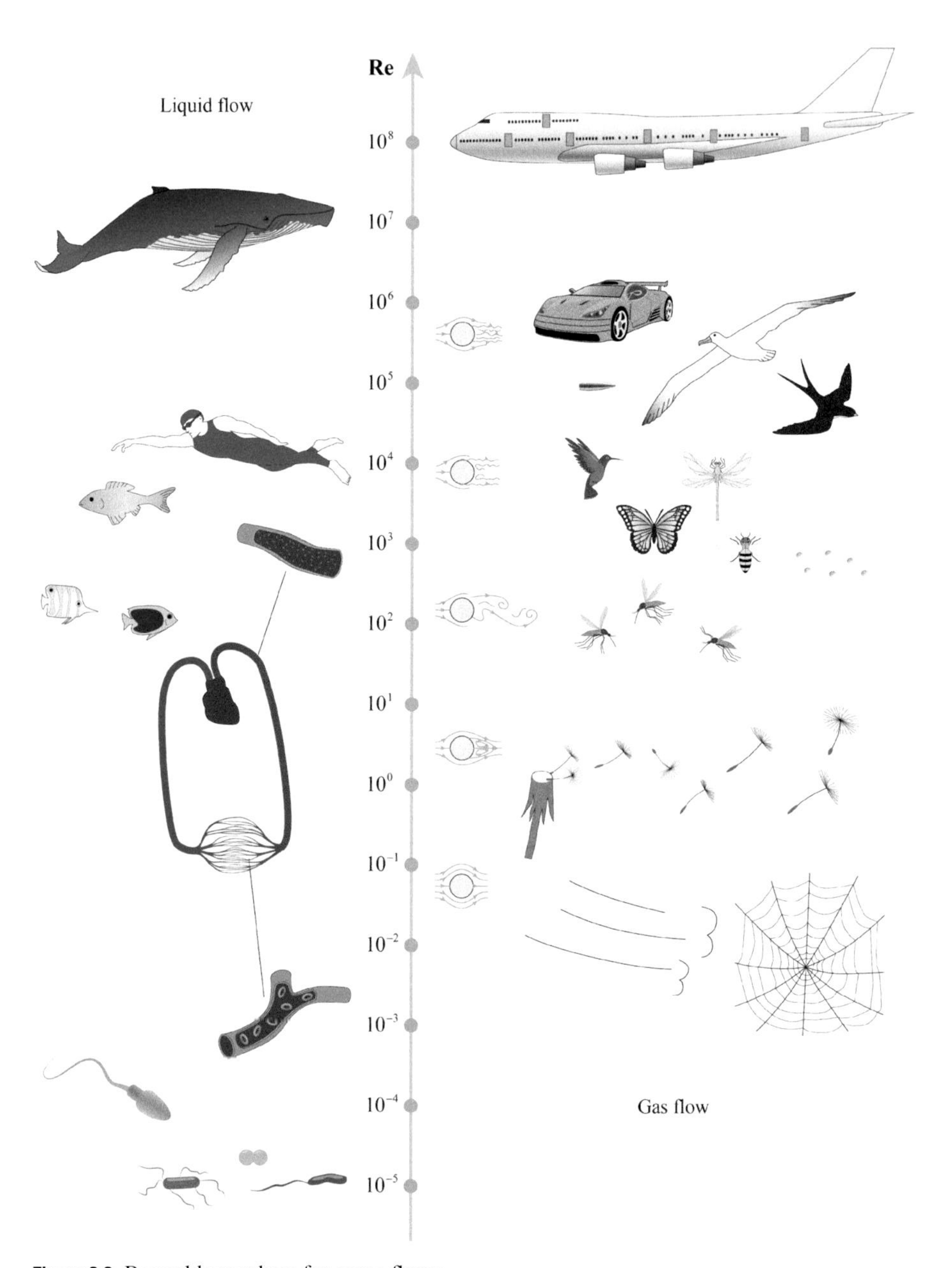

Figure 8.2 Reynolds numbers for some flows.

8.2.2 Mach Number

The Mach number is

$$Ma = \frac{V}{a}, \tag{8.2}$$

where V is the velocity and a is the local speed of sound.

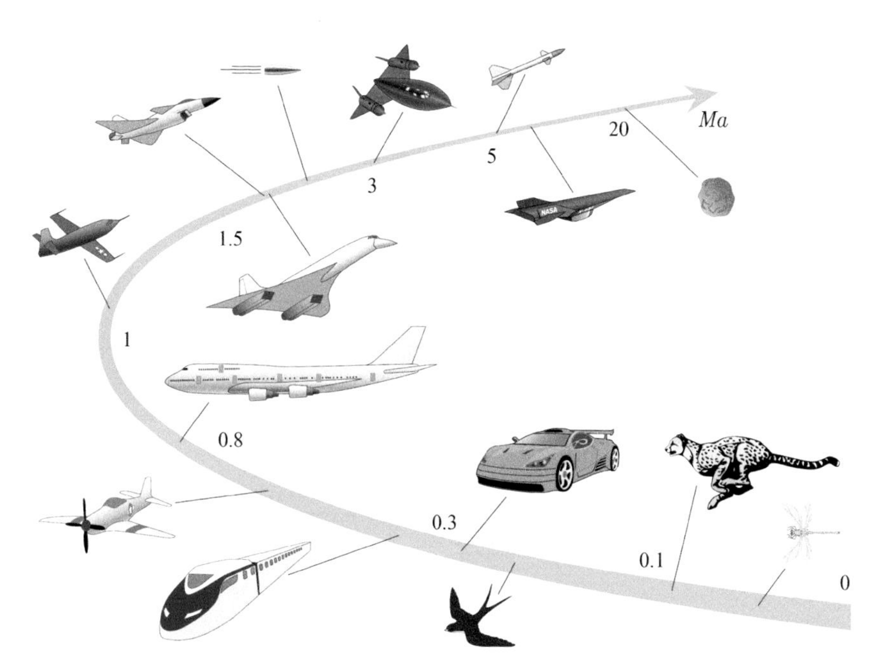

Figure 8.3 Mach numbers of typical objects in motion.

Similar to the importance of the Reynolds number in fluid mechanics, the Mach number is probably the most famous dimensionless number in aerodynamics. This number, named after the Austrian physicist Ernst Mach, provides a physical meaning of the ratio of the speed of an object to the local speed of sound. In fluid mechanics, the Mach number represents the ratio of inertial forces to elastic forces. The larger the Mach number, the smaller the modulus of elasticity of a fluid with respect to the inertial force.

At a very small Mach number, a gas has a modulus of elasticity so large that it can almost be regarded as a rigid body. Then, the gas is usually considered incompressible, and a change in velocity only produces a differential pressure. As the Mach number increases further, a change in velocity also produces an elastic force by compressing the gas. In transonic and supersonic flows the Mach number is the most important factor determining the flow state. For engineers who work with high Mach number flows a common problem they face is the interactions between shock wave and boundary layer, where both elastic and viscous forces affect the flow. If model experiments are needed for this type of flow, the best option is to have both the Mach and Reynolds numbers match the actual situation.

Figure 8.3 shows the Mach numbers of typical objects in motion. In daily life, typical flow velocities are far below the speed of sound, and the Mach number has little influence on the flow. Consequently, model experiments mostly take the Reynolds number into account. Compressibility effects become progressively more important for high-speed flows passing an aircraft or a racing car.

8.2.3 Strouhal Number

The Strouhal number is

$$St = \frac{fL}{V}, \tag{8.3}$$

where f is the frequency of periodic flow; L is the characteristic length; and V is the fluid velocity.

For unsteady, oscillating flows, the Strouhal number is a dimensionless value used to measure the degree of oscillation. It represents the ratio of (local or unsteady) inertial forces to (convective or steady) inertial forces. The higher the Strouhal number, the greater the intensity of oscillation. When a fluid flows around a bluff object, periodic vortex shedding, known as a Karman vortex street, is usually generated in the wake region. Over a certain Reynolds number range the vortex sheds periodically. Then, a sound can be heard and the object appears to sing. Vincenc Strouhal (1850–1922) defined the Strouhal number in his study of wires singing in the wind.

For the flow around a circular cylinder over a wide Reynolds number range, the Strouhal number defined by the vortex shedding frequency is almost constant (i.e., $St \approx 0.21$). Thus, the vortex shedding frequency f can be calculated from the fluid velocity and cylinder diameter. For the flow around a wire 5 mm in diameter, the Reynolds number is 3,000 on a Beaufort wind scale of 5 (wind speeds ~10 m/s), and a Karman vortex street will be generated in the wake region. According to Equation (8.3), the vortex shedding frequency can be calculated as 420 Hz, which is audible to human ears. At the same Reynolds number, the vortex shedding frequency around a tree 50 cm in diameter at a wind speed of 10 m/s is 4.2 Hz, which is not audible. For a chimney with a diameter of 5 m, wind speeds of at least 476 m/s are needed for audible sound ($f > 20$ Hz). Clearly, there are no wind speeds that high on Earth. Even if there were, a supersonic flow with complex wave interactions would occur, which does not obey the aforementioned law.

8.2.4 Froude Number

The Froude number is

$$Fr = \frac{V}{\sqrt{gL}}, \tag{8.4}$$

where V is the fluid velocity; L is the characteristic length; and g is the gravitational acceleration.

The Froude number represents the ratio of inertia forces to gravitational forces. This number is generally considered only when dealing with the motion associated with the free surface of liquids in a gravitational field. It was defined by the British scientist William Froude (1810–1879) to predict the drag of boats.

The Froude number is less commonly used because the gravitational forces are negligible in most fluid dynamics problems. Let us look at an example. For the flow through a corner at a radius of 10 mm, if the fluid velocity is 20 m/s, its centrifugal acceleration is:

$$a = \frac{V^2}{r} = \frac{20^2}{0.01} = 40{,}000\,\mathrm{m/s^2} \approx 4{,}000g.$$

In this case, the inertial force is far greater than the gravitational force, which can be completely neglected.

Since acceleration is independent of fluid density, the above conclusion can be applied to both gases and liquids. However, in practical engineering problems we may clearly feel that the influence of the gravitational force on liquid motion is far greater than that on gas motion. The reason is that the inertia forces are easily balanced by the differential pressure forces for a gas but not so easily balanced for a liquid. For air and water, the freestream dynamic pressures of a 20 m/s flow are:

$$\text{air}: \frac{\rho V^2}{2} = 245\,\mathrm{Pa} \approx 0.002\,\mathrm{atm},$$

$$\text{water}: \frac{\rho V^2}{2} = 2\times 10^5\,\mathrm{Pa} \approx 2\,\mathrm{atm}.$$

It is very common for air to flow at this speed, but a large differential pressure is needed to have water flow at the same speed. Driven by the same differential pressure, gases move at high speeds while liquids move at low speeds. The previous analysis on the centrifugal acceleration demonstrates that the gravitational force is important only at small fluid velocities. Therefore, common sense tells us that the gravitational force has a significant influence on the water flow but a negligible one on the airflow. The fact is that water normally flows at low speed. For a high-speed water flow, such as a water jet cutter, gravitational forces do not need to be considered.

Although the gravitational effect is noticeable for low-speed airflow, it is still much smaller than the viscous effect at low Reynolds numbers. Therefore, for a flowing gas we generally do not take the gravitational force into account. However, the gravitational effect is not negligible if a change in temperature leads to a nonuniform density. For example, gravitational force is what drives natural convection; then, another dimensionless number, the Grashof number, needs to be considered.

8.2.5 Euler Number

The Euler number is

$$Eu = \frac{p}{\rho V^2}, \tag{8.5}$$

where p is the fluid pressure; ρ is the fluid density; and V is the fluid velocity.

The Euler number represents the ratio of pressure forces to inertial forces. In reality, it is the differential pressure, rather than the absolute pressure, that influences the motion of the fluid the most. Therefore, a modified Euler number is defined as:

$$Eu = \frac{\Delta p}{\rho V^2}. \tag{8.5a}$$

This dimensionless number is often expressed in another form, namely the pressure coefficient:

$$C_p = \frac{\Delta p}{\rho V^2/2}.$$

For an incompressible flow the modified Euler number represents the ratio of the pressure difference between two points to the dynamic pressure head. According to Bernoulli's equation, the modified Euler number also represents the degree of acceleration of a fluid. For a steady, inviscid flow with no body forces or surface tension forces, or a flow whose state is determined only by differential pressure, the pressure coefficient depends solely on the geometry of the flow field. Taking the ideal flow past a circular cylinder as an example, the surface pressure coefficient is only related to the geometric angle:

$$C_p = \frac{p - p_\infty}{\rho V_\infty{}^2/2} = 1 - 4\sin^2\theta.$$

8.2.6 Weber Number

The Weber number is

$$We = \frac{\rho V^2 L}{\sigma}, \tag{8.6}$$

where ρ is the fluid density; V is the fluid velocity; L is the characteristic length; and σ is the surface tension coefficient of the liquid.

The Weber number is a dimensionless parameter representing the ratio of inertial forces to surface tension forces. Surface tension is a phenomenon occurring on the surface of a liquid. For small-scale flow problems such as capillary phenomena, soap bubbles, and droplet breakup, the effect of surface tension, which may be equal or greater than the local inertial force, is worth considering. The smaller the Weber number, the more important the surface tension. When the Weber number is far greater than 1, the effect of surface tension is negligible.

Raindrops are roughly spherical due to the surface tension of water. However, the effect of surface tension on the water flow in a large waterfall is not worth considering. The breakup of droplets in a high-speed airflow is also closely related to the Weber number, which is usually the key to efficient combustion in an engine combustion chamber.

8.3 Governing Equations in Dimensionless Form

The various dimensionless numbers discussed in Section 8.2 can be obtained either through experiments or dimensional analysis. In this section, we will derive the most important factors that affect the flow field from the governing equations. Compared to dimensional analysis, this is a more rigorous method, and its physical meaning is clearer. Fluid motion is described by the fundamental governing equations of fluid dynamics. For the sake of simplicity, only the Navier–Stokes (N-S) equations for an incompressible, two-dimensional flow are analyzed here. The governing equation in the x direction is expressed as:

$$\rho\frac{\partial u}{\partial t}+\rho u\frac{\partial u}{\partial x}+\rho v\frac{\partial u}{\partial y}=\rho f_x-\frac{\partial p}{\partial x}+\mu\left(\frac{\partial^2 u}{\partial x^2}+\frac{\partial^2 u}{\partial y^2}\right).$$

In a wind tunnel experiment the prototype test conditions are duplicated in the scaled model at the "same" location in order to create similar flows. The "same" location is just a relative concept in nature. Consider an airfoil, for example. The flow separation on the upper surface occurs downstream of its leading edge by 30% of the chord length. Thirty percent means $x^*=x/c=0.3$, where x is the distance from the separation point to the leading edge, c is the chord length, and x^* is the dimensionless length scale.

The governing equations in dimensionless coordinates can be used to describe the flows of similar geometry but different size. Furthermore, these equations can describe the flowing fluid under various boundary conditions, providing a better generality.

Take V, p_0, L, and τ as the reference quantities of velocity, pressure, length, and time, respectively. The dimensionless quantities are expressed as:

$$u^*=\frac{u}{V},\ \ v^*=\frac{v}{V},\ \ p^*=\frac{p}{p_0},\ \ x^*=\frac{x}{L},\ \ y^*=\frac{y}{L},\ \ t^*=\frac{t}{\tau}.$$

Thus, each physical quantity can be expressed as:

$$u=u^*V,\ \ v=v^*V,\ \ p=p^*p_0,\ \ x=x^*L,\ \ y=y^*L,\ \ t=t^*\tau.$$

Substituting these parameters into the governing equations, we obtain:

$$\left[\frac{\rho V}{\tau}\right]\frac{\partial u^*}{\partial t^*}+\left[\frac{\rho V^2}{L}\right]\left(u^*\frac{\partial u^*}{\partial x^*}+v^*\frac{\partial u^*}{\partial y^*}\right)=[\rho f_x]-\left[\frac{p_0}{L}\right]\frac{\partial p^*}{\partial x^*}+\left[\frac{\mu V}{L^2}\right]\left(\frac{\partial^2 u^*}{\partial x^{*2}}+\frac{\partial^2 u^*}{\partial y^{*2}}\right).$$

Each term within square brackets in this formula is measured in newtons per cubic meter. These terms represent the forces acting on the fluid and have the following meanings:

$\left[\frac{\rho V}{\tau}\right]$: local inertial force (unsteady inertial force);

$\left[\frac{\rho V^2}{L}\right]$: convective inertial force (steady inertial force);

$[\rho f_x]$: body force;

$\left[\frac{p_0}{L}\right]$: pressure force; and

$\left[\frac{\mu V}{L^2}\right]$: viscous force.

By dividing all the terms in the above equation by the convective inertial force $\rho V^2/L$, and substituting the body force with gravity, we obtain:

$$\left[\frac{L}{\tau V}\right]\frac{\partial u^*}{\partial t^*}+\left(u^*\frac{\partial u^*}{\partial x^*}+v^*\frac{\partial u^*}{\partial y^*}\right)=\left[\frac{gL}{V^2}\right]-\left[\frac{p_0}{\rho V^2}\right]\frac{\partial p^*}{\partial x^*}+\left[\frac{\mu}{\rho VL}\right]\left(\frac{\partial^2 u^*}{\partial x^{*2}}+\frac{\partial^2 u^*}{\partial y^{*2}}\right).$$

As can be seen, the terms within square brackets correspond to some dimensionless number mentioned above:

$$St\frac{\partial u^*}{\partial t^*}+\left(u^*\frac{\partial u^*}{\partial x^*}+v^*\frac{\partial u^*}{\partial y^*}\right)=\frac{1}{Fr^2}-Eu\frac{\partial p^*}{\partial x^*}+\frac{1}{Re}\left(\frac{\partial^2 u^*}{\partial x^{*2}}+\frac{\partial^2 u^*}{\partial y^{*2}}\right). \tag{8.7}$$

There are four dimensionless numbers that determine the state of an incompressible flow: Euler number (Eu), Reynolds number (Re), Froude number (Fr), and Strouhal number (St). Assuming that the flow is steady and the gravity is negligible, the formula can be simplified as:

$$\left(u^*\frac{\partial u^*}{\partial x^*}+v^*\frac{\partial u^*}{\partial y^*}\right)=-Eu\frac{\partial p^*}{\partial x^*}+\frac{1}{Re}\left(\frac{\partial^2 u^*}{\partial x^{*2}}+\frac{\partial^2 u^*}{\partial y^{*2}}\right). \tag{8.8}$$

That is, the flow state is only determined by the Euler and the Reynolds numbers. Assuming that the viscous force is negligible, the formula can be simplified as:

$$Eu\frac{\partial p^*}{\partial x^*}+\left(u^*\frac{\partial u^*}{\partial x^*}+v^*\frac{\partial u^*}{\partial y^*}\right)=0. \tag{8.9}$$

That is, the flow state is only determined by the Euler number. Since the Euler number can be expressed as the pressure coefficient, Equation (8.9) states that two flow fields with similar geometry and identical boundary conditions have identical pressure distributions. In other words, the pressure field is only determined by its kinematic characteristics – that is, the characteristics of potential flow.

The Mach number is not discussed in the above analysis because the elastic forces are not present in the governing equations for an incompressible flow. For an adiabatic, isentropic, and compressible flow, a case that was analyzed in Chapter 7, the pressure distributions are only determined by two dimensionless numbers: the Mach number and the specific heat ratio.

8.4 Flow Modeling and Analysis

Similar geometry and identical dimensionless numbers are necessary in order to have similar flows. This is actually impossible because only a prototype could be such a model. The key to replacing the prototype with the model is to concentrate on the major factors and ignore the minor ones. That is to say, to require that only some dimensionless numbers be identical. Next, we analyze the selected dimensionless numbers for three specific problems.

8.4.1 Low-Speed Incompressible Flow

In a low-speed incompressible flow, the flow state is highly influenced by the Reynolds number but negligibly influenced by the Mach number. The properties of incompressible flows can usually be described as a function of its Reynolds number. As long as the Reynolds number remains the same, the conclusions of a model experiment can be applied to an actual flow. Since there are three independent variables – V, L, and ν ($\nu = \mu/\rho$) – in the expression of the Reynolds number for model experiments, a number of options are available.

If the prototype is too small to be accurately measured (for example, the lift of insect wings), the experiment can be done with a scale model that is larger than the actual size of the prototype. The Reynolds number is kept constant by reducing the velocity. Conversely, if the actual flow field is too large to properly simulate in a wind tunnel (for example, the aerodynamic forces acting on a skyscraper), we need to work with a smaller model and a higher fluid velocity. In addition, a water tunnel can be used to visualize the air flowing over an aircraft as long as the Reynolds number is kept the same and the gravitational effect excluded.

The above method poses no problem in theory, but there is no easy way to implement it. For example, limited by the size of a wind tunnel's test section, a 400-meter skyscraper is scaled to a 0.4 m model. In order to ensure an identical Reynolds number, the wind speed in the wind tunnel needs to be 10 km/s to simulate an actual wind speed of 10 m/s! Even if a wind tunnel speed of 10 km/s could be attained, the flow past the model would have become hypersonic. Then, the compressibility effects and the changes in air properties would render the two flows totally dissimilar. Then, how can flow similarity be guaranteed?

In high Reynolds number flows, the turbulent eddy viscosity is dominant, while the molecular viscosity can almost be ignored. Therefore, if the Reynolds number is greater than a critical value that ensures a fully developed turbulent flow, the actual and model flows are considered to be approximately similar. The critical Reynolds number typically lies in the range of 10^5–10^7, known as the self-modeling zone. If the Reynolds number is less than some critical value (such as 2,300 in pipe flow) and the flow is fully laminar, the flow at one Reynolds number might be inferred from that at another Reynolds number based on some laminar theory. Therefore, it is possible to convert the results between different Reynolds numbers in this case.

At Reynolds numbers at which the flow may or may not be laminar, achieving flow similarity is most troublesome. Consider, for example, the measurement of lift for an airfoil mounted in a wind tunnel. Although the Reynolds number using chord length as the characteristic length is large enough, the local Reynolds number near the leading edge is bound to be low. If the Reynolds number is not kept the same, and a lower flow speed is used, the laminar-to-turbulent transition takes place at some point downstream of the actual transition position. Then, an artificial transition approach is usually adopted. According to the Reynolds number for the flow over a prototype, we estimate (or measure) the transition position. Artificial disturbances are introduced at the transition position over the model to promote transition-to-turbulence in the boundary layer, such as adding a layer of fine sand with the thickness of the order of the local boundary layer. Figure 8.4 shows how to apply the artificial transition to a scaled airfoil in order to maintain the same surface pressure distribution at different Reynolds numbers.

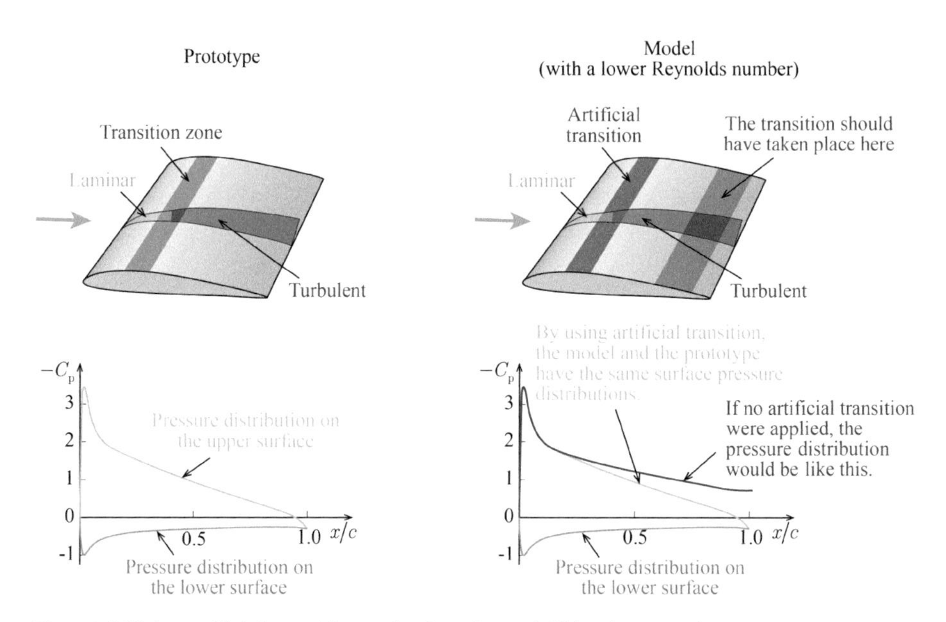

Figure 8.4 Using artificial transition technology in model blowing experiments.

8.4.2 High-Speed Compressible Flow

Compressibility is usually the most important factor in aerodynamics experiments. Especially when a flow is transonic or supersonic, there could be shock waves in the flow, generating discontinuities. These phenomena cannot be simulated by low-speed experiments.

However, it is not entirely impossible to simulate a supersonic airflow by using a low-speed experiment. Sometimes we can simulate shock waves with water surface waves, and do some qualitative research on wave propagation and interference, but this does not satisfy the complete flow similarity. It is possible to generate a shock wave in a low-speed flow as long as the speed of sound through the given fluid is sufficiently low. In the 1940s and 1950s, NASA performed a large number of supersonic compressor experiments. At that time, the problems of strength and vibration characteristics of the blades under high-speed rotation had not been solved. The solution was to use Freon as a working substance. Sound travels much slower through Freon than through air, so the low-speed rotation in Freon can be used to simulate the high-speed rotation in air. Nowadays, of course, this approach is no longer practical. Freon is so notorious for destroying the ozone layer that it is no longer used even for refrigeration.

The Mach number describes the importance of compressibility effects for high-speed subsonic flows. The characteristics such as lift and drag obtained from low-speed model experiments are significantly different from those obtained at high-speed flow conditions. For example, when a low-speed and large-scale flow is used to simulate a high-speed and small-scale flow, the Reynolds numbers for the two flows are

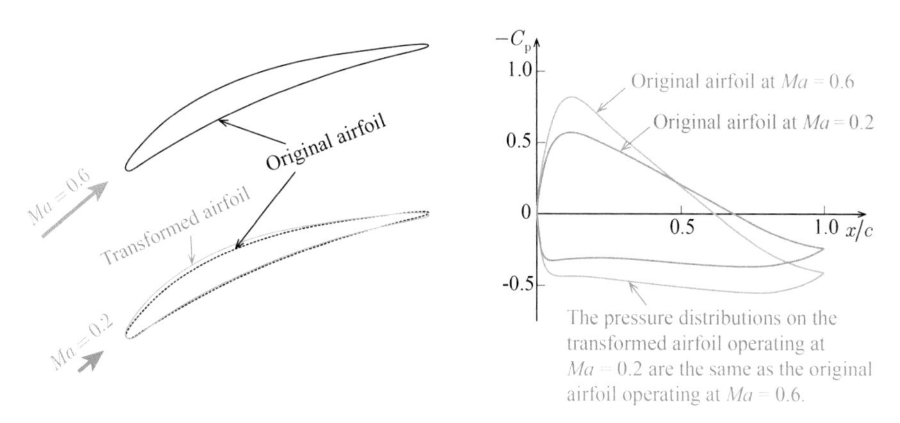

Figure 8.5 The high–low speed similarity transformation is applied to simulate low-speed flows.

kept the same. However, the Mach numbers are not the same. It is necessary and possible to conduct a high-speed to low-speed similarity transformation. A solution to this problem, called the Prandtl–Glauert transformation rule, was first proposed by Prandtl and published by Hermann Glauert (1892–1934) after systematic research.

An alternative method is to redesign the shape of an airfoil so that the pressure coefficient distributions along it are kept the same for both low and high speeds. The shape of the model designed in this way is different from that of the prototype. However, since the pressure distributions along the model and prototype remain the same, the conclusion from the low-speed flow past the model can be applied directly to the high-speed flow past the prototype. For the vanes on a fan, the thickness and curvature of the low-speed model will be greater than those of the high-speed prototype, as shown in Figure 8.5.

If both high-to-low speed transformation and artificial transition are adopted simultaneously, both Reynolds and Mach numbers for the flows past the model and prototype can be different, but the modified Euler number representing the ratio of differential pressure forces to inertial forces must be kept constant. This is because the pressure gradient is the principal driving force in most flows.

8.4.3 A Real-Life Example: A Milk Drop

Figure 8.6 shows the splash of a milk drop. Let us examine the dimensionless numbers needed to describe this type of flow.

First, this flow is driven by gravity and forms a surface wave on the liquid surface, so gravity is an important determinant. That is to say, the flow is related to the Froude number, Fr.

Second, the splattered milk droplets tend to have spherical shapes under the action of surface tension, which is connected to the Weber number, We.

Third, the liquid's shape is constrained by both viscous and inertial forces, which points to the Reynolds number, Re.

Figure 8.6 The splashing of milk.

Fourth, the splashing of milk is an unsteady process associated with the Strouhal number, *St*.

We can see that, even for such a seemingly simple flow, a correct simulation requires the simultaneous consideration of at least four dimensionless numbers: *Fr*, *We*, *Re*, and *St*. Many modern movies use animation to vividly simulate real scenes, in which fluid mechanics plays an important role. Using computational fluid dynamics software to simulate real flows, coupled with post-production and other techniques, stunning waves and hurricanes can be created. In addition, real-world photography also often resorts to similarity theory, by using a small pool to mimic a sea, for example. Whether animation or scenery, the final image effects need to comply with common sense to be realistic. Therefore, it is crucial to maintain flow similarity. The cartoon *Thomas and Friends* provides an example: The smoke from the train chimney is often laminar or low Reynolds number turbulent flow, which is different from the smoke from a full-scale train. It can be speculated that the model used in filming was relatively small.

Expanded Knowledge

Flows at Extremely Low Reynolds Numbers

The flow of a viscous fluid at extremely low Reynolds numbers ($Re \ll 1$), where the inertial forces are negligible compared to the viscous forces, is quite different from the typically observed flow at high Reynolds numbers. This kind of flow is called creeping flow or Stokes flow.

The N-S equations state that a fluid element in equilibrium is generally subjected to four types of forces: gravitational, differential pressure, viscous, and inertial. At extremely low Reynolds numbers, both gravitational and inertial forces can be ignored, and there is a balance of differential pressure and viscous forces. We are generally familiar with the situation of ignoring the gravitational or the viscous forces, but what does it mean to ignore the inertial forces?

If the inertial forces can be ignored, it means that the object has no inertia – that is, there is no acceleration. Then, a net force is necessary to keep an object moving with constant speed in such a fluid; the object will stop immediately when the net force vanishes, which no longer conforms to Newton's law. A more rigorous way to put it is that *when the Reynolds number is far smaller than 1, both gravitational and*

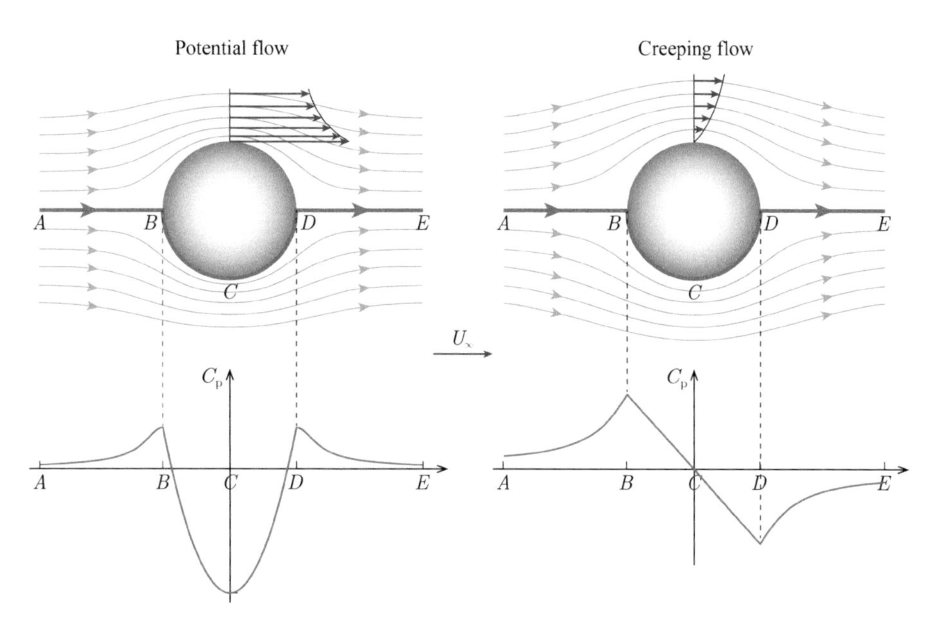

Figure 8.7 The streamlines for potential and creeping flows past a stationary sphere.

inertia forces can be ignored, and the viscous forces and differential pressure forces throughout the flow field are always equal in magnitude but opposite in direction, so there is no acceleration or deceleration in the flow field. It is like trying to push a heavy piece of furniture over rough ground: if we push it hard, the object moves; in the absence of a net force, the object stops immediately. The same is true for the movement of the microorganisms in liquids, which must keep swimming or wriggling to maintain speed.

In a creeping flow there is no flow separation, since the inertial forces are negligible. The flow past a cylinder looks very similar to the potential flow, which is symmetrical on both sides of the cylinder. However, the two flows are different. Figure 8.7 compares the streamlines for creeping and potential flows past a stationary sphere. As can be seen, these two types of flow are different in streamlines, velocity, and pressure distributions. In reality we cannot make a potential flow, so a creeping flow is often shown in the classroom to help students understand a potential flow.

Questions

8.1 Do some investigation on the altitude test bed for aircraft engines. Think about questions such as why it costs so much to build such a test bed, and why ordinary test beds cannot meet the test requirements.

8.2 Learn how to determine the length scale in Reynolds number calculations for atmospheric motions in meteorology.

9 Analysis of Some Flow Phenomena

After stirring a cup of tea, the water ascends at the center and descends at the outside, driving the tea leaves toward the center of the bottom.

In this chapter, 25 carefully selected flow-related phenomena are analyzed, with the purpose of consolidating the understanding of the subject and strengthening the ability to link theory with practice.

9.1 What Are the Shapes of Objects in Outer Space? Properties of Fluids

We are used to living on Earth and take a lot of things for granted. From the perspective of mechanics, the environment on Earth is, however, not common. There is gravity generated by universal gravitation, and atmospheric pressure caused by the atmosphere weighing down on our planet's surface. Objects in outer space, in contrast, experience a greater freedom.

A solid on Earth experiences both gravity and atmospheric pressure, which create stresses within it. If it is suddenly relocated to outer space, such internal stresses will be released as gravity and atmospheric pressure disappear. However, the solid is practically incompressible and will maintain its shape, the change in stress not being sufficient to break it down.

A liquid, such as water, has no definite shape; however, this is true only on Earth. A fluid at rest can only experience normal stresses since it cannot generate shear stresses. Without a container, water always spreads on the ground as a result of continuous shear deformation under the action of gravity. There is no gravity in outer space, and therefore surface tension plays a dominant role in pulling the water droplets into spherical shapes. Therefore, water in outer space should be perfectly spherical.

However, such a pure mechanical experiment is not easy to conduct. If liquid water is suddenly placed in the vacuum of outer space, its boiling point will be significantly lowered so that the water will immediately boil and vaporize at its current temperature. Some of the water will change from liquid to gaseous phase, while the rest of the water will freeze due to the temperature drop. In a weightless environment, the boiling water manifests itself as small bubbles, first appearing in the interior. Smaller bubbles contact each other and then coalesce into larger ones. What is the final shape of the water? Most of the water vapor escaped into the surrounding vacuum, leaving one or several pieces of broken ice shell. There might be some water vapor trapped in the remaining ice.

Therefore, a small amount of water cannot stay liquid in outer space. On the other hand, a huge amount of water can be kept together by universal gravitation to maintain a certain pressure and temperature inside, forming a spherical liquid planet. If the planet is rotating, it becomes an oblate spheroid.

If a certain amount of air initially at atmospheric pressure is suddenly relocated to outer space, it will rapidly expand until its pressure reduces to ambient pressure, which is close to zero. If we consider air as an ideal gas, since the work done on/by the ideal gas is zero in free expansion, its internal energy remains unchanged. Finally, the gas temperature may decrease to the cosmic background temperature through radiation.

9.2 Upside-Down Cup of Water: Incompressibility of Liquids

The upside-down cup of water experiment is often used to demonstrate the existence of atmospheric pressure. Put a cardboard sheet (or plastic board) over the mouth of a cup fully filled with water. Hold the cup and the cardboard, and turn them upside-down. Slowly release the hand that holds the cardboard. The cardboard will not fall down and the water will not spill out. The explanation usually given is that the atmospheric air pushing up from underneath the cardboard is strong enough to overcome the weight of the water pushing down on it. Since atmospheric pressure can support approximately a 10 m high column of water, it is no problem to hold up a cup of water.

This explanation seems reasonable, but it is actually problematic because the water in the cup has 1 atm within itself, hence the downward force exerted by the water on the cardboard should be equal to the weight of the water plus 1 bar. Therefore, the argument that atmospheric pressure supports the cardboard and water from falling simply is not accurate.

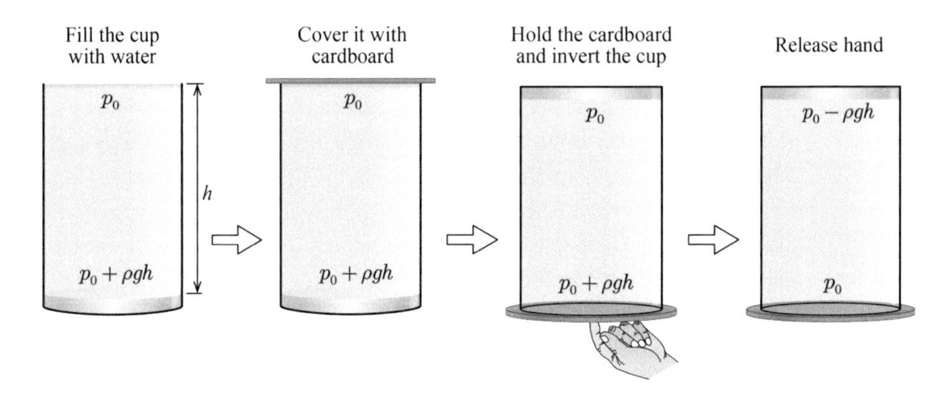

Figure 9.1 A change in water pressure in the upside-down cup.

For the sake of simplicity, assume that both the cylindrical cup and the cardboard do not deform, the cardboard is flat and weightless, and there are no air bubbles in the cup.

Figure 9.1 shows the whole process of the upside-down cup of water experiment. After we fill the cup with water, both the air and water at the mouth of the cup experience atmospheric pressure p_0, and the pressure at the bottom of the cup is $p_0 + \rho gh$. Gently place the cardboard on the mouth of the cup. The water pressure in the cup remains unchanged since the cardboard does not exert any force on the water. Hold the cardboard and turn the cup upside-down. As long as the cardboard and the cup do not deform, the water pressure in the cup would remain unchanged. However, the water pressure at the bottom of the cup becomes p_0 and the water pressure at the mouth of the cup becomes $p_0 + \rho gh$. Subsequently, slowly take away the hand that was holding the cardboard. One would expect the cardboard to fall down since the pressure on its upper surface is greater than that on its lower surface. Contrary to that, the cardboard stays covering the cup and the water does not spill out. According to force analysis, the pressure on the upper and lower surfaces of the weightless cardboard should be equal. Consequently, the water pressure at the mouth of the cup should become the atmospheric pressure p_0, and the water pressure at the bottom of the cup should become $p_0 - \rho gh$. In other words, the water pressure in the cup decreases after you release your hand. How does this happen?

After you overturn the cup and take your hand away, the cardboard would have to move downward a small amount away from the rim of the cup due to the unbalanced pressure on its upper and lower surfaces. As shown in Figure 9.2, the water fills the gap between the cardboard and the rim of the cup without spilling out, which indicates that the water has slightly expanded. Water is almost incompressible, so a tiny expansion will cause a large drop in pressure. Therefore, the water pressure at the mouth of the cup decreases to the atmospheric pressure as the cardboard drops an imperceptible distance.

According to this interpretation, both the water pressure and the air pressure at the gap shown in Figure 9.2 are equal to the atmospheric pressure. To keep water from

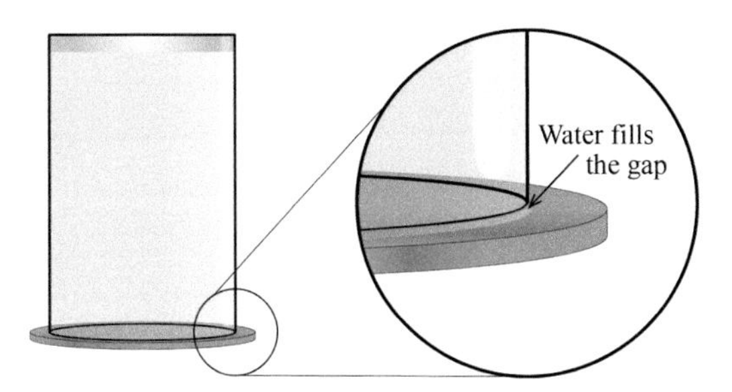

Figure 9.2 The gap between the cardboard and the edge of the cup is filled with water.

spilling out, the gap must be small enough for the surface tension of water to work, otherwise this balance of pressures will be broken and air will enter the cup, leading to the failure of this experiment. A calculation based on the compressibility of water concludes that a 0.1 μm gap is sufficient to achieve the required pressure drop for an average-sized cup. Within such a small gap, the surface tension of water is strong enough.

Many influencing factors have been ignored in the above analysis, making the interpretation not entirely in accord with the facts. An obvious objection is that, since we can clearly observe the gap between the cardboard and the rim of the cup, this gap is obviously larger than 0.1 μm. Next, let us examine some factors that have been ignored.

First, if the compressibility of water is to be considered, so is the deformation of the cup. Since the water pressure inside the cup is largely lower than the atmospheric pressure, the cup experiences an inward pressure, causing its volume to decrease a little. In addition, the squeezing of the hand holding the cup also decreases the volume of the cup. Part of the water is pushing out of the cup, resulting in the actual gap being larger than its theoretical value.

Second, gases from the air are dissolved in the water. A decrease in water pressure releases a small amount of gas to create tiny bubbles that occupy some space in the cup. Since air is more compressible than water, the same pressure drop causes a larger expansion of bubbles so that more water is pushed out of the cup. This is another factor which causes the actual gap to be much larger than the theoretical value. Actually, the experiment can still work even if there are large amounts of bubbles in the cup or the cup is half full. As long as a small amount of water covers the mouth of the cup and the cardboard sticks to it, the water will stay in the cup. It is just a little more difficult to conduct the experiment. All in all, the key to a successful demonstration is that the gap should not be too large.

To conduct a successful demonstration, a cardboard sheet is preferable to a glass one, since it is lighter and easier to deform. The deformation of the cardboard can increase the volume of the cup with the gap left unchanged (the cardboard bulges at the center, increasing the volume within the cup). During the experiment, some teachers may "cheat" by deliberately applying an inward force on the cardboard to make it concave (the water spills out a little), as shown in Figure 9.3. Due to the rebound

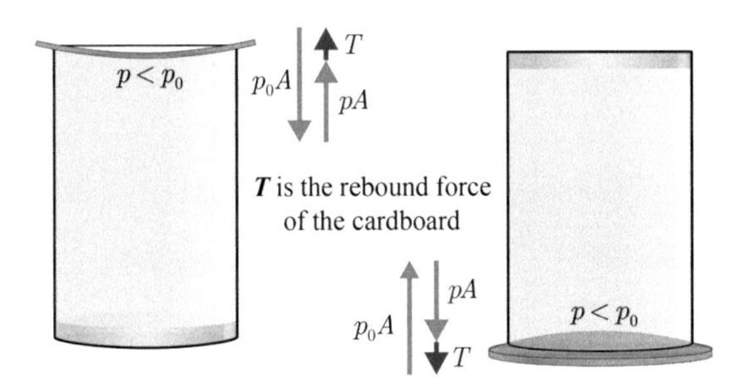

Figure 9.3 Pushing the cardboard concave inward makes the experiment easier to succeed.

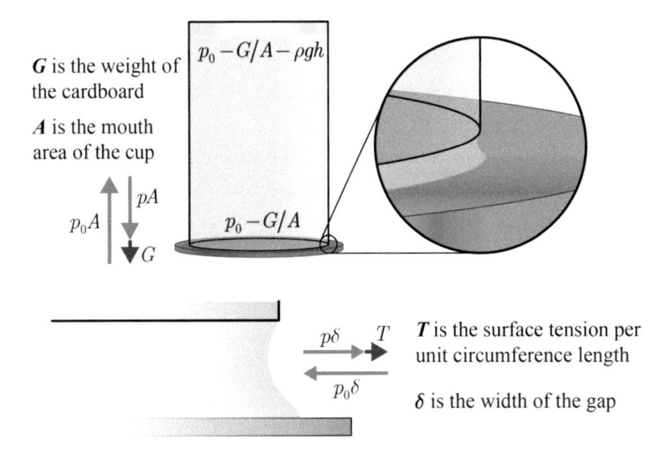

Figure 9.4 The weight of the cover plate is balanced by the surface tension of water at the gap.

force of the cardboard, the water pressure at the mouth of the cup will be slightly lower than the atmospheric pressure, and the experiment is more likely to work. As an added bonus, the concave cardboard is hard to slide from side to side after the cup is turned, which also helps stabilize the water–air interface between the cardboard and the rim of the cup.

Considering the weight of the cardboard sheet, the water pressure at the mouth of the cup should be slightly lower than the atmospheric pressure. If the weight of the cardboard is G, the water pressure at the mouth of the cup should be $p = p_0 - G/A$, as shown in Figure 9.4. The pressure difference between water and atmosphere is balanced by the surface tension of the water. The water surface at the gap must curve inward, causing the resultant force of water pressure and surface tension to equal the force of atmospheric pressure. That is to say, the weight of the cardboard is directly supported by the pressure difference inside and outside the cup, which needs to be balanced by the surface tension of the water at the gap. Therefore, the water in the cup can be heavy, but the cover plate should not be too heavy. If a heavy object such as a thick glass or a metal sheet are used as the cover plate, the demonstration is less likely to succeed.

9.3 Air Blockage: Compressibility of Gases

The air blockage phenomenon occurs when one or more air bubbles enter the liquid flow in a pipe and the flow is blocked. For example, the air in brake lines, which is usually the result of a leak in the brake line or the precipitation of overheated vapor after long-term use, is likely to cause brake failure; or air bubbles trapped within our main blood vessels may result in insufficient blood supply and even death.

Some books claim that air blockage occurs because the air in brake lines cannot transfer pressure. This statement is not accurate, since both liquid and gas can transfer pressure. A car brake system is illustrated in Figure 9.5. When there are no air bubbles in brake lines, the force exerted by the foot on the brake pedal is transferred to the brake calipers through the brake fluid; when there are air bubbles in brake lines, the force exerted on the brake pedal can theoretically still be transferred to the brake calipers. So, what causes a brake failure?

Actually, it is displacement rather than pressure that is transferred in brake lines. When we step on the brake, the brake fluid moves a certain distance and pushes the brake calipers to provide a clamping action on the brake disk. If there is a big air bubble in the brake line, the brake fluid downstream of the air bubble moves a much shorter distance than that upstream of the bubbles because the volume of the air is reduced in compression. Therefore, the same pedal displacement produces much less brake calipers displacement, resulting in brake failure. In this case, the driver feels weaker feedback from the brake pedal because the compressed gases create a limited increase in pressure. Therefore, it does make sense to say that the pressure cannot be transferred if there are air bubbles in brake lines. However, the root cause is that the compressibility of air bubbles makes it hard to establish high pressure.

Therefore, the occurrence of air blockage depends on the mechanical pattern of the power end. If displacement is a finite value (e.g., brake pedal and ventricular contraction), the air bubbles indeed cause a blockage in the flowing liquid. However, if the power end is a constant-pressure boundary condition, the air bubbles do not necessarily

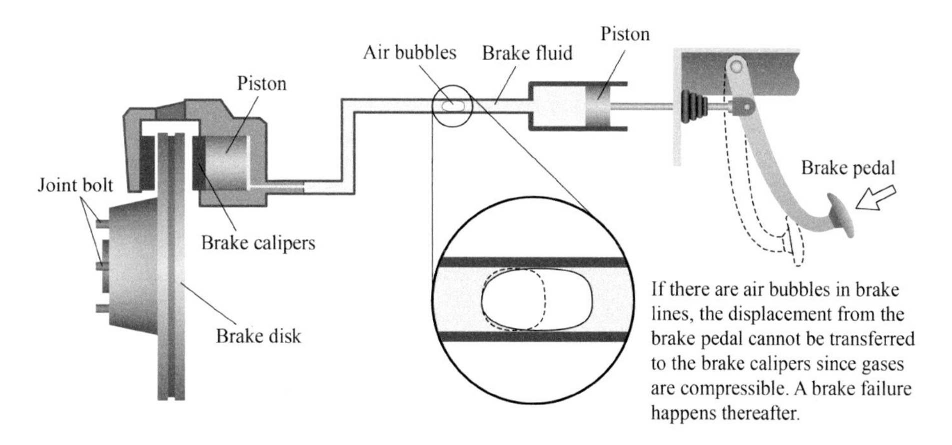

Figure 9.5 Basic brake system schematics and the principle of air embolism.

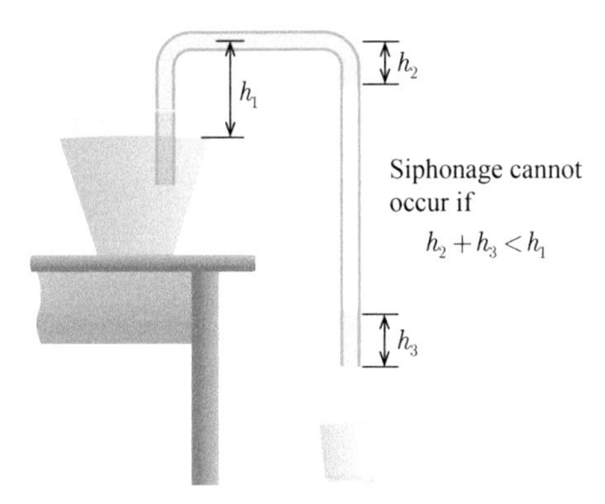

Figure 9.6 Air blockage during siphonage.

block the passage of liquid. For example, a large amount of air contained in the tap water pipe rarely leads to blockage. The same is true for the flow in a heating pipe system.

In addition, air blockage can also occur in siphonage pipes, as shown in Figure 9.6. This is because the gas in the siphonage pipe weighs less than the liquid it replaces.

9.4 How Balloons Create Thrust: Momentum Theorem

When an inflated rubber balloon is released, the air escapes from its opening and the balloon takes off in the opposite direction. This flow phenomenon has the same principles as rockets.

There are generally two explanations for the thrust of an inflated balloon. The law of conservation of momentum states that the momentum of the balloon and air are conserved, while Newton's third law of motion tells us that the deflating balloon accelerates due to the reaction force of the escaping air.

Both of these explanations are correct, and actually correspond to the integral and differential approaches in mechanics, respectively. As shown in Figure 9.7, the balloon and the air inside it are chosen as a control volume. As the balloon moves through the air, there will be an opposite drag force from the surrounding air. If the drag force is equal to the thrust, the balloon will move at a constant velocity; if the drag force is less than the thrust, the balloon will accelerate, with an inertial force imposed upon the control volume.

The thrust does not explicitly exist in the above momentum integral approach. If we study the forces acting on the balloon itself and exclude the air inside it, the thrust can be explicitly observed in the momentum differential approach.

The air pressure inside the inflated balloon is greater than the atmospheric pressure outside. As the air escapes through the opening, the stationary air inside the balloon has the same pressure everywhere, but near the opening, pressure decreases to the atmospheric pressure due to the increase of local velocity. For every single force of the distributed pressure force illustrated with a dark blue arrow on the inner surface of the balloon in Figure 9.8, there is an opposite force on the other inner side of the balloon, except on the surface of the balloon opposite the opening (the leftmost one

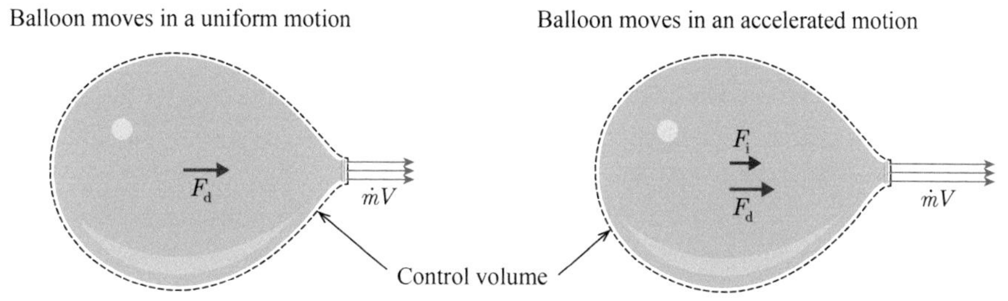

Balloon moves in a uniform motion

F_d $\dot{m}V$

Control volume

Air drag F_d is the only horizontal external force acting on the control volume, that is

$$F_d = \dot{m}V$$

Balloon moves in an accelerated motion

F_i F_d $\dot{m}V$

The horizontal external forces acting on the control volume include air drag F_d and inertia force F_i, that is

$$F_d + F_i = \dot{m}V$$

Figure 9.7 A balloon and the air inside it are chosen as a control volume to analyze the external forces exerted on it.

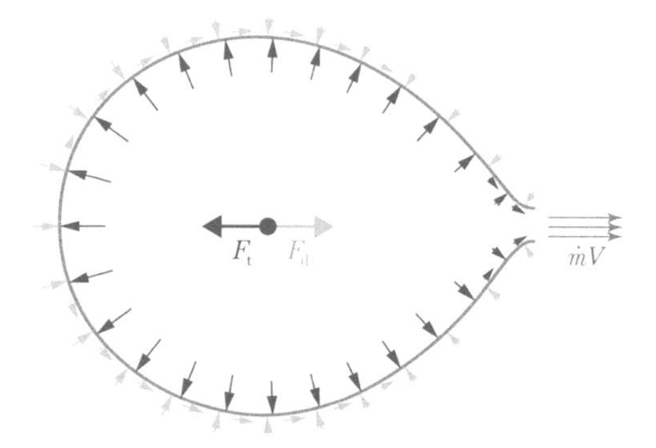

If the balloon moves at a uniform speed, the thrust, F_t, is equal to the drag, F_d. The thrust mainly comes from the pressure difference inside the balloon, and the drag comes from the sum of pressure difference and shear stress outside the balloon.

Figure 9.8 Force field analysis on a deflating balloon.

in Figure 9.8). The pressure difference between this surface and the opening creates thrust that propels the balloon forward. Figure 9.8 shows all the forces exerted on the balloon. The thrust is mainly created by the imbalanced pressure forces inside the balloon, and there is negligible shear force near the opening. The imbalanced pressure forces outside the balloon, accompanied by the minor contribution of shear forces, create drag that hinders the balloon.

In short, the laws of motion in relation to the balloon can be summarized as follows:

(1) The high pressure inside the balloon pushes the air out via the opening.
(2) Since the pressure inside the balloon is equal to the atmospheric pressure at the opening, the inner surface of the balloon opposite to the opening has no opposing force, which creates a net forward force on the balloon, known as thrust.
(3) If the balloon moves at a uniform velocity, there is a balance between thrust and drag.

9.5 Thrust of a Water Rocket: Independent of Working Substance

A water rocket is a kind of toy using the compressibility of air and the momentum of water to produce thrust. It is commonly seen in extracurricular scientific and technological works contests. A water rocket has air inside a plastic bottle with some water added. More air is pumped into the bottle to establish high pressure. When the compressed air reaches a certain pressure, the water is expelled from the bottle through the opening at the bottom, so the rocket (bottle) is pushed upward at high speed. Its propulsion principle is the same as that of a deflating balloon, but can last longer.

Different types of water rockets have been designed to compete for the world altitude record. Experiments have proved that a bottle filled to approximately one-third with water flies the highest, as illustrated in Figure 9.9. A bottle full of water obviously does not work, because the compression rate of water is so low that almost no water will be expelled from the opening to push the bottle upward. A bottle full of air will fly upward, but not very high.

One theory is that water, being about 1,000 times heavier than air, produces more thrust than air at the same exhaust velocity. This explanation seems plausible, but is not the answer to why both air and water are needed for a water rocket. Next, let us calculate the thrust produced by the expelled water and air, respectively.

Let p_0 be the atmospheric pressure, and p_1 is the static pressure of the fluid inside the bottle. If we ignore the weight of water and assume the bottle diameter to be much larger than the opening diameter, Bernoulli's equation gives the exhaust velocity at the opening:

$$V_1 = \sqrt{\frac{2(p_1 - p_0)}{\rho}} = \sqrt{\frac{2p_{1g}}{\rho}},$$

where p_{1g} is the gauge pressure of the fluid inside the bottle.

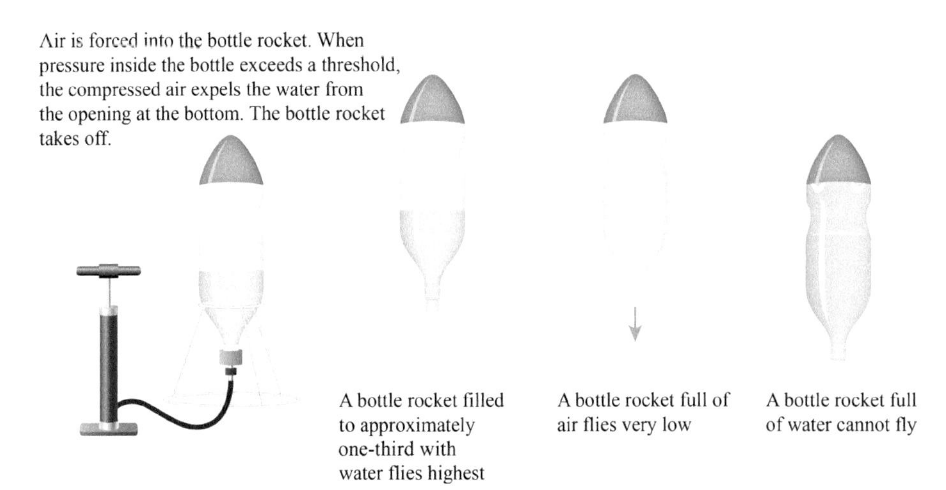

Figure 9.9 A schematic diagram of a bottle rocket.

The thrust exerted by the expelled water or air on the bottle is equal to its rate of change of momentum:

$$T = \dot{m}V_1 = \rho A V_1^2 = \rho A \frac{2p_{1g}}{\rho} = 2p_{1g}A.$$

Notice that the amount of thrust depends only on the opening area and the internal gauge pressure, and is independent of the working substance. Thus, the thrust is identical whether we use air or water as the working substance. Although the denser water seems to provide a larger momentum, the exhaust velocity will be lower under the same gauge pressure. Since the bottle full of air is lighter, the bottle with only air actually produces higher initial acceleration.

Why can a bottle rocket with water as a working substance fly higher? The reason is that, although air or water initially produce the same thrust, the amount of work done by the working substance determines the flight altitude. In other words, the bottle rocket needs not only high thrust, but also long working time. For example, if the gauge pressure of the air inside the bottle is 0.3 atm, the exhaust velocity of air can reach more than 200 m/s, so the bottle full of air will be empty very quickly. By contrast, for the same condition, the exhaust velocity of water is less than 8 m/s, resulting in a much longer time for the water to be consumed. Therefore, the bottle rocket using water as a working substance can fly much higher. Meanwhile, the compressibility of air is also essential to expel the water out at a relatively steady speed.

9.6 Turbojet Engine Thrust: On Which Components?

Most modern civil and military aircraft are powered by jet engines. In a turbojet engine, air is drawn into the compressor, and the pressurized air is discharged from the compressor into a combustion chamber, where the high-pressure air and fuel mixture is ignited to produce a high-energy gas stream. The gas expands in a turbine to produce work that drives the compressor, and is then exhausted out of the nozzle. The momentum theorem states that the air passing through the engine has to be accelerated in order to produce forward thrust to the engine.

Similar to the thrust produced in a deflating balloon, the thrust of the turbojet engine is actually the sum of the pressure and viscous forces acting on various surfaces of all the engine components. We need only to analyze pressure forces, which are far larger than viscous forces. According to the characteristics of pressure, only the backward-facing surface is subjected to forward thrust.

Figure 9.10 schematically shows the thrust allocation among the components of a turbojet engine. The compressor is used to pressurize the air, so the pressure on the backward-facing surfaces of the compressor blades is always larger than that on the forward-facing surfaces, thus giving a forward force on the compressor. The pressure on the inner wall of the combustion chamber is approximately the same everywhere due to the relatively low air velocity. However, the backward-facing wall area is obviously larger than the forward-facing one (just like in a balloon), thus giving a forward force on the combustion chamber. Unlike the compressor, the pressure on

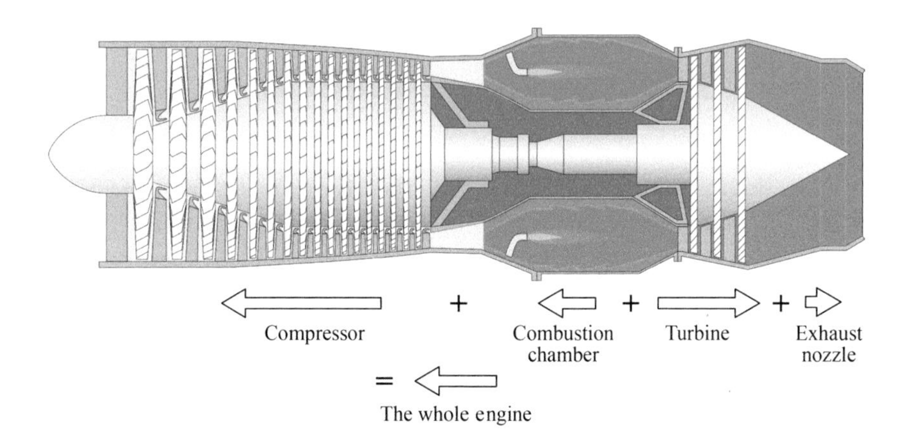

Figure 9.10 Thrust components of a turbojet engine.

the forward-facing surfaces of the turbine blades is larger than that on the backward-facing surfaces, thus producing a backward force on the turbine. If the exhaust nozzle is converging, the exhaust gas exerts a backward force on it. However, an exhaust cone is installed in the center of the exhaust nozzle on which the exhaust gas exerts a forward force, so the resultant force on the nozzle can be forward.

To sum up, in a turbojet engine, most of the thrust is generated at the compressor and combustion chamber. Part of the thrust can also be generated by the exhaust nozzle, while the force generated by the turbine is completely directed backward. Of course, the turbine is indispensable to the engine, because it not only outputs power to drive the compressor, but also provides the compressor with appropriate exhaust conditions. These two functions work together to ensure a large forward force on the compressor.

Large passenger aircraft are always powered by large bypass ratio turbofan engines in which the thrust is chiefly generated by the fan. Meanwhile, the resultant force acting on the compressor, combustion chamber, turbine, and exhaust nozzle is still directed forward, thus producing part of the thrust. Low-speed passenger aircraft are powered by turboprop engines in which the thrust is almost entirely generated by the propeller, while the forces acting on the other engine components essentially cancel out.

9.7 Total Pressure and Its Measurement: Not a Property of Fluids

Total pressure is defined as the pressure of a flowing fluid that has been isentropically decelerated to zero velocity, or as the sum of the static and dynamic pressures. The static pressure is the actual pressure of a gas flow, while the dynamic pressure is a defined property of a gas flow, which is the increased pressure when a flow is decelerated to zero velocity.

If a pressure sensor moves with the fluid at the same speed, the measured value refers to the static pressure. In this case, there is no concept of total pressure, or the

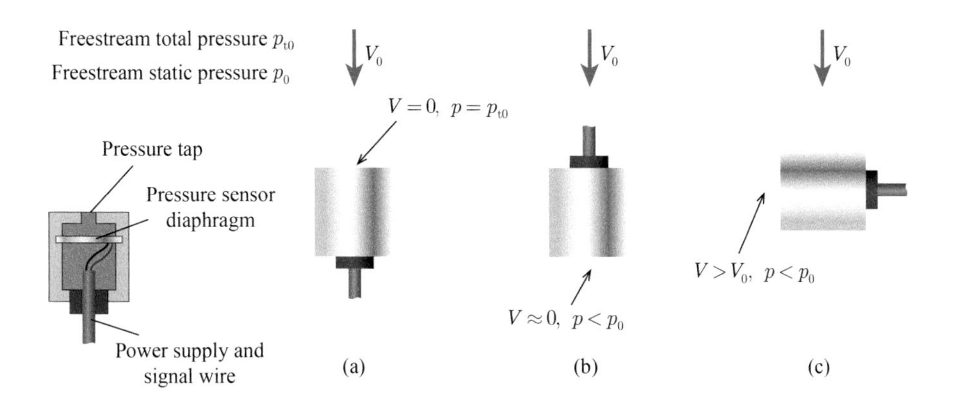

Figure 9.11 Pressure measured by a pressure sensor mounted in the airflow.

total pressure equals the static pressure, since the relative velocity is zero. The most reasonable way to measure the static pressure of a flowing fluid is to move with it, which is, however, generally not easy to implement.

As a fluid flows past the pressure sensor, the fluid is disturbed, forming a boundary layer or separation zone on the surface of the pressure sensor. The pressure in the boundary layer remains constant along the normal direction, and the pressure in the separation zone also remains basically constant. Therefore, the mainstream velocity near the pressure tap essentially determines the pressure measured by the sensor. If the pressure tap faces against the freestream, the flow velocity near it is close to zero, as shown in Figure 9.11(a). If the pressure tap faces away from the freestream, there is a separation zone near it where the air velocity is also close to zero, as shown in Figure 9.11(b). If the pressure tap is parallel to the freestream, the air velocity near it is usually higher than the freestream velocity, as shown in Figure 9.11(c). Next, let us analyze the pressure measured by the sensor under each of these three conditions.

The case shown in Figure 9.11(a) is relatively simple. Because the deceleration of the flow is very close to an isentropic adiabatic process, the pressure measured by the sensor basically equals the total pressure of the freestream. It should be noted that, if the freestream is supersonic, a shock wave will form in front of the sensor. Then, the sensor measures the total pressure downstream of the shock wave, which needs to be converted into the actual total pressure of the incoming flow.

For the flow shown in Figure 9.11(b), although the local velocity is close to zero, the pressure measured by the sensor is much lower than the freestream total pressure; actually, the measured pressure is even lower than the freestream static pressure. The reason is the following: A separation zone forms on the leeward side of the object, where viscous effects cannot be ignored and the pressure distributions cannot be explained by Bernoulli's principle. The pressure remains approximately the same in the separated zone as that at the separation point due to the very low velocity. For flow past a blunt body, the separation usually occurs just behind the point of maximum velocity, where the velocity is higher than the freestream velocity. Therefore, the pressure at the separation point is lower than the freestream static pressure.

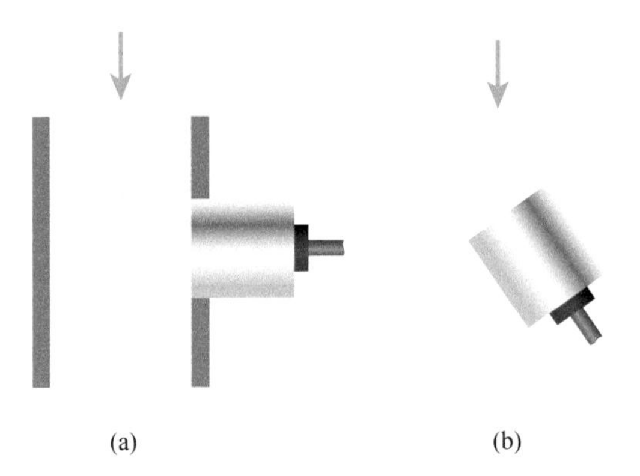

Figure 9.12 Two methods to measure the freestream static pressure.

For the case in Figure 9.11(c), the pressure measured by the pressure sensor is lower than the freestream static pressure. This is because the mainstream disturbed by the pressure sensor is accelerated beside the side surface.

To measure the freestream static pressure in a stationary coordinate system, two typical methods are used. One is flush-mounted sensors, which do not introduce disturbances to the freestream. For a channel flow with constant cross-section area, the measured pressure equals the freestream static pressure if the flow acceleration due to the increase of displacement thickness is not considered, as shown in Figure 9.12(a). The other method introduces disturbances to the freestream, but ensures that the velocity at the pressure tap equals the freestream velocity – for example, by mounting the pressure tap at some angle to the direction of the freestream, as shown in Figure 9.12(b).

To sum up, a pressure sensor can only measure the static pressure of a fluid, rather than the total pressure. Total pressure is actually the static pressure measured when a moving fluid is isentropically decelerated to zero velocity. In other words, static pressure is a real property of a fluid, while the total and dynamic pressure are imaginary properties that exist as definitions. The total and dynamic pressure change with the selected coordinate system.

Generally, we explain lift by using a coordinate system fixed on an airplane. This is equivalent to the case in which the airplane is mounted in a wind tunnel and air is blown to it. In this case, the total pressure used in the analysis should be called the total pressure relative to the airplane. We generally conclude that the high pressure on the nose of the airplane is due to the stagnation of the freestream. This familiar interpretation is based on the premise that the coordinate system is fixed on the airplane. For a stationary observer on the ground, the pressure on the windward side of the airplane is higher than the atmospheric pressure due to the work done by the airplane on the freestream. If the pressure at some location of the airplane surface is equal to the atmospheric pressure, this is equivalent to saying that the air at that location is not disturbed (the local mainstream velocity equals the freestream velocity). If the pressure is less than the atmospheric pressure, this corresponds to the fact that the air is accelerated.

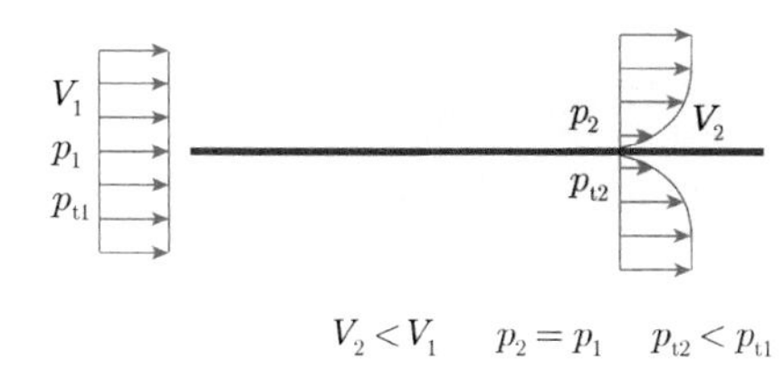

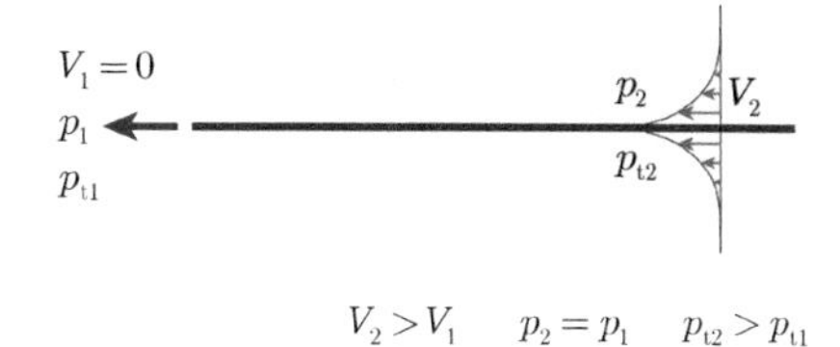

Figure 9.13 Fluid parameters within the boundary layer over a flat plate in two defined coordinate system.

As an incompressible flow decelerates, the pressure rise is entirely the result of the push work done by the ambient fluid on the decelerated fluid. As a compressible flow decelerates, both push work and compression work occur, so the dynamic pressure calculated by the compressible formula is larger than that calculated by the incompressible one.

If the fluid is accelerated by viscous shear forces rather than differential pressure, the situation is somewhat different. If subjected to shear forces only, the fluid is being dragged rather than pushed, which does not squeeze the fluid. So, the pressure inside the fluid remains unchanged. In other words, the viscous shear forces can increase only the kinetic energy rather than the pressure potential energy of the fluid. For an incompressible flow, the friction-induced temperature rise does not affect the pressure. In that case, the viscous forces that drag the fluid to move will not cause any pressure change.

Figure 9.13 shows the pressure and velocity distributions in the boundary layer over a zero-thickness flat plate. The figure on the left shows a uniform flow past a stationary flat plate. That on the right shows a zero-thickness flat plate moving through a stationary fluid.

The case on the left has been analyzed in detail by using the boundary layer theory. The pressure remains the same throughout the flow field. The velocity within the boundary layer is less than the mainstream velocity. The total pressure varies from the mainstream to the wall in the same way that the velocity (relevant to the dynamic pressure) varies. The viscous effects inside the fluid result in irreversible conversion of mechanical energy into internal energy, causing a decrease in the total pressure within the boundary layer. For the case on the right, only the viscous forces are imposed on the fluids by the flat plate, so the pressure remains the same throughout the flow field. The fluid in the vicinity of the flat plate is dragged by the viscous forces and accelerates from rest. The total pressure still varies from the mainstream to the wall in the same way as the velocity does. The work done by the viscous forces on the fluid causes an increase in the fluid's kinetic energy, thereby increasing the total pressure within the boundary layer. The left and right figures are actually the expression of the same flow in different reference frames. Static pressure, as a property of the fluid, is reference-frame-independent, so the static pressure distributions in the two figures are the same.

9.8 Why Does a Converging Flow Accelerate? Balance of Basic Laws

A subsonic converging flow increases in velocity and decreases in static pressure based on the continuity equation and Bernoulli's equation. However, one may ask: If the wall of a converging pipe can be regarded as an obstacle to the fluid, shouldn't the wall slow down the fluid rather than speed it up?

The motion of a fluid must obey Newton's second law, which says that the acceleration of an object depends on the amount of force applied on it. From this point of view, the pressure drop is the cause, and the acceleration is the effect. In fact, the prescribed inlet and outlet pressure boundary conditions determine the velocity distributions in a pipe flow. Next, we give a specific example to analyze the velocity and pressure variations in a pipe flow.

Figure 9.14 shows the flow model. Assume that there are two sufficiently large vessels filled with air at different pressures. Three different types of pipes with constant, converging, and diverging cross-sectional areas are installed between the two vessels and cut-off by valves. When the valve suddenly opens, the air starts flowing. As long as the size of the pressure vessel is far larger than that of the pipe diameter, the flow can reach a steady state after a certain period of time. Next, let us observe how a steady flow is achieved, and what it looks like. For the sake of simplicity, it is assumed that the flow is inviscid and gravity is ignored. In addition, the pressure difference between the two vessels is not large, causing the steady flow in the pipe to be subsonic. The conclusion based on this model is also applicable to the case in which gas stored under pressure in a closed vessel is discharged into the atmosphere.

First, let us discuss the flow through a constant cross-section pipe. This case actually describes a similar process to a shock tube. For details, please refer to Tip 7.2 in Chapter 7. When the valve suddenly opens, the air at the high/low pressure interface starts

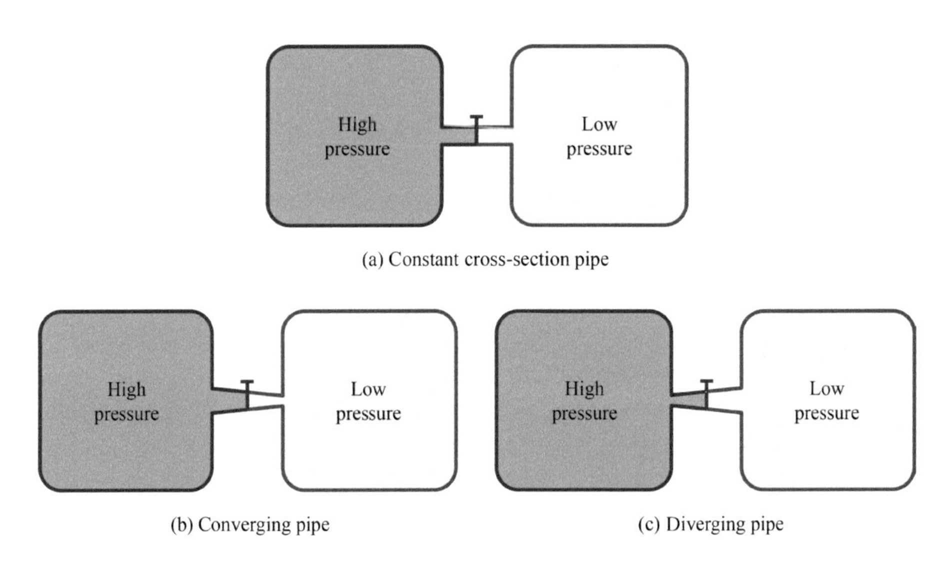

Figure 9.14 Three types of pipes with constant inlet and outlet pressure sources.

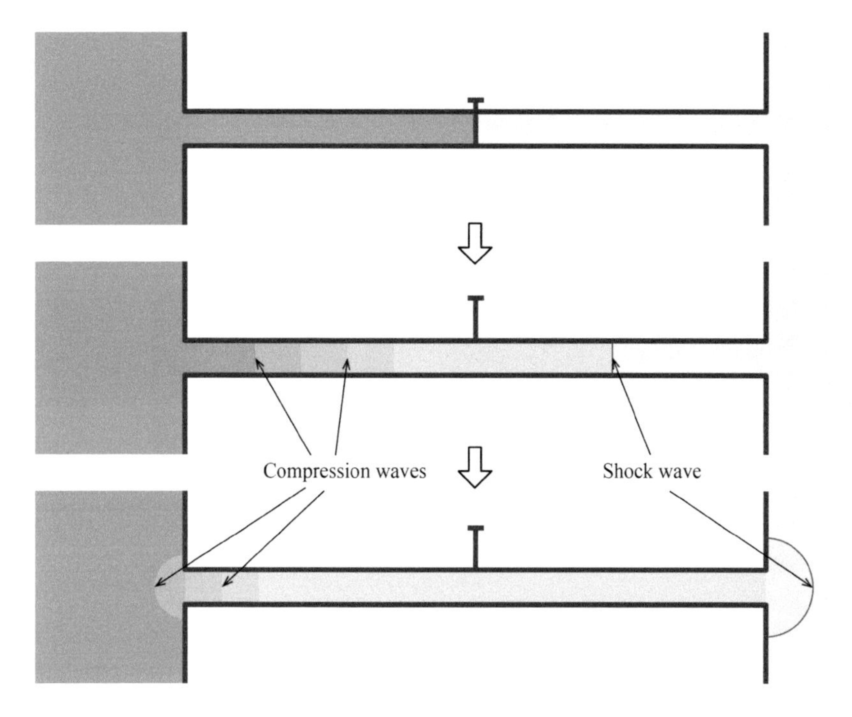

Figure 9.15 The fluid flows through a constant cross-section pipe after the valve suddenly opens.

moving toward the right under the action of differential pressure. At the same time, a strong compression wave (i.e., shock wave) originates from the interface and moves faster than the contact interface in the same direction, forming a discontinuous pressure interface, which separates the pipe into a high-pressure (left) and a low-pressure (right) region. A series of expansion waves originate from the valve and propagates to the left, forming a transitional zone which is separated into high-pressure (left) and low-pressure (right) regions. The shock wave propagates into the low-pressure chamber at the right outlet of the pipe and dissipates through expansion. Subsequently, the high-pressure gas behind the shock wave decreases in pressure. The expansion waves propagate into the high-pressure chamber at the left inlet of the pipe, thereby decreasing the pressure in a zone around the inlet. Figure 9.15 illustrates this process.

Across the expansion waves, the pressure on the upstream side of the valve decreases. After the shock wave sweeps, the pressure on the downstream side of the valve quickly returns to its original value. Figure 9.16(a) shows the pressure distribution of the finally steady flow. In the constant cross-section pipe, the air is in a uniform flow. At the right outlet of the pipe, the gas flows into the low-pressure chamber as a jet. The static pressure inside the jet is equal to the pressure in the low-pressure chamber. The pressure at the inlet of the pipe is equal to that at the outlet, which is equal to the pressure in the low-pressure chamber. A low-pressure zone forms in the high-pressure chamber, corresponding to a large-scale converging flow.

The processes toward final flow stabilization in the converging and diverging pipes are similar to those in the constant cross-section pipe. The final steady flows of them

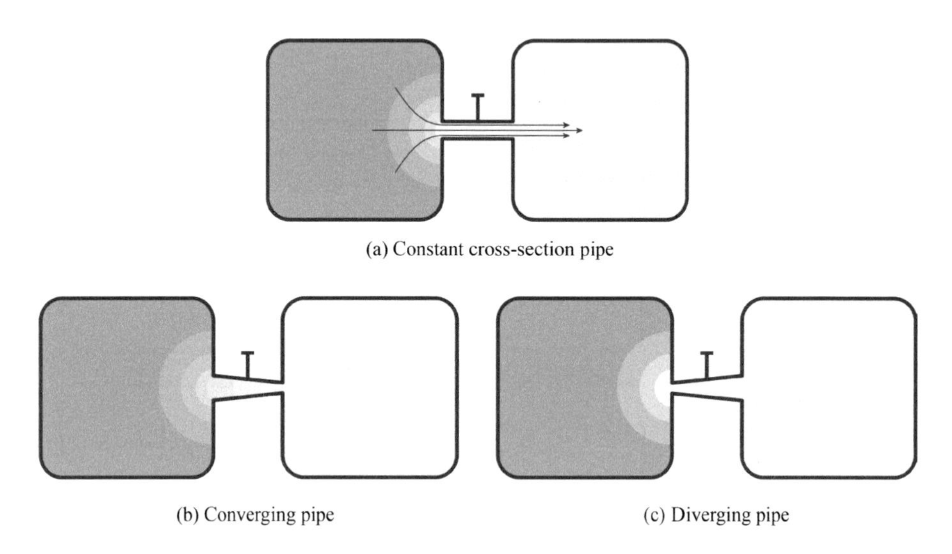

Figure 9.16 Final steady flow state in three types of pipes.

are shown in Figure 9.16(b) and (c). In the three cases, the outlet pressure is always equal to the pressure in the low-pressure chamber, and the converging flow inside the high-pressure chamber always generates a low-pressure zone upstream of the inlet. The difference is that for the converging pipe, the pressure at the inlet does not fully decrease to equal the pressure in the low-pressure chamber, and further decrease happens in the converging pipe; for the diverging pipe, the pressure at the inlet decreases to a value lower than the pressure in the low-pressure chamber, and there is a pressure increase in the diverging pipe.

Regardless of whether the pipe is converging or diverging, a jet forms at the pipe outlet where the static pressure is always equal to the pressure in the low-pressure chamber, and the total pressure is always equal to the pressure in the high-pressure chamber. Therefore, the outlet velocity is the same for the three cases, and the inlet velocity should obey the continuity equation. We have found that a converging pipe has a slower inlet velocity than a diverging pipe. According to Bernoulli's principle, the inlet pressure is high for the converging pipe and low for the diverging one. Therefore, we have reached the conclusion that the pressure in a converging pipe drops streamwise, and the pressure in a diverging pipe rises streamwise.

Next, let us discuss only the flow through a converging pipe. Compared with the constant cross-section pipe, at every streamwise location the pressure in the converging pipe is higher and the velocity is lower. Apparently, this higher pressure and lower velocity is the result of the pipe wall. In other words, the converging wall applies force on the fluid, making its pressure higher and its velocity lower.

Since the wall of the converging pipe acts as an obstacle to the flow, why does the fluid accelerate from the inlet to the outlet? This acceleration is caused by the pressure drop between inlet and outlet. The converging pipe, chosen as the control volume, is isolated from its surroundings for force field analysis, as shown in Figure 9.17. Since the momentum outflow rate is greater than the momentum inflow rate, the control

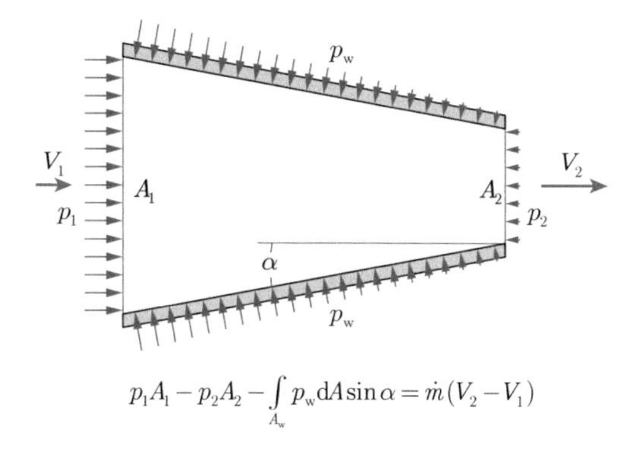

Figure 9.17 Subsonic flow through a converging pipe.

volume must be subjected to a net force whose direction is consistent with the direction of the flow. This force can only come from the differential pressure. The control volume has three control surfaces: the inlet, outlet, and conical surfaces. The forces by the conical and outlet surfaces act in the opposite direction to the flow direction, impeding the flow, while only the pressure force through the inlet surface is pushing the fluid to move, which is the only source of fluid acceleration.

It is worth noticing that for an unsteady flow through a diverging pipe, it is possible for the inlet pressure to be higher than the outlet pressure at some instant. For every type of pipe in the above example, when the valve suddenly opens, the inlet pressure is always equal to the pressure in the high-pressure chamber, and the outlet pressure is always equal to the pressure in the low-pressure chamber. However, this flow state is not balanced. For a steady flow, the inflow rate must match the outflow rate. The steady flow is a balanced state, and the flow pattern depends on boundary conditions. The present examples assume a constant-pressure inlet/outlet condition. If the two vessels are not infinite, the pressures inside them eventually tend to be equal and the fluid will be at rest everywhere. Whether converging or diverging, the pressure will be the same everywhere.

9.9 Impulsive Force and Stagnation Pressure: Relationship between the Momentum Equation and Bernoulli's Equation

When a perfectly elastic ball strikes a rigid wall perpendicularly, it will bounce back with the same magnitude of velocity, by the law of conservation of momentum and conservation of energy. From the perspective of force field analysis, the ball must be subjected to a force opposite to the direction of its initial motion, known as impulsive force. This force must act for a certain period of time in order to produce a change of momentum, which can be expressed as:

$$m\Delta V = \int_0^{\Delta t} F\mathrm{d}t,$$

where Δt is the time during which the ball stays in contact with the wall. As can be imagined, the contact time is generally very short and the force between them is large.

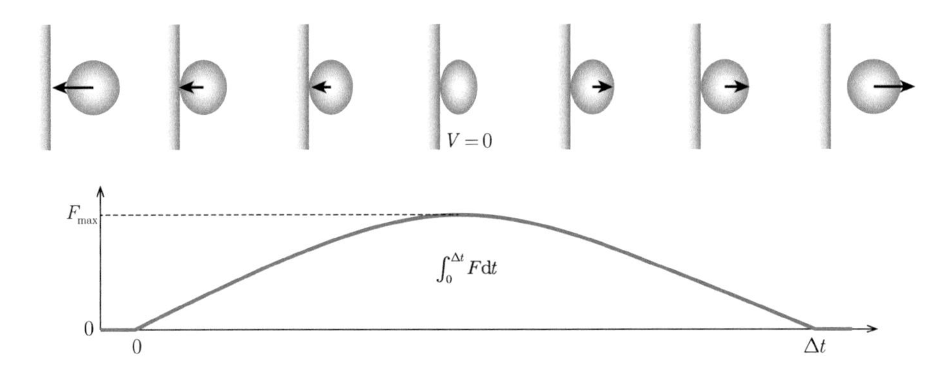

Figure 9.18 The change in velocity and impulsive force during the collision between elastic ball and rigid wall.

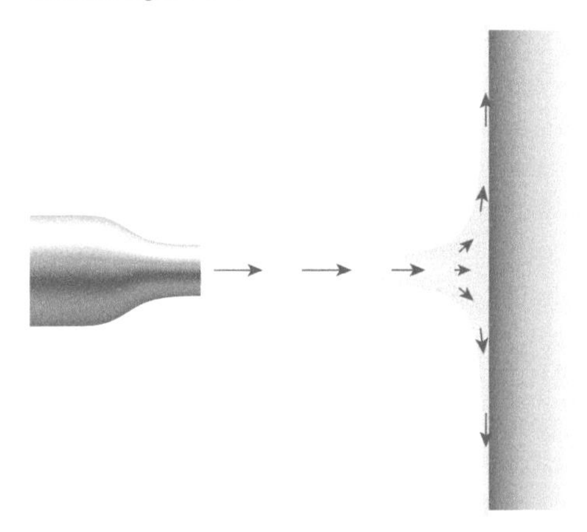

Figure 9.19 A liquid jet that impinges on solid surfaces.

This is why collisions often cause damage. From the perspective of the ball, the impulsive force relates to the elastic force exerted on the wall by the ball, corresponding to its deformation. A simple estimation based on mechanics concludes that when the two objects collide, the impulsive force increases from the instant of contact to the instant when the ball attains maximum deformation, and then decreases until the instant of separation. Figure 9.18 illustrates the deformation of the ball over the above-mentioned stages, and the change of magnitude of the impulsive force over time.

Next, let us discuss the impulsive force exerted by a jet of fluid. A fluid hitting a wall is obviously not an elastic collision, but comes closer to a plastic deformation. As shown in Figure 9.19, a jet of water hits the wall vertically, and spreads out in all directions along the wall. In the horizontal direction, the momentum of the jet per unit time is completely converted into the impulsive force on the wall:

$$F = \dot{m}V = \rho A V^2 ,$$

where A is the jet cross-sectional area and V is the velocity of the undisturbed jet.

As we know, the pressure produced by decelerating the fluid from velocity V to zero is equal to the dynamic pressure of the fluid:

$$p_{\mathrm{d}} = \frac{1}{2}\rho V^2 .$$

If the impulse force exerted by the jet on the wall is calculated by multiplying the dynamic pressure of the jet times its cross-sectional area, we have:

$$F = p_{\mathrm{d}} A = \frac{1}{2}\rho A V^2 .$$

This value is only half of the above result given by the momentum theorem, so, it is obviously not correct.

This is because the action area of the jet on the wall is not equal to but much larger than the jet cross-sectional area. The action area and corresponding pressure distributions on it (see Figure 6.29) are required to calculate the impulsive force exerted by the jet on the wall, expressed as:

$$F = \int_{A_0} p \mathrm{d}A ,$$

where A_0 is the total area of contact between jet and wall.

It is not easy to solve such a problem, because the area of contact and corresponding pressure distributions on it are hard to calculate. The problem is somewhat similar to the collision between a solid ball and a wall. Both are easier to solve with the momentum equation, but it is much harder to analyze the instantaneous forces between the two objects.

Next, let us analyze another problem involving impulsive force: the deflated balloon or the water rocket. As discussed earlier, these two examples can be explained by applying the momentum theorem or the differential pressure. According to the momentum theorem, the thrust created by the balloon is:

$$T = \dot{m}V = \rho A V^2 = \rho A \frac{2(p_{\mathrm{t}} - p_{\mathrm{a}})}{\rho} = 2(p_{\mathrm{t}} - p_{\mathrm{a}})A.$$

According to the differential pressure model shown in Figure 9.8, if the air pressure inside the balloon (except for the opening) is thought to be its total pressure, the thrust can be calculated as

$$T = (p_{\mathrm{t}} - p_{\mathrm{a}})A,$$

which is only half of the actual thrust.

The cause of this error is that the air pressure inside the balloon is not everywhere equal to its total pressure. In the vicinity of the opening, the air velocity obviously cannot be ignored. As velocity increases, static pressure decreases. Therefore, the outward forces exerted by the air inside the balloon near the opening is lower than total pressure, thereby producing a part of additional thrust.

9.10 Pressure of Jet Flow: A Pressure-Dominated Flow

In most fluid mechanics problems, it is generally assumed that the static pressure of a jet flow is equal to the ambient pressure. Such a condition is not a hypothesis, but

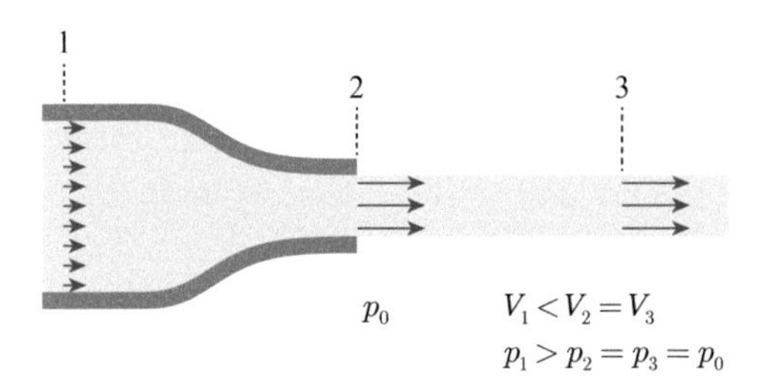

Figure 9.20 Velocity and pressure profiles inside a jet of ideal fluid.

When jet pressure is lower than ambient pressure, the jet decelerates due to the differential pressure.

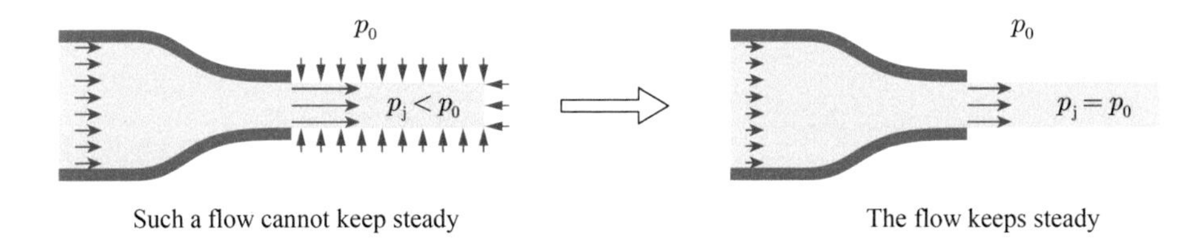

Figure 9.21 Flow pattern when the exhaust pressure is lower than the ambient pressure.

a fact. In subsonic flow, this condition is basically accurate; in supersonic flow, the static pressure inside the jet may be different from the ambient pressure. Here, we consider only a subsonic incompressible flow.

A jet flow of ideal fluid moves in uniform linear motion if the static pressure of the jet flow is the same as the ambient pressure, and the ambient fluid remains stationary, as shown in Figure 9.20. What happens if the static pressure of the jet flow is not equal to the ambient pressure?

If the pressure of the jet flow is lower than the ambient pressure, the jet will be decelerated and squeezed into a narrower shape due to the compression from high-pressure ambient fluids, as shown in Figure 9.21. Therefore, the selected part of the jet should seemingly decrease in both length and diameter, which never actually happens for an incompressible flow. The actual situation is as follows: The jet flow is squeezed inward by the high-pressure ambient fluids on both sides and downstream. To maintain the selected part of the jet unchanged in volume, the fluid upstream of the nozzle decelerates until the exhaust pressure is increased to match the ambient pressure. In extreme cases, the exhaust velocity is reduced to zero, but the exhaust pressure is still lower than the ambient pressure. Then, the ambient fluid flows back into the nozzle, and the nozzle becomes a suction port.

If the exhaust pressure is higher than the ambient pressure, the exhaust jet will expand outwards in all directions, reflected as an increase in streamwise velocity and transverse diameter. The fluid elements in the jet should seemingly expand, but this never actually happens for an incompressible flow. The actual situation may be described as follows: The exhaust jet tends to expand, causing the static pressure inside the jet to decrease. To maintain the selected part of the jet unchanged in

When jet pressure is higher than ambient pressure, the jet accelerates due to the differential pressure.

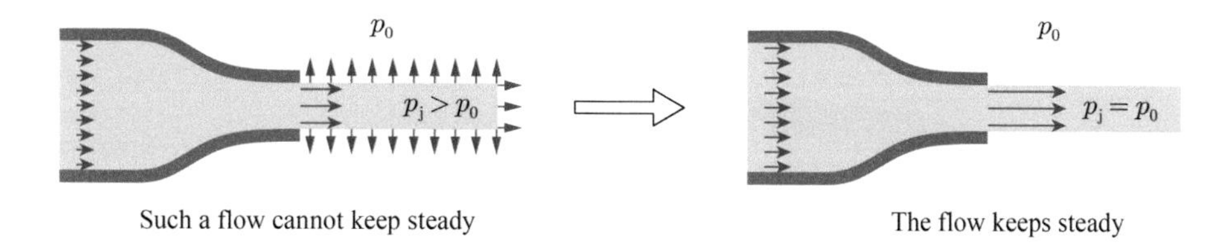

Figure 9.22 Flow pattern when the exhaust pressure is higher than the ambient pressure.

volume, the fluid upstream of the nozzle accelerates to fill the "gap" inside the jet until the exhaust pressure drops to match the ambient pressure. Then, a balanced jet flow forms, as shown in Figure 9.22.

As long as the exhaust pressure is higher than the ambient pressure, the flow always accelerates throughout the nozzle. Once the exhaust velocity reaches the speed of sound, the exhaust pressure cannot be reduced any further. In this case, the ambient pressure cannot affect the flow inside the nozzle. A series of expansion waves form downstream of the nozzle outlet to reduce the exhaust pressure, and this is not a subsonic jet problem anymore.

To sum up, for inviscid and incompressible flows, the exhaust pressure must be equal to the ambient pressure. For subsonic compressible flows, the above analysis becomes slightly more complicated, but the conclusion remains the same. For supersonic compressible flows, the downstream pressure disturbances cannot propagate upstream into the nozzle, and the ambient pressure does not affect the static pressure inside the jet. Therefore, the exhaust and ambient pressures need not be equal. A wave system of shock, expansion, and compression waves forms downstream of the nozzle outlet to achieve the balance between the exhaust and ambient pressures.

For an actual flow, the ambient fluid is entrained into the jet due to viscous effects, the jet no longer maintains a completely parallel flow, and there will be a slight difference between the pressure inside the jet and the ambient pressure. The ambient fluid is accelerated from rest to be entrained into the jet flow. Based on Bernoulli's principle, its pressure decreases. The pressure of the jet is still equal to the pressure of the surrounding fluid, but slightly lower than the pressure of the far field.

9.11 Faucet Flow Control: Total Pressure Determines Jet Speed

Valves, such as a faucet, are used to control the flow rate in a pipe. Since the valve inside a faucet is not located at the pipe outlet, the outlet area remains unchanged as the faucet valve is being closed. Thus, the decrease in flow rate should be caused by the decrease in outlet velocity rather than area. This phenomenon is consistent with our own experience; the outlet velocity indeed decreases as we throttle down a faucet.

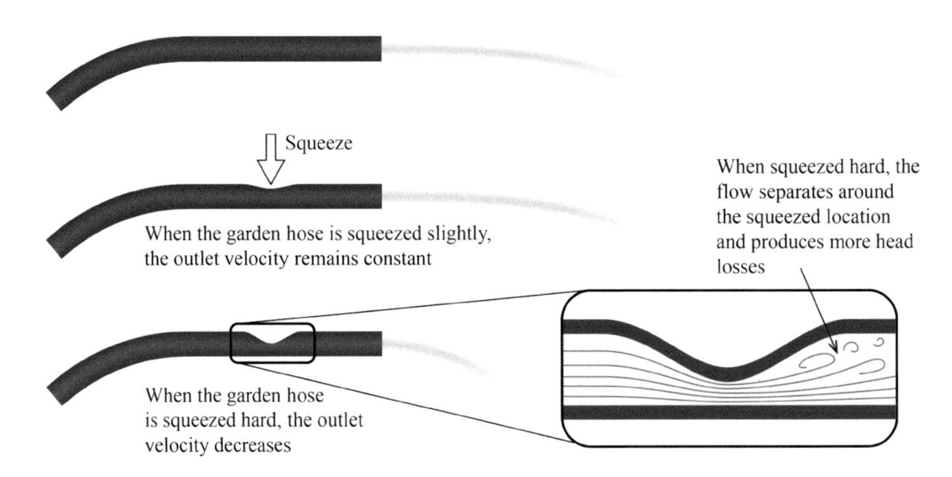

Figure 9.23 The variation in outlet velocity when you squeeze the middle of a garden hose with different forces.

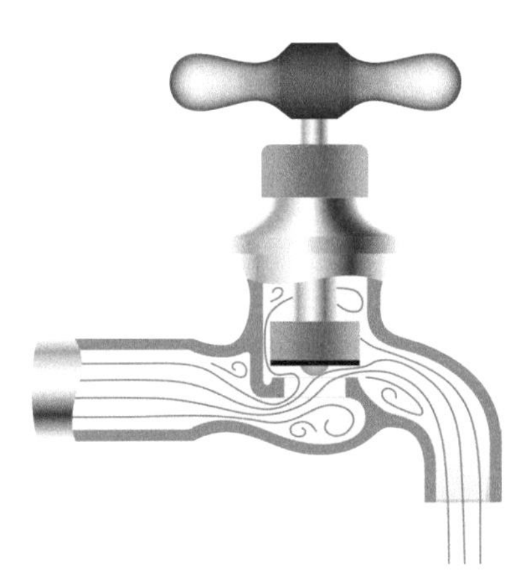

Figure 9.24 Flow inside a faucet.

According to Bernoulli's equation, the outlet velocity is determined by the following formula:

$$V = \sqrt{2(p_t - p_a)/\rho},$$

where p_t is the water total pressure, p_a is the atmospheric pressure, and ρ is the water density.

The passage area of the valve affects only the local velocity, while the total flow rate through the pipe may remain unchanged. According to the continuity equation $A_1V_1 = A_2V_2$, the outlet velocity can be independent of the valve passage area. To conduct such an experiment, slightly squeeze the middle of a garden hose used for watering lawns, and its outlet velocity indeed remains unchanged. However, if you squeeze the hose hard, the outlet velocity will decrease. This is because large local head loss occurs due to the flow separation around the squeezed location, thus causing the outlet total pressure to decrease. Figure 9.23 shows the variation in outlet velocity when the garden hose is squeezed with different forces.

The faucet controls the outlet velocity by adjusting the local head loss. Figure 9.24 shows the internal structure of a traditional faucet. The fluid flowing through the inner chamber of the faucet has to experience contraction, expansion, and turning. Flow separation occurs due to the adverse pressure gradients or sudden turning, accompanied by a large total head loss that reduces the outlet total pressure. Since the static pressure of the jet must be equal to the ambient pressure, it is the outlet total pressure that determines the outlet velocity. Valves are designed to minimize head loss in the fully opened position, and generate head loss in some half-closed positions. It is better to let the flow rate and the degree of closure obey some definite rule (for example, a linear relationship), which can be achieved by optimizing the internal flow path. For example, ball valves can help regulate the flow rate better than brake valves, and needle valves allow more precise flow rate control.

9.12 Squeeze the Outlet of a Hose to Increase Velocity: Total Pressure Determines Jet Speed

We all know by experience that squeezing the outlet of a garden hose will increase jet speed, as shown in Figure 9.25. There are popular science books that explain this phenomenon by the continuity of flow. The arguments are that if the flow rate is held constant, the outlet velocity increases as the outlet area decreases. This seemingly reasonable explanation does not really stand up to scrutiny.

If the flow rate is held constant, squeezing the outlet can indeed increase the outlet velocity. However, there is no reason for the flow rate to remain the same in actual situations. If we reduce the outlet area to one-half of its original value, the outlet

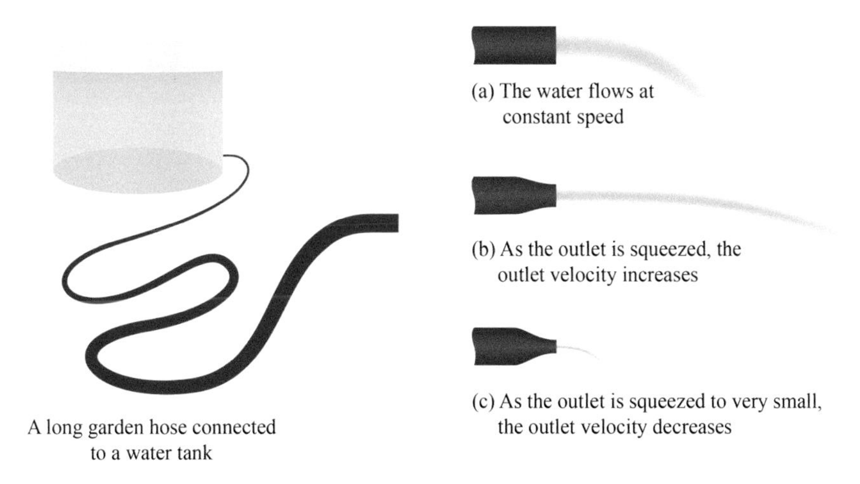

Figure 9.25 The outlet velocity changes as we squeeze the garden hose.

velocity will increase, but not to twice as much. When the outlet area decreases to close to zero, would the outlet velocity go to infinity? Of course not. In fact, the tighter we squeeze the outlet, the smaller the flow rate. So, what is the real reason for the increase in outlet velocity?

For an incompressible flow, the exhaust velocity of a jet is completely determined by outlet total pressure and ambient pressure. The outlet velocity can be calculated using Bernoulli's equation:

$$V = \sqrt{2(p_t - p_a)/\rho}\,.$$

Since the ambient pressure and water density remain unchanged, the increase in outlet velocity indicates that the total pressure of water inside the hose increased. Normally, the garden hose is connected to a constant-pressure water supply device, such as a water faucet or water tank. Assuming that the level of water in the tank is held constant, Bernoulli's equation states that the outlet total pressure should be a fixed value:

$$p_t = p_0 + \rho g h\,.$$

It seems that the outlet velocity should remain constant, independently of the outlet area.

Obviously, the theoretical conclusion is inconsistent with the actual situation. It is clearly inappropriate to apply Bernoulli's equation to explain the flow along the garden hose, because the head loss along the pipe is so significant that the effect of viscosity cannot be ignored.

The principle of this phenomenon is as follows: The outlet velocity is determined by the local total pressure, which equals the total pressure at source minus the pressure drop along the pipe. As the outlet is squeezed, the reduction in outlet area leads to a decrease in flow rate throughout the piping system. Subsequently, the reduced flow rate decreases the velocity everywhere in the pipe, which lessens the pressure drop along the pipe. Therefore, squeezing the outlet increases the total pressure of the water at the outlet, thus causing an increase in outlet velocity.

The above analysis may be summarized as follows: the outlet area decreases → the flow rate decreases → the head loss along the pipe decreases → the outlet velocity increases → part of the flow rate recovers. Of course, the changes in the above parameters occur simultaneously during the whole process. The outlet velocity depends on the pressure drop along the pipe, which in turn depends on the flow velocity.

Assume that the length of the water pipe is 10 m, the inner diameter is 2 cm, and the freestream gauge pressure is 0.5 atm. Using Equation 6.28(a) for loss coefficients for a turbulent pipe flow, we can estimate the matching curve, as shown in Figure 9.26. In this diagram, the outlet velocity is plotted as the vertical ordinate, while the volume flow rate is plotted as the horizontal ordinate. The curve shows the operating characteristics of the whole pipeline, and each straight line corresponds to the operating characteristics of an outlet with specific diameters. The intersection between curve and straight line corresponds to an operating point. As we squeeze the outlet, the operating point moves to the upper-left along the curve. As the outlet area tends to zero, the outlet velocity approaches its maximum, which is the attainable velocity

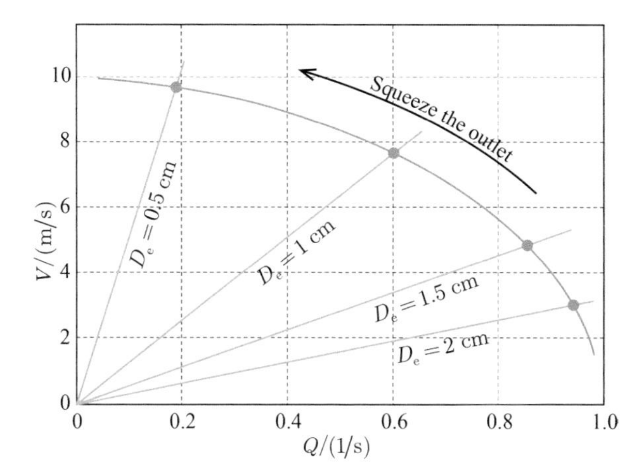

Figure 9.26 The combined effects of pipe length and outlet area.

by the water flowing through the pipe without head loss. According to Bernoulli's equation, this maximum outlet velocity is calculated as:

$$V_{\max} = \sqrt{\frac{2 \cdot \Delta p}{\rho}} = \sqrt{\frac{2 \times 0.5 \times 101{,}325}{1{,}000}} \approx 10\ \mathrm{m/s}\,.$$

It is a prerequisite for this phenomenon that the pipe should be very long. If the pipe is so short that the head loss along the pipe is almost negligible, the change in outlet area will not affect the outlet velocity. As we squeeze the outlet, the outlet velocity remains unchanged.

If the outlet is squeezed to a very small area, a large local head loss will occur in the converging portion near the outlet. Then, the outlet velocity decreases, as shown in Figure 9.25(c), instead of approaching the maximum velocity as shown in Figure 9.26.

9.13 Suction and Blow: Pressure-Dominated Flows

It is a common experience that suction and blow have significantly different effects even with the same flow rate. You can easily blow out a candle from tens of centimeters away, but if you try to suck it out from the same distance, the flame will be totally motionless. Standing in front of a fan, you can feel a strong airflow; standing behind it, you can barely feel any airflow. Here is the reason for this phenomenon: the air blown by the fan flows in one direction, characterized as a jet flow; while the air sucked by the fan flows in from all directions, characterized by large spherical area with low flow velocity. Next, let us take a closer look at the significant differences between blow and suction flow.

The blowing fluid forms a jet. Neglecting the effect of viscosity, the static pressure inside the jet is equal to the ambient pressure, and the fluid travels at a constant velocity in a straight line. Even though the effect of viscosity is considered, the jet can still travel a very long distance. For a high Reynolds number flow, the jet flow can be detected at

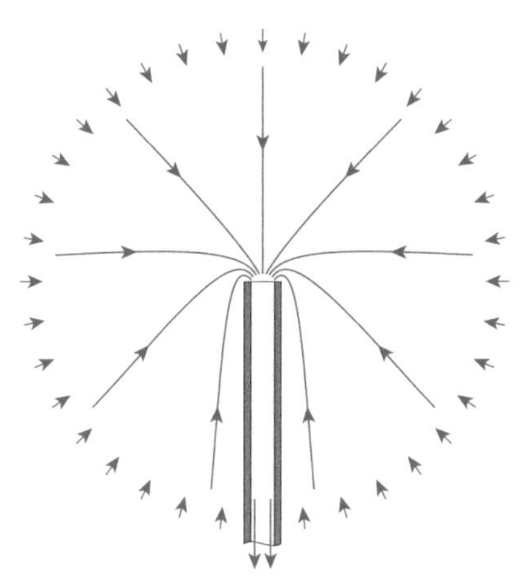

Figure 9.27 Flow pattern during inhalation.

more than 100 times the diameters downstream of the jet orifice. On the other hand, when sucking air into an orifice, a low-pressure zone forms near the orifice. Naturally, the surrounding fluid will be accelerated toward the low-pressure zone. Since subsonic flow decreases in pressure when the flow is converging, this low-pressure zone can always remain, which is consistent with the actual suction. On the other hand, a supersonic flow increases in pressure when the flow is converging; no matter how hard you try, the suction flow outside the orifice cannot be accelerated to supersonic speed.

Figure 9.27 shows the flow pattern near a pipe orifice during suction. Since the fluid flows from all directions, the outlet of the control volume is taken at the orifice, and the inlet of the control volume is taken at a spherical surface excluding the cross-section of the pipe. The schematic diagram shows that for the control volume, the inlet area is far larger than the outlet area. According to the continuity of flow, the velocity at a short distance from the pipe orifice is negligibly small. This is why the distance at which suction has an effect is much smaller than that of blowing.

Some water jet-propelled marine animals, such as the nautilus, have only one opening (a funnel orifice) for jet propulsion, and can propel themselves by intermittently sucking and ejecting seawater. When the nautilus sucks water, the water flows into its funnel from all directions, and the change in momentum along the direction of motion is so small as to produce only a negligible negative thrust; when it ejects water, the water is expelled in one direction, producing a large positive thrust. Therefore, the nautilus obtains a sufficient positive thrust by intermittently sucking and ejecting seawater.

9.14 Wind Near Buildings: Complex Three-Dimensional Unsteady Flow

Near a tall building there is always a place where the wind speed is higher than the wind in an open field. People living in South China name this draft "Longtang wind" – "Longtang" means "alley between buildings."

Some books explain this phenomenon using the continuity of flow. If the air flows from a wide channel into a narrow one, its velocity will increase. There is perhaps some truth in that explanation, but the causes of drafts are complex. Moreover, the acceleration of a fluid depends only on the forces acting on it, so it is more reasonable to explain the draft from the perspective of force field analysis.

This kind of problem is studied in a branch of civil engineering. Figure 9.28 shows the flow patterns of wind passing through buildings with different cross-sectional shapes in wind tunnel experiments. It is obvious that some of these three-dimensional flows cannot be explained with a simple one-dimensional theory. However, no matter how flow patterns change, the basic principle is the same. Because in this type of airflow the viscous forces are negligible, the acceleration and deceleration of airflow are basically caused by differential pressure forces. The steady flow characteristics state that the wind speed is low on the windward and leeward sides of a building, while on the flank sides of the building it is high.

Estimated from an ideal flow around a cylinder, the maximum wind speed on either side of a building can be twice as high as that on the ground. Therefore, if there is a wind of Beaufort force 4 on the ground, there may be a wind of Beaufort force 7 on the flank sides of a building. Since most buildings are neither streamlined nor circular, flow separations occur everywhere when the air flows past them. Due to the unsteady

Figure 9.28 Wind flow patterns around buildings (redrawn based on Gandemer J. Discomfort due to wind near buildings: aerodynamic concepts. NBS, 1978).

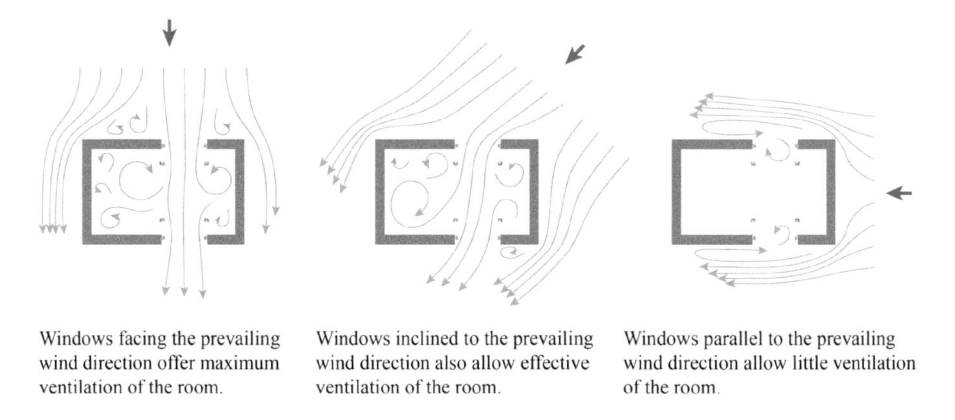

Figure 9.29 The ventilation through the windows on two opposite sides of a slab-type building.

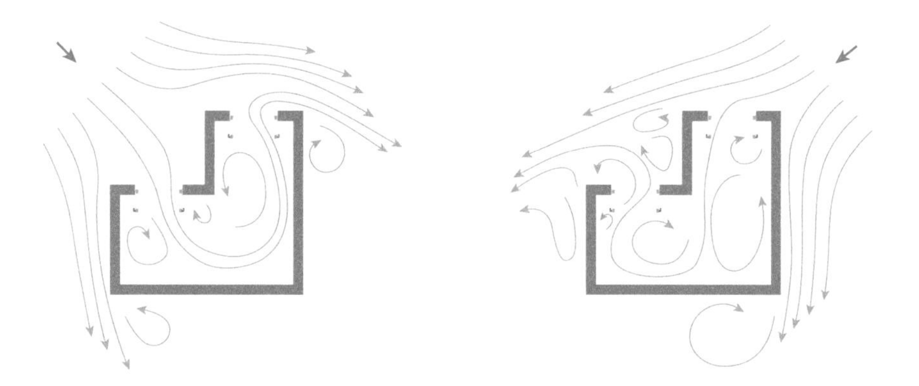

Figure 9.30 The rooms in a tower-type building also allow adequate ventilation with appropriate wind directions.

characteristics of separated flows, the instantaneous wind speeds near tall buildings may partially exceed Beaufort force 7.

Building design architects pay close attention to making use of natural ventilation. Adequate ventilation can generally be obtained for a slab-type building with its front and rear windows opened. Figure 9.29 shows the ventilation through the windows on two opposite sides of a slab-type building in different wind directions. In most cases, these two windows allow effective ventilation of the rooms. It is only when the air moves completely parallel to the two windows that the ventilation is inadequate.

Figure 9.30 shows the ventilation through two windows on the same side of a tower-type building. If both the layout and the wind direction are appropriate, it is possible for the two windows to ensure adequate ventilation. However, if the two windows are installed on the same plane, the ventilation will be poor, and can rely only on an unsteady flow – that is, a gust.

9.15 Coandă Effect: Viscous Effect is Indispensable

The Coandă effect refers to flow phenomena in which a flow stays attached to a nearby surface even when the surface curves away from the flow. Figure 9.31 schematically demonstrates the Coandă effect of a water jet. If you hold the back of a spoon vertically in touch with a thin stream of water from a faucet, the water will move toward and finally flow along the curved back of the spoon rather than fall vertically. This flow phenomenon is quite important because it is related to producing lift and increasing pressure in diverging pipes. The Coandă effect is closely associated to the effects of fluid viscosity. It can be argued that the Coandă effect cannot take place without viscosity. For an ideal inviscid flow, even if the wall surface curves away from the jet direction, the fluid can still flow along its original direction of motion, forming a "dead water zone" between the fluid and the wall surface. In this case, the pressure remains the same everywhere, and the mainstream does not have to attach itself to the wall surface, as shown in Figure 9.32(b).

Since there is no reason for a jet of inviscid fluid to follow an adjacent curved surface, viscosity should be a necessary condition for the Coandă effect. Now, assuming that a flow is initially inviscid and suddenly becomes viscous some time later, then the

Figure 9.31 Coandă effect.

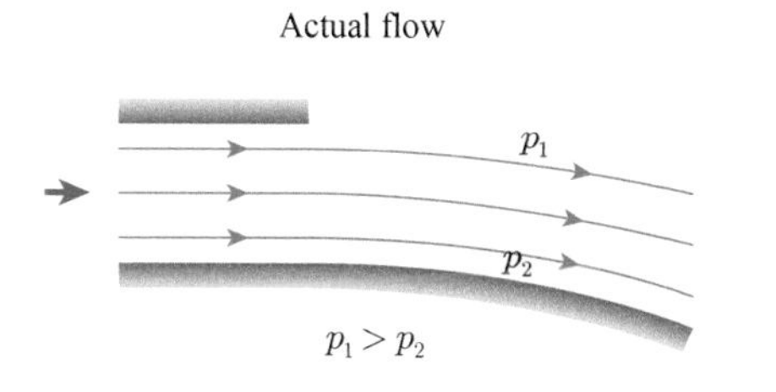

(a) Fluid moves over the curved wall surface due to Coandă effect

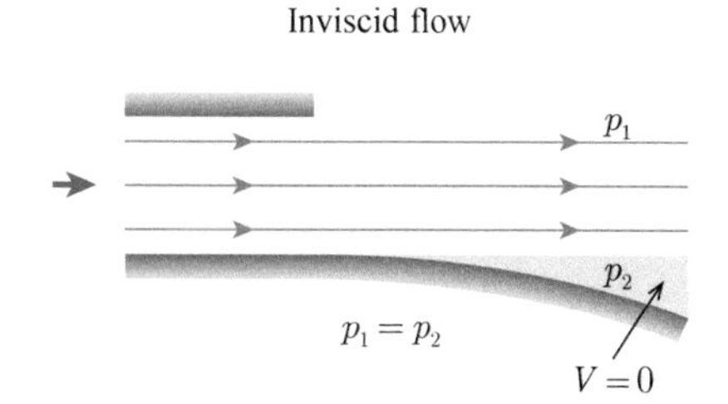

(b) Fluid subjected to no external force moves in a straight line.

Figure 9.32 No Coandă effect for ideal flow.

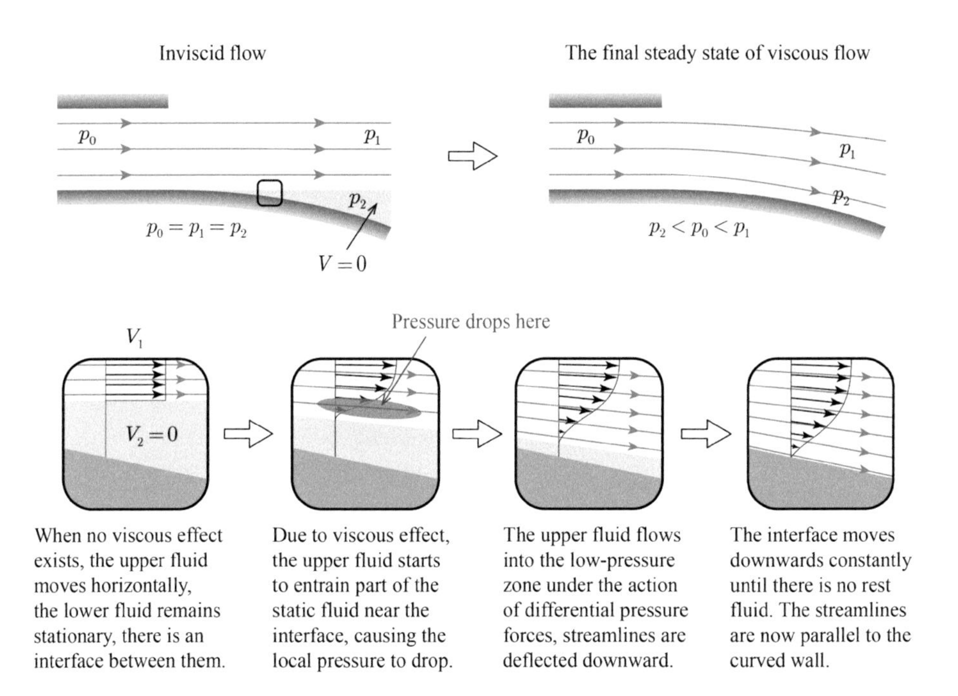

Figure 9.33 Generation mechanism of the Coandă effect.

fluid in the upper layer of the dead water zone will soon be entrained into the mainstream by viscous stress so that a zone of lower pressure develops. Due to the pressure gradient, the fluids in both the dead water zone (lower) and the mainstream (upper) are deflected into the shear layer to fill this low-pressure zone, as shown in Figure 9.33. As the mainstream is pulled toward the wall surface, its downstream pressure increases due to the enlargement in cross-sectional area. The increased pressure further compresses the mainstream and deflects it laterally. If the wall surface is slightly curved, the mainstream can adhere to the surface; if the wall surface is sharply curved, the downstream boundary layer separates from the surface, and the mainstream no longer stays attached to the surface after separation.

To be rigorous, the "viscosity" mentioned here refers to not only the molecular viscosity, but also the turbulence viscosity. That is to say, for high Reynolds number flow with molecular viscosity negligible, the Coandă effect still exists since turbulent entrainment drives the dead water zone to move.

The Coandă effect can also be explained as follows: A free jet of fluid entrains and mixes with its surrounding medium as it leaves a nozzle. If an obstacle is placed close to the jet, the jet still tends to "entrain" the obstacle. Since the obstacle cannot be moved, the jet attaches itself to the obstacle by flowing along its surface.

For the case shown in Figure 9.31 there is surface tension at the interface between the water stream and the ambient air, and in this layer of flow there is very high pulling force. Strong adhesive forces also exist between the water and the solid wall.

Figure 9.34 The surface tension of water plays a decisive role.

When the water touches the back of a spoon, the water will be attached to the curved surface intensely. The water flowing along the curved surface can be deflected by a considerable angle without separation. If the spoon is replaced by a cylindrical cup, the water can still adhere to the curved surface for a long distance, even after flowing 90° around the cylindrically curved surface. The flow pattern is shown in Figure 9.34. It is clear that differential pressure forces due to viscosity alone are not sufficient to deflect the water by such a large angle; it is the surface tension of water that plays a decisive role here.

9.16 Shape of a Raindrop: Surface Tension and Pressure Distribution

Most cartoonists typically draw raindrops in the shape of teardrops, as shown in Figure 9.35, which does not conform with reality. In fact, waterdrops are only shaped like teardrops when they are about to fall off an object. Figure 9.36 shows the shape of a water drop falling from a leaf. Before the water drop falls from the leaf, its upper part is pulled into a tip shape. After it leaves the leaf, the water drop quickly forms into a roughly spherical shape due to the surface tension of water.

So do raindrops have a spherical shape? Answering this question requires a careful analysis of the forces exerted on the raindrops. A falling raindrop is mostly subjected to three types of force: gravity, surface tension, and aerodynamic forces. Only raindrops falling freely in vacuum can have a spherical shape because they are solely subjected to surface tension. Raindrops falling in air are not in a free-fall movement, but are in uniform motion with a constant final speed. With transverse wind not considered, the upward force of air drag and the downward force of gravity are equal.

The air drag consists of friction drag and pressure drag. For airflow past a sphere, the pressure drag dominates. Figure 9.37 shows the pressure distribution on the surface of the sphere in airflow. The red arrows pointing inwards into the ball indicate

Figure 9.35 The shape of raindrops in various pictures.

Figure 9.36 Falling raindrops.

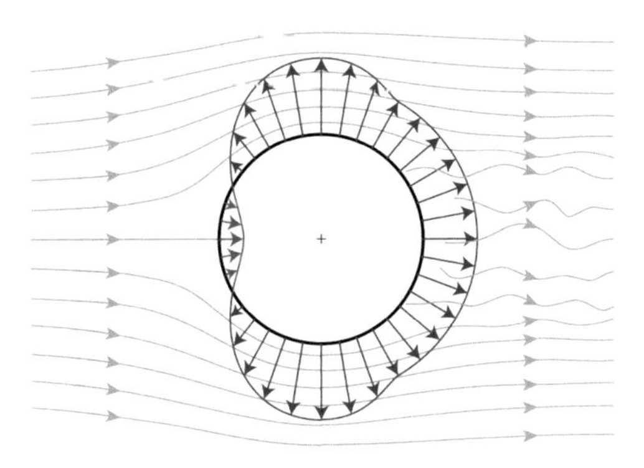

The red arrows indicate that the wall pressure is higher than the freestream pressure.

The blue arrows indicate that the wall pressure is lower than the freestream pressure.

Figure 9.37 The pressure distribution on the surface of the sphere in airflow.

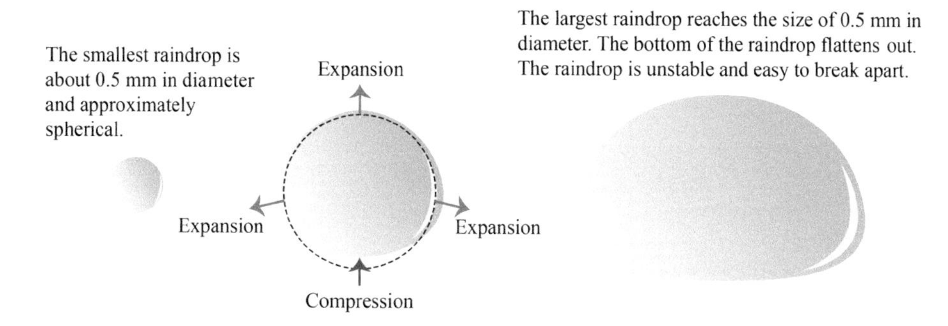

Figure 9.38 Shapes and causes of raindrops of different sizes (expansion and compression).

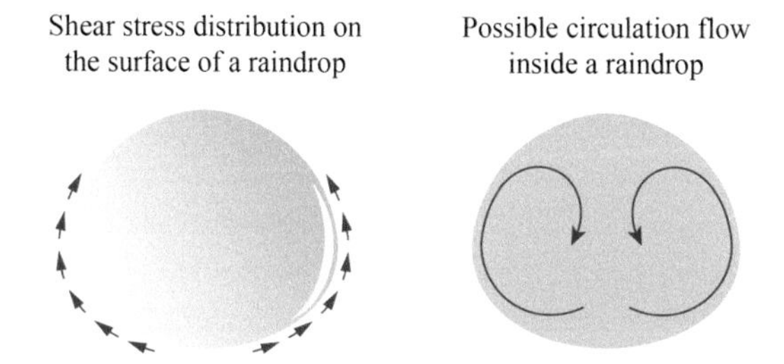

Figure 9.39 Shear stresses may cause circulation flow inside a raindrop.

that the local pressure is higher than the atmospheric pressure, while the blue arrows pointing outwards mean that the local pressure is lower than the atmospheric pressure. It can be seen that the front of the sphere is under pressure action, while both sides and the rear are under "suction" action. The surface forces determine the degree to which raindrops deviate from spheres.

The smaller the raindrop, the greater the surface tension and the closer to a perfect sphere it will become; the larger the raindrop, the more dominant the differential pressure and the more deviated from a perfect sphere it will become. Figure 9.38 shows the shape of raindrops of different sizes. As can be seen, the large raindrops cannot remain spherical, but become more like a hamburger – flattened on the bottom and with a curved dome top. Once the size of a raindrop gets too large, it will eventually break apart into smaller raindrops due to surface instability. (You can pour a cup of water from a certain height to observe the instability of this large mass of water.)

The viscous shear forces exerted by the ambient air may form ripples on the surface of raindrops, affecting the stability of water droplets. Under certain conditions, the shear force may also cause the water inside the droplet to continuously circulate. As shown in Figure 9.39, the water is dragged by the shear forces to flow along the surface from the front to the rear; it flows into the droplet at the rear, and eventually it flows out of the front again. This statement on water circulation is only the conjecture of the author.

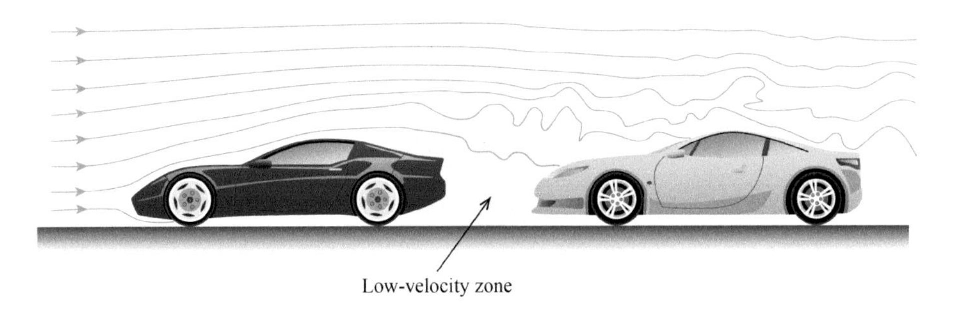

Figure 9.40 Vacuum effects of racing cars.

9.17 Vacuum Effects in Racing Cars Related to Incoming Flow Velocity

In car racing, we often observe the following phenomenon: A car suddenly moves over to one side and speeds up to overtake the car in front of it, which cannot avoid being overtaken. This is a common way for cars to overtake in motor racing. The driver of the rear car uses the rear vacuum produced by the front car to pass ahead of it.

Figure 9.40 shows the flow past two cars. The rear car rides in the wake of the front car. We know that the drag force exerted on an object is directly proportional to the square of the incoming flow velocity. Compared to the incoming flow velocity of the front car, that of the rear car is lower; therefore the drag force exerted on the rear car is much smaller. Furthermore, the pressure in the wake of the front car is lower than the atmospheric pressure, which also reduces the pressure drag exerted on the rear car. However, the pressure reduction is a relatively secondary factor compared to the decreased incoming flow velocity.

In a bicycle race, cyclists ride in single file and take turns at the lead. The air drag that the trailing cyclists need to overcome is estimated to be less than 60% of that exerted on the lead rider.

9.18 Larger in Size, Longer in Range: Scale Effect

A rifle bullet typically has an initial speed ranging from about 800 m/s up to about 1200 m/s. If the barrel is elevated at a certain angle, the obliquely fired bullet can travel a distance of up to 5 km. A cannonball has a muzzle velocity comparable to that of a bullet, but its shooting range can reach 20 km. What is the reason for this?

According to the analysis of oblique projectile motion, the range of an object launched with an initial velocity of 1,000 m/s at an angle of 45° to the horizontal is calculated to be 51 km. We are clearly aware that the distance traveled by a rifle bullet is far smaller than this theoretical value. The distance traveled by a cannonball is greater, but still far smaller than 51 km. Figure 9.41 compares the ballistic trajectories of three kinds of projectiles. It should be obvious that the difference between the actual shooting range and the theoretical value is due to air drag. What is the reason for the noticeable difference between a rifle bullet and a cannonball?

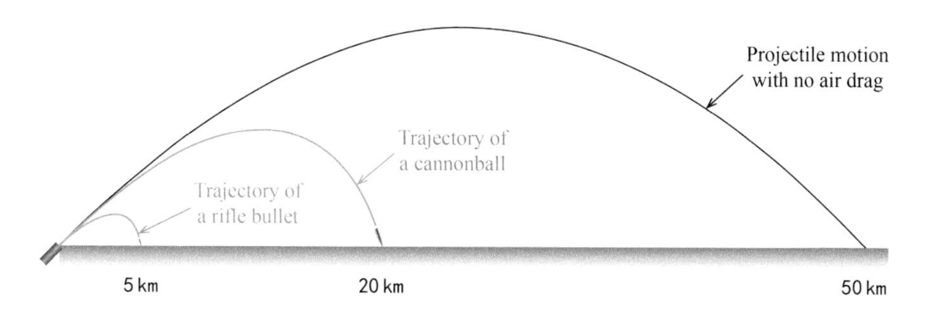

Figure 9.41 Ballistic trajectories of the projectiles both with and without air drag with initial velocity of 1 km/s at an angle of 45° to the horizontal.

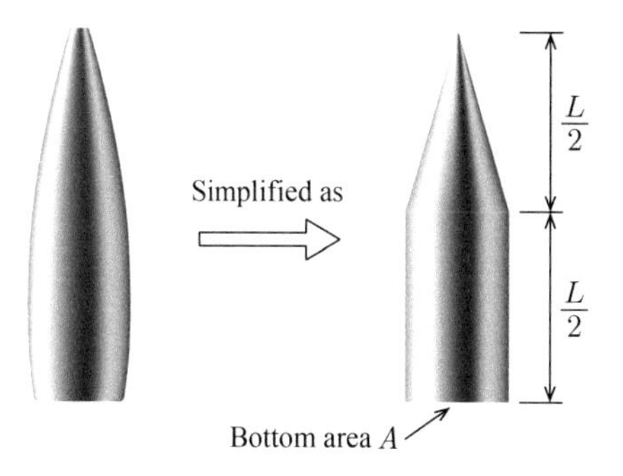

Figure 9.42 The simplified shape of a bullet.

There is a significant difference in the Reynolds numbers for the bullet and shell, which are the same in shape but totally different in size. There are two major effects of this number on air drag: its effect on the surface flow pattern causes the friction drag to change; and that on the separation point produces a change in the pressure drag. Both the rifle bullet and the cannonball travel very fast. The calculated Reynolds numbers are so large that the boundary layers over the surfaces of the two projectiles are virtually turbulent, and the separation points are approximately the same. The supersonic bullet and the cannonball suffer the same shock drag, which depends only on the shape of the projectile and the freestream velocity. Therefore, the drag coefficient remains basically the same for the two projectiles.

We denote the drag coefficients for both of the projectiles by C_D, the windward area by A, and the volume by B. For the sake of simplicity, both projectiles are assumed to be homogeneous in material, with the same density of ρ_{bullet}. Let the density of air be ρ_{air}, then the air drag with velocity V is:

$$D = C_D \cdot A \cdot \frac{1}{2}\rho_{\text{air}} V^2 .$$

There is a fixed relationship between the volume and windward area of an object for a given shape. Now, if the bullet is simply viewed as a cone on top of a cylinder, as shown in Figure 9.42, the relationship between the volume and the windward area is:

$$B = A \cdot \frac{L}{2} + \frac{1}{3}\left(A \cdot \frac{L}{2}\right) = \frac{2}{3} AL .$$

Thus, the mass of the bullet or shell is:

$$m = \rho_{\text{bullet}} B = \frac{2}{3} \rho_{\text{bullet}} AL \, .$$

The acceleration due to air drag is:

$$a = \frac{D}{m} = \frac{C_D \cdot A \cdot \frac{1}{2} \rho_{\text{air}} V^2}{\frac{2}{3} \rho_{\text{bullet}} AL} = \frac{3}{4} C_D \cdot \frac{\rho_{\text{air}}}{\rho_{\text{bullet}}} \cdot \frac{V^2}{L} \, .$$

We can now discuss the effect of bullet and cannonball sizes (i.e., L). If a cannonball is n times the size of a bullet, its acceleration due to air drag is only $1/n$ of that of the bullet. In general, the size of a cannonball is dozens of times larger than that of a bullet, and the air drag effects on the trajectories of the two projectiles are quite different.

In reality, there is another important reason why a cannonball travels much further than a bullet. The cannonball has not only a longer shooting range but also a higher shooting altitude. We know that flying at a higher altitude typically means flying through thinner air, and the smaller air drag results in a faster traveling cannonball with a longer trajectory. However, the prerequisite for this is still that the drag acting on the cannonball be sufficiently smaller to allow it to fly higher.

As we have seen, the smaller the object, the greater the influence of air drag. When two iron balls of different size fall freely through the air, the bigger one will land first, because the air drag has a stronger effect on the smaller one. The body density of ants is similar to that of humans, but regardless of how high they are falling from, they will not fall to their deaths. This is also because small objects are greatly affected by air drag. Another factor is that the motion of an ant is characterized by a small Reynolds number, which results in a large drag coefficient.

Scale has an effect not only on fluid mechanics, but also on solid mechanics. Popular science books often claim that ants are stronger than elephants, which is not fair. Smaller animals usually have greater strength relative to their weight, and the same is true for people. Let us take a look at weightlifting world records: The men's 56 kg snatch record is 138 kg, while the men's 105 kg snatch record is 200 kg. A lightweight athlete lifts 2.5 times his body weight, while the lift to body weight ratio of a heavyweight athlete is only 1.9.

The reason why small creatures have greater strength is that strength is determined by the cross-sectional area of the muscle, which is proportional to the square of the scale, while weight is proportional to the cube of the scale. This analysis is the same as that of air drag mentioned above.

In addition, the scale effect is also biologically significant. For example, small animals need to eat more high-energy foods to maintain a constant body temperature. This phenomenon is relevant to heat transfer problems. The smaller the animal, the relatively larger its skin surface area, which causes faster heat dissipation.

9.19 Meandering of Rivers: Pressure-Dominated Channel Vortex

It is not surprising that rivers meander through flatlands, since water flows along a local gradient that can vary at different locations. However, even an originally straight river will become sinuous with time, as shown in Figure 9.43. What is the reason behind this?

A river channel increases in bend curvature continuously until the river cuts through a meander neck to shorten its course, leaving behind detached oxbow lakes in the abandoned channel. After that, the straight river channel continues bending over time. The high curvature of such a river channel influences the flow distribution, causing outer bank erosion and inner channel blockage, creating another curve, and then another, and another … until eventually a winding, snake-like river forms. This natural change in the river course is due to the vortices generated by the water flowing through a curved channel. This type of vortex flow is usually known as a channel vortex.

Figure 9.44 shows the river flowing through a curved channel. When the river bends, the pressure on the outer bank is always higher than that on the inner bank, in accordance with centrifugal force. For both inner and outer banks, the water surface in contact with the atmosphere is subjected to atmospheric pressure, which cannot provide the pressure-gradient force directed inward. When the river enters a curved channel from a straight one, part of the water moves toward the outer bank at the surface, so the water surface near the outer bank has a higher altitude. Since the pressure of the water beneath the water surface increases with increasing depth, the higher water surface elevation along the outer bank – therefore, the higher water pressure – generates a centripetal force in order to turn the river. In the boundary layer closest to

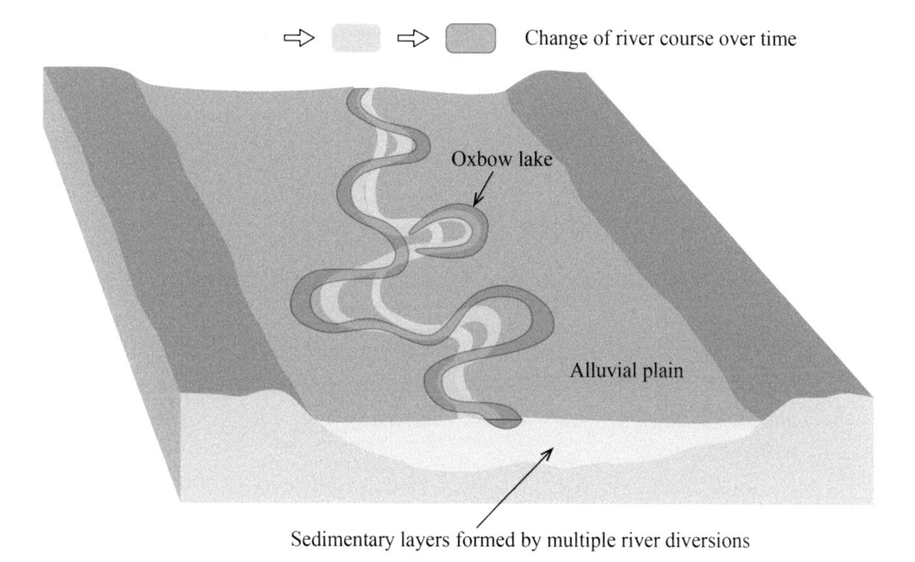

Figure 9.43 A river changes its course over time.

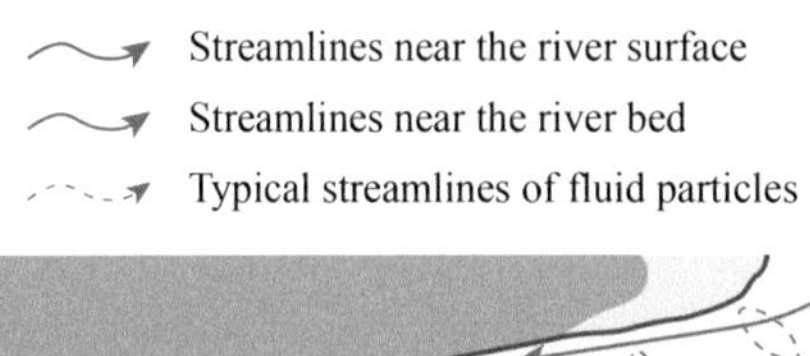

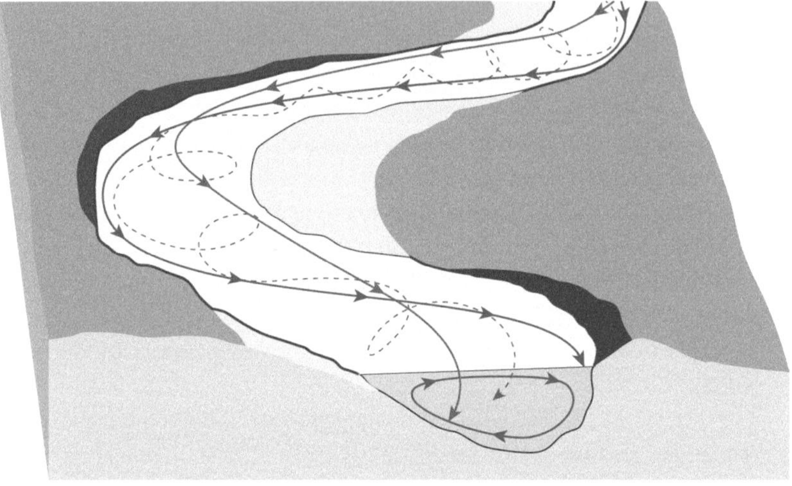

Figure 9.44 River in a curved channel and formation of channel vortex.

the bottom, the velocity of the water is slower, which requires less centripetal force to turn. The boundary layer flow at the bottom must thus follow a smaller radius than the main flow.

Hence, when the water in a river flows through a bend, it moves toward the outer bank at the surface and toward the inner bank at the bottom. This will generate a swirl movement of the water on the channel cross-section, known as a channel vortex. The water near the surface flows over the outer bank at high speeds and causes more soil erosion, while the water near the bottom transports sediments and deposits them on the inner bank. As time goes by, the channel migrates in the direction of the outer bank and increases its bending. This is why rivers in floodplains naturally tend to have sinuous channels and change course over time.

9.20 Tea Leaves Gather in the Middle of the Cup: Another Channel Vortex

When stirring half a cup of tea with loose tea at the bottom, the tea leaves will soon gather at the center of the bottom of the cup. This is a confusing phenomenon: Why are the tea leaves, which are heavier than water, not thrown to the side wall by centrifugal forces?

Actually, this is also a problem of vortex motion with viscosity effects included, just like channel vortices occurring in riverbank erosion. It was Einstein who provided one of the earliest explanations of this phenomenon; he also explained the erosion at the outside portion of a river bend.

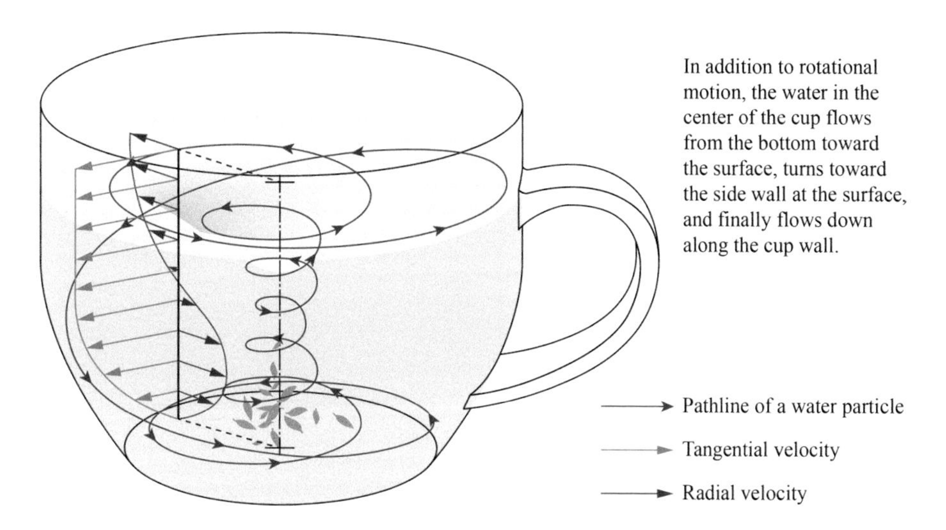

Figure 9.45 Three-dimensional swirling flow in a teacup.

When the water is stirred by a spoon, a boundary layer forms at the bottom of the cup. The angular velocity, and therefore the centrifugal force, will be smaller near the bottom than higher up near the surface. However, the centripetal force generated by the differential pressure between the outside and inside of the cup bottom is determined by the height of the water surface. This differential pressure is exactly suitable for the rotating water at half the cup depth, but too large for that in the bottom boundary layer. Therefore, the inward-directed differential pressure force pushes the water near the bottom from the side wall toward the center along a spiral line.

The water at the center of the cup flows from the bottom toward the surface. Subsequently, it turns toward the side wall at the surface, and finally flows from the surface toward the bottom near the side wall. Figure 9.45 shows the three-dimensional vortex flow structure. As the rotational speed decreases, the ascending water in the center of the cup cannot overcome the gravity of the tea leaves, which are left to gather at the center of the bottom.

9.21 Iron Ox Moves Upstream: Pressure-Dominated Horseshoe Vortex

There is a traditional story in China: One year, the Yellow River flushed away an iron ox statue. Days later, the local people conducted a search operation downriver, but no trace of the statue could be found. Finally, the iron ox turned up in the upper reaches of the river. The explanation in the story for this event is that the water could not flush away the very heavy iron ox, but removed mud and sand in front of it until it toppled into the pit ahead. As time went by, the iron ox rolled upstream along the riverbed.

In reality, the iron ox would move only several meters upstream, with several turnovers at most. Figure 9.46 shows a rock rolling upstream along the riverbed. Let us discuss how the water removes the sediments from under the base of the rock.

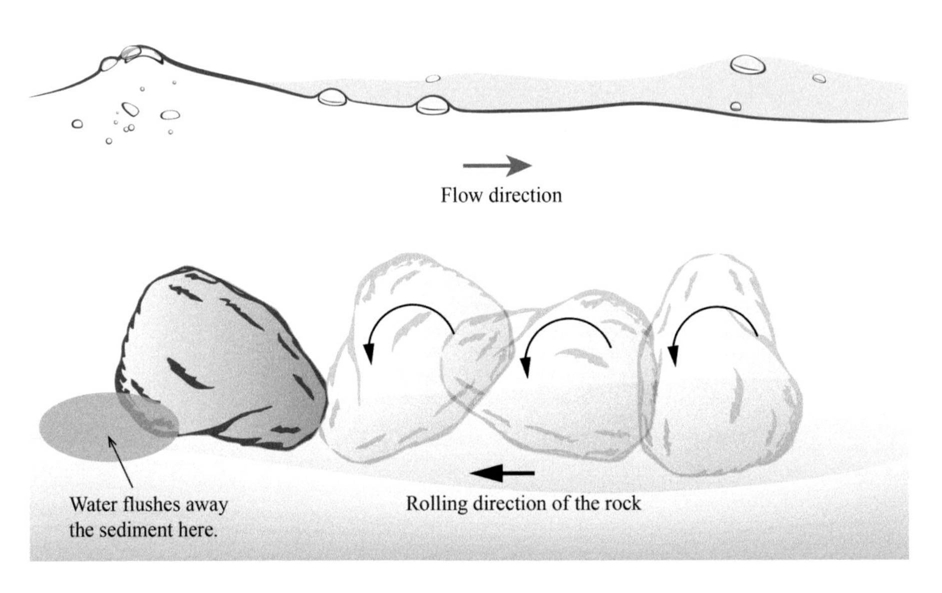

Figure 9.46 A rock rolling upstream along the riverbed.

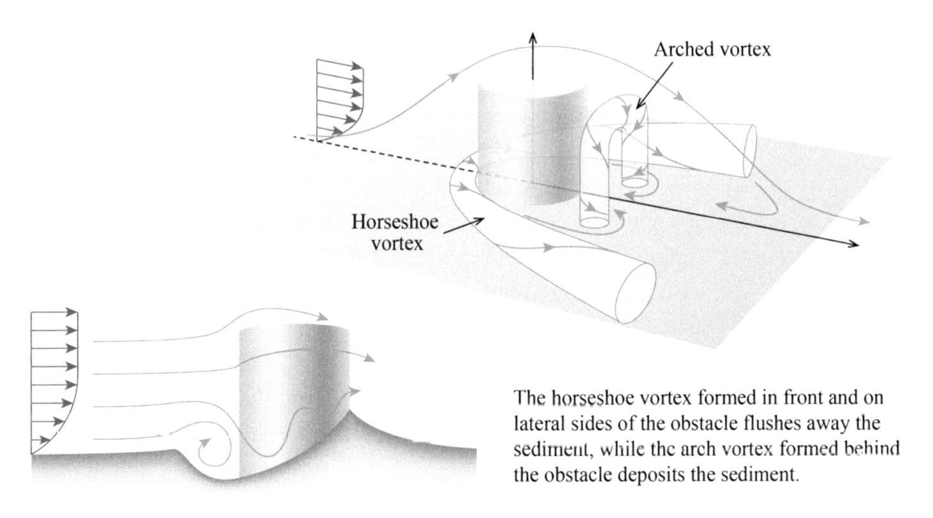

Figure 9.47 Flow over an obstacle lying on the riverbed and sediment transport by water.

As shown in Figure 9.47, each layer of water flows horizontally, a net force caused by hydrostatic pressure is balanced by gravity, and a boundary layer forms over the riverbed. An obstacle lying on the riverbed will produce a stagnation region upstream of the obstacle. The pressure at the stagnation region is the sum of the freestream static and dynamic pressures. The water near the riverbed is also stagnated, but its dynamic pressure is small, resulting in a smaller pressure rise. The closer the water is to the riverbed, the smaller the pressure rise after stagnation. At the obstacle's leading edge, pressure is not balanced along the surface of the obstacle. The high-pressure fluid in the upper layer will squeeze the low-pressure fluid in the lower one. As a result, the freestream upstream of the obstacle is deflected downward.

Due to the adverse pressure gradients caused by the obstacle, the boundary layer separates to generate a vortex near the leading edge. The two legs of the vortex extend downstream and wrap around the obstacle, as shown in Figure 9.47. Since the vortex has a shape vaguely reminiscent of a horseshoe, it is called a horseshoe vortex. Such a vortex tends to form in the typical flow around surface-mounted obstacles.

Now let us look at the horseshoe vortex created by the iron ox lying on the riverbed: The water near the riverbed flows away from the iron ox. That is to say, the water is constantly flushing out the sediments at the base of the iron ox and moving them downstream, especially the sediments in front of the statue; this is why a pit forms on the upstream side of it.

Moreover, a horseshoe vortex always forms near the ground as the wind blows through big trees and tall buildings. People living in places where heavy snow is common may have seen the effect of such vortices. If it snows on a windy day, the snow will be kept from settling onto the ground near the windward side of isolated buildings, but it will deposit on the leeward side, often forming a snow bank of considerable height. The major reason for this is the horseshoe vortex.

9.22 Pressure Change by a Passing Train: Not Just Bernoulli's Equation

Caution is recommended when standing near the edge of a railway platform, because when a train passes by at high speed a standing person may get "sucked" toward the train. A common explanation is that the air close to the train is dragged and moves faster than the surrounding air. Bernoulli's principle states that the faster-moving air has lower pressure than the surrounding stationary air, so the differential pressure pushes the person toward the train, as shown in Figure 9.48.

However, this explanation is not correct. Let us review the conditions for applying Bernoulli's equation: along a streamline, steady, inviscid, and incompressible. In this example, if Bernoulli's equation is to be applied between points A and B, only the last condition – incompressibility – is satisfied (for a qualitative analysis, Bernoulli's

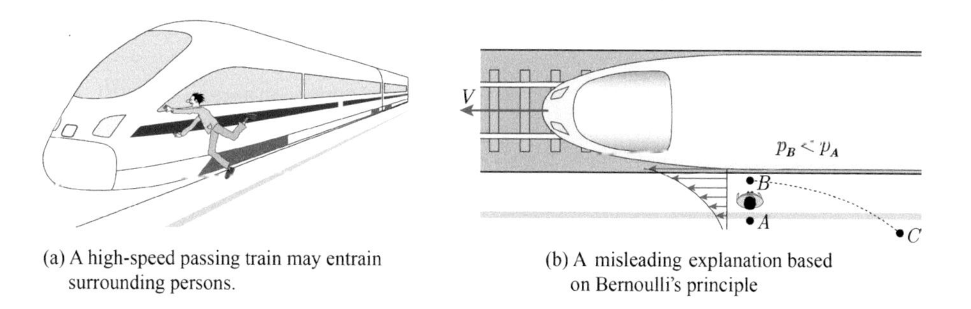

(a) A high-speed passing train may entrain surrounding persons.

(b) A misleading explanation based on Bernoulli's principle

Figure 9.48 A person gets "sucked" toward a train and its explanation based on Bernoulli's principle.

principle "faster-moving fluid is at lower pressure" can be applied to problems of both incompressible and compressible fluid flow). Let's analyze the other three conditions.

First, it is obvious that points A and B are not in the same streamline, which, however, does not affect the conclusion. Because a remote point C has the same atmospheric pressure as point A, applying Bernoulli's equation along C–B is equivalent to applying it along A–B, as shown in Figure 9.48(b)

Second, the airflow caused by the train is obviously unsteady. The train does work on the originally stationary air. The air in front of the train is pushed by the train to accelerate, and with increased pressure. That is so say, the airflow does not satisfy Bernoulli's equation. On this point, see Figure 4.8 in Chapter 4, which shows the air velocity and pressure distributions in front of a moving sphere.

Third, the description "The air around the train is dragged and moves faster than the surrounding air" means that the air at point B is dragged along with the train through viscous forces, generating a type of viscous flow to which Bernoulli's equation cannot be applied. If the air is dragged by viscous forces, its velocity will increase while air pressure will remain unchanged. On this point, we can refer to Figure 9.13, which shows the velocity and pressure distributions of the airflow past a zero-thickness plate at zero angle of attack. Moreover, the air forms boundary layers to the side of the train, while the typical nominal thickness of the boundary layer is so small that the person ordinarily stands outside it.

In fluid mechanics, an unsteady flow is usually transformed into a steady one through an appropriate coordinate transformation. The equivalent steady case is that air is blown through a stationary train model. Then, the only factor preventing the application of Bernoulli's equation is viscosity. Figure 9.49(a) shows the velocity and pressure distributions around the train with the train as reference frame. As can be seen, the boundary layer is very thin and only the mainstream (inviscid flow) needs to be analyzed. A high-pressure area (in red) caused by air deceleration forms on the windward surface of the train nose, and a low-pressure zone (in blue) caused by air acceleration forms around and a bit behind the train nose. In other areas, the pressure is practically equal to the atmospheric pressure. By means of a coordinate transformation, we obtain the velocity distribution relative to a reference frame fixed on the ground, as shown in Figure 9.49(b). The air flows away from the train nose, forming a region around it with velocities greater than zero (deeper red color denotes velocity magnitude).

Now let us analyze the aerodynamic forces acting on a person standing beside the train. The airflow the person feels is shown in Figure 9.49(b), while the static pressure felt is coordinate-independent, as shown in Figure 9.49(a). Figure 9.50 displays the superimposition of the velocity and static pressure distributions. Obviously, the aerodynamic forces acting on people standing at P_2 and P_3 are the greatest. The aerodynamic force consists of differential pressure and viscous force. For a high Reynolds number flow around a bluff body such as a human body, the viscous forces are negligible, and only the differential pressure forces are taken into account. When no one is standing on the platform, the pressure gradient caused by the fast-moving train already generates a differential pressure force, called "differential static

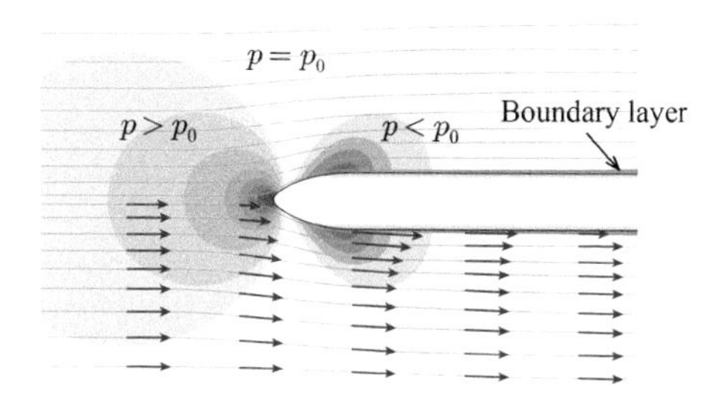

(a) Streamlines and velocity and pressure distributions relative to a train fixed reference frame

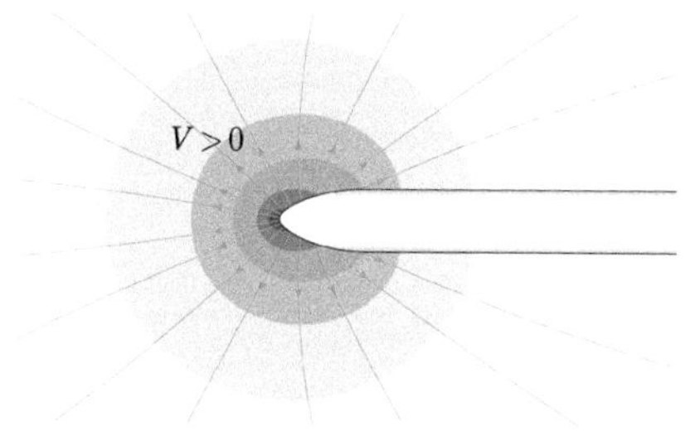

(b) Streamlines and velocity distributions relative to a ground fixed reference frame

Figure 9.49 Velocity and pressure distributions near the train in two coordinate systems.

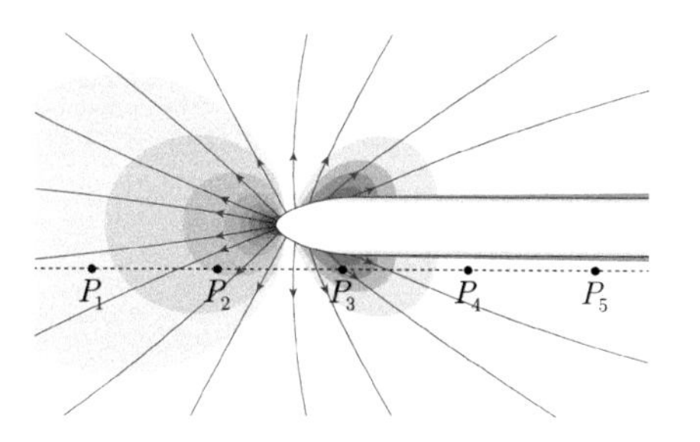

It is estimated that the aerodynamic forces acting on the human body are composed of two parts: the force generated by static pressure difference, F_p, and the force generated by velocity, F_v.

Assuming the train moves at a speed of 280 km/h, the distance between the human body and the side face of the train is 0.5 m.

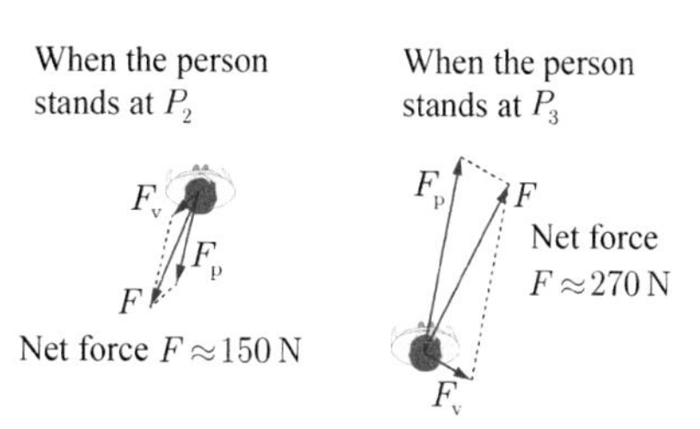

Figure 9.50 Aerodynamic forces acting on the human body standing at five different positions.

pressure force" here. The human body disturbs the local velocity field, generating an additional differential pressure force, called "differential dynamic pressure force" here. By a rough estimate, the total differential pressure force can be treated as the sum of these two forces.

It is assumed that the train moves at a speed of 280 km/h, the distance between the standing person and the side face of the train is 0.5 m, the windward area of the human body is 0.7 m^2, and the drag coefficient is 0.4. According to numerical simulation results, other flow parameters at P_2 and P_3 can be estimated as follows:

differential pressure between the front and back of the human body at P_2: 200 Pa;
differential pressure between the front and back of the human body at P_3: 400 Pa;
air velocity relative to the human body at P_2: 10 m/s;
air velocity relative to the human body at P_3: 18 m/s.

The differential static pressure force is the product of the differential pressure times the area, while the differential dynamic pressure is the product of the dynamic pressure times the area times the wind drag coefficient. Thus, we obtain:

differential static pressure force at P_2: 140 N;
differential static pressure force at P_3: 280 N;
differential dynamic pressure force at P_2: 17 N;
differential dynamic pressure force at P_3: 56 N.

The resultant force is the vector sum of the differential static pressure force plus the differential dynamic pressure force. Finally, we estimate that the aerodynamic force exerted on the person standing at P_2 is about 150 N, pointing away from the train; the aerodynamic force exerted on the person standing at P_3 is about 270 N, pointing toward the train.

It follows that the aerodynamic force is indeed large enough to "suck" the human body toward the train at P_3. However, this force is not the result of the air around the train being dragged and moving faster than the surrounding air, but is largely caused by the low-pressure zone around the train nose.

Taking the train as the reference frame, the formation of this low-pressure zone can be explained by Bernoulli's principle, but the acceleration of the airflow has nothing to do with the viscous forces from the fast-moving train. Moreover, as shown in Figure 9.50, the velocity relative to the ground at P_3 points backward, implying that the train nose squeezes the air upstream of it. The air acceleration is caused by a "give way" effect, or differential pressure forces.

The aerodynamic forces exerted on the person standing at P_5 are negligible. That is to say, the side face of the train does not pose a threat to the person standing half a meter away from the train nose. At the rear of the train, the person may stand inside the thick boundary layer over the side face of the train. Then, the airflow may exert a thrust on the human body, pointing along the direction the train is heading.

To sum up, when a fast-moving train passes a person standing too close to a railway track, the person will be pushed away first and then pulled toward the train nose. As the train body passes by, the aerodynamic forces acting on the human body are negligible. When the rear part of the train passes by, the aerodynamic forces acting on the human body are directed along the heading direction of the train.

9.23 How Lift Is Created: The Coandă Effect is the Key

There are two ways to explain why an airfoil produces a lift, corresponding to the integral and the differential approaches. The interpretation based on the integral approach is the most rigorous and the least controversial, while that adopting the differential approach is not easy to express rigorously and is the subject of some debate. Here, both of these approaches are introduced. In addition, there are also approaches to explain how the lift is created based on potential flow and vortex dynamics, which are considered nonintuitive and will not be discussed here.

The interpretation based on the integral approach may be regarded as the combination of Newton's second and third laws of motion. In short, if the airfoil is angled correctly, the airflow over it is deflected downward, thus generating an upward lift on the airfoil. According to the momentum theorem or Newton's second law, the air

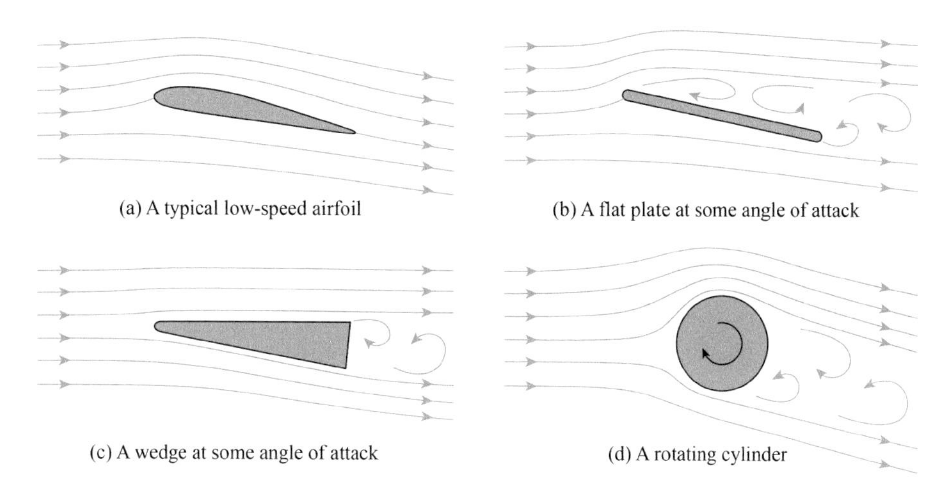

Figure 9.51 Several types of objects that can generate lift.

leaving the trailing edge of the airfoil is given a downward component of momentum, which requires the airfoil to exert a downward force on the air. From the viewpoint of Newton's third law, an equal and opposite force from the air pushes the airfoil upward and hence provides the aerodynamic lift.

Figure 9.51 shows several types of objects that can deflect air downward to create lift. Figure 9.51(a) shows the airflow over a typical low-speed airfoil at some angle of attack. The downward deflection of the air is mainly due to the Coandă effect on the upper airfoil surface. Figure 9.51(b) shows the airflow over a flat plate at some angle of attack. Although air is separated on the upper surface, the plate can still deflect air downward. Figure 9.51(c) shows the airflow parallel to the upper surface of a wedge. The lower surface of the wedge has an important role to deflect the air downward. Figure 9.51(d) shows the airflow over a rotating cylinder. The same principle can also be used for curve shooting in football games.

However, many find the interpretation based on the integral approach nonintuitive. Since an airfoil in flight experiences an upward lift, there must, therefore, be a low-pressure zone on the upper airfoil surface, or a high-pressure zone on the lower airfoil surface, or both. The explanation based on the pressure distributions between upper and lower surfaces seems more convincing, and has been adopted by popular science books. Next, let us analyze the difference in pressure across the airfoil to explain the lift – that is, the differential approach.

A popular explanation is the following: The airfoils are curved, convex on the upper surface and flat on the lower one. The flow over the upper surface, which is longer in path than the lower surface, travels faster in order to reach the trailing edge simultaneously. According to Bernoulli's equation, the higher velocity produces a lower pressure on the upper surface, thus creating a net upward force. This analysis seems reasonable, but is actually not correct. It is easy to find a counterexample. As shown in Figure 9.52, an airfoil with a wavy upper surface cannot produce lift although its upper surface is obviously longer than its lower one.

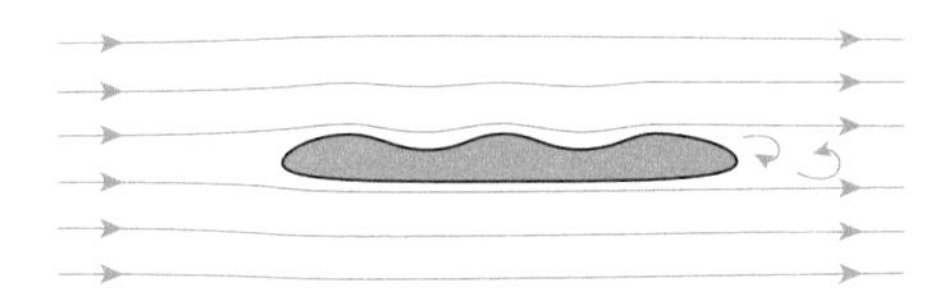

Figure 9.52 An airfoil with a wavy upper surface.

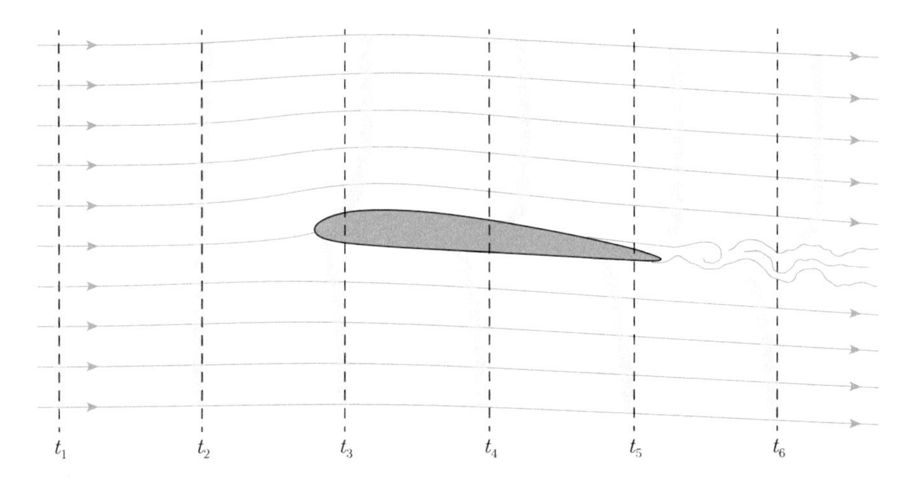

Figure 9.53 Air flowing over the upper and lower surfaces of the airfoil does not reach the trailing edge simultaneously.

A fatal flaw in the above argument is that the airflow over the upper and lower surfaces of the airfoil was assumed to reach the trailing edge at the same time. However, experimental results demonstrate that the airflow over the upper surface actually reaches the trailing edge earlier. Interested readers can search and watch videos on flow display experiments. Figure 9.53 shows the schematic diagram based on a video. As can be seen, the fluid moves in parallel and uniformly at instant t_1. Before approaching the leading edge of the airfoil at instant t_2, the upper half of the fluid has slightly accelerated, while the lower half has slightly decelerated. At instant t_3, the fluid above the airfoil continues to accelerate except for that within the boundary layer, while the fluid below the airfoil further decelerates. This difference in velocity continues until the fluid above the airfoil leaves the trailing edge first, while the fluid below the airfoil has no chance to catch up.

The argument that the airflows over the upper and lower surfaces reach the trailing edge simultaneously may be based on the following interpretation: The undisturbed air away from a flying aircraft obviously remains at rest. Therefore, the undisturbed air above and below the airfoil moves at the same velocity as the flight speed if the airfoil is taken as the frame of reference. However, the premise of this interpretation is that the airfoil disturbs only the fluid near it. In fact, the range of the disturbed fluid is far greater than what some people might intuitively feel. Figure 9.54 shows the airflow past the airfoil over a wider range. The streamlines, which are located multiple times the chord length away from the upper and lower surfaces, have been disturbed. The air further away from the airfoil can be deemed as moving at the same velocity.

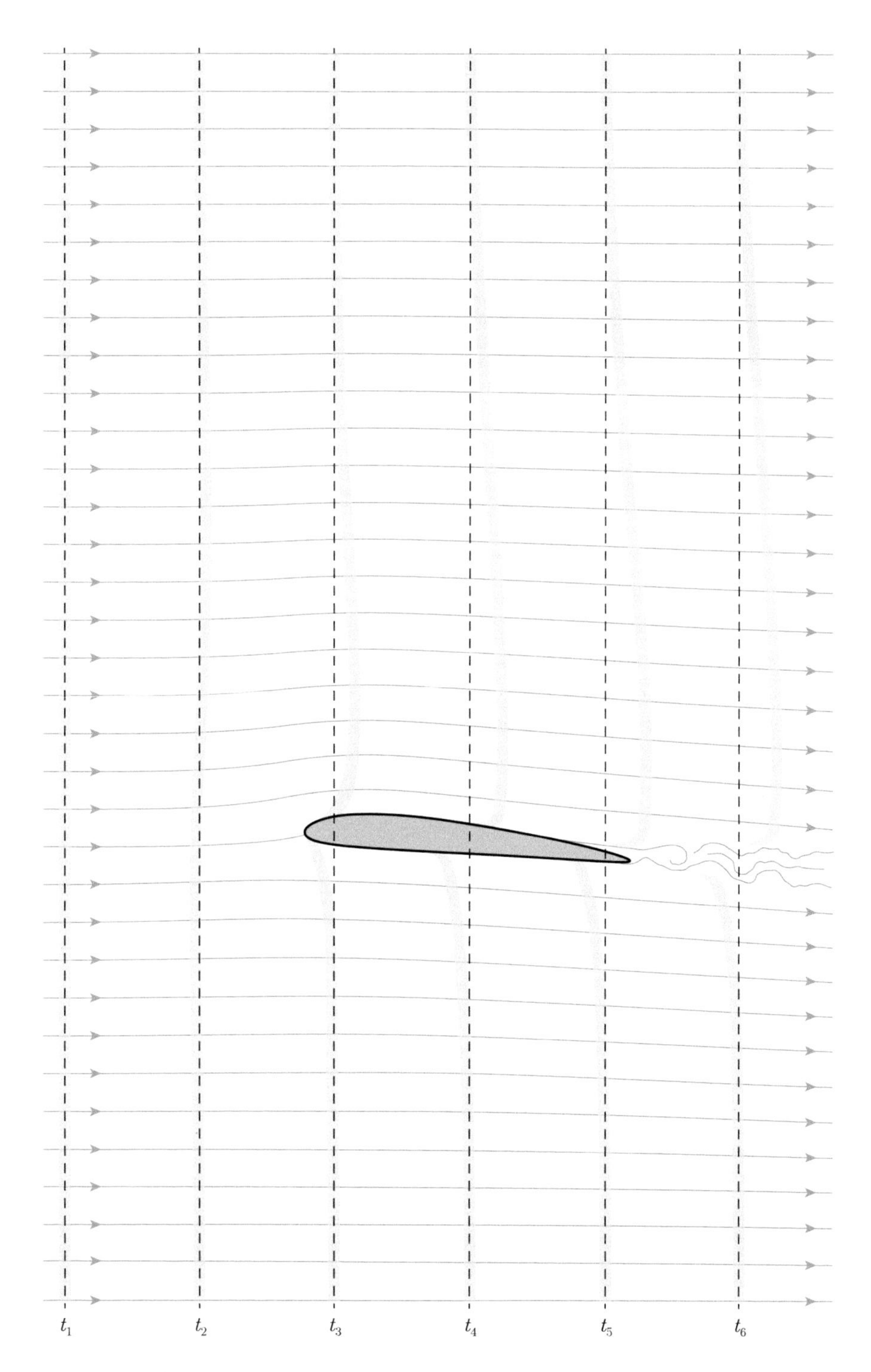

Figure 9.54 Influence range of the upper and lower surfaces of an airfoil.

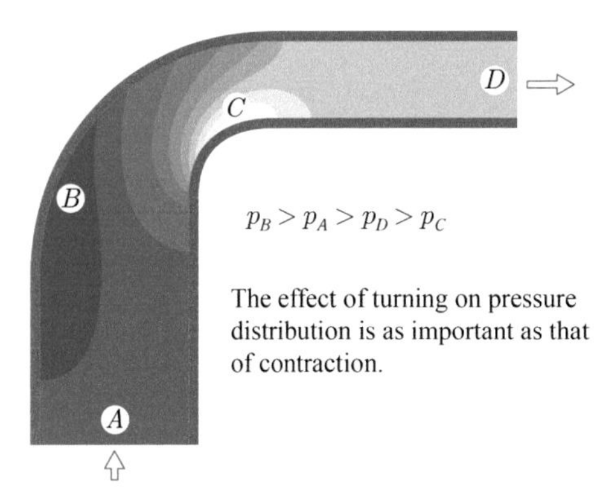

Figure 9.55 Flow through successive turning and contraction.

Let us return to the explanation of lift. The argument that the airflows over the upper and lower surfaces reach the trailing edge simultaneously is clearly incorrect to explain the lower pressure on the upper airfoil surface. There is another explanation based on the continuity equation. Its basic idea is that the convex upper surface reduces the cross-sectional area of the airflow over it, resulting in higher velocities and lower pressure on the upper airfoil surface. This explanation, although it seems logical, is not rigorous. First, the two-dimensional flow of fluid through a contraction is not bound to cause a decrease in pressure, because a contraction in structure does not necessarily produce an acceleration in two-dimensional flow. Take a two-dimensional flow through successive turning and contraction, as shown in Figure 9.55, for example. The pressure at point B is higher than that at point A because a centripetal force is required to make the fluid turn in a curving path. Furthermore, the continuity equation states that the mass flow rate remains the same over any section of pipe, which is the result of steady flow, not the cause of velocity change. The change in fluid velocity is directly related to the external forces acting on the fluid. A driving force produces an acceleration, while an impeding force produces a deceleration. Neglecting viscous forces, the driving forces include differential pressure and gravitational forces. According to Newton's law, it is the decrease in pressure that causes the fluid velocity to increase.

The lift is caused by the low-pressure zone above the airfoil. The oncoming air gets sucked into this low-pressure zone and accelerates along the upper surface. How is the low-pressure zone generated? It is because the airflow along the convex upper surface requires a centripetal force which has to be created by the difference in pressure. Far away from the airfoil, the pressure is atmospheric. Therefore, the pressure on the upper airfoil surface must be lower than the atmospheric pressure. As shown in Figure 9.55, the very low pressure at point C is not caused by the contraction of the channel, but by the large local streamline curvature which requires a significant centripetal force. The flow over the upper airfoil surface, especially near the leading edge, is like that through point C in Figure 9.55. As the air flows along the convex upper surface, the local low-pressure zone is the requirement of centripetal force, or the result of centrifugal force. For the reason why the airflow remains attached to the convex upper surface rather than traveling in a straight line, please refer to the Coandă effect discussed in Section 9.15.

In summary, it is not a simple matter to explain the lift rigorously and persuasively. The above analysis may be summarized as follows: Due to the Coandă effect, the airflow sticks to the upper surface and follows the curved airfoil. A low-pressure zone forms under the action of the centrifugal force, which corresponds to the differential approach. At the same time, the low-pressure zone "sucks" the upper air downward. The air that has passed over the airfoil deflects downward to provide lift, which corresponds to the integral approach. In short, the low-pressure zone above the upper airfoil surface is the source of lift.

The above discussion focuses only on the upper airfoil surface. For general airfoils, the upper surface is far more important than the lower one since the lift is mostly produced by the low-pressure zone above the upper surface. As shown in Figure 9.54, the influence range of the upper surface on the airflow is significantly larger than that of the lower one, and the upper surface deflects more air downward. Because it is less important, the lower face is often covered with various equipment (engines, auxiliary fuel tanks, weapons, detection devices, etc.), while the upper surface must be kept undisturbed.

We now examine the airflow around a ball. As shown in Figure 9.37, only a small part of the windward area of the ball experiences higher pressure than the freestream pressure, while most of the pressure experienced by the ball surface is lower than the freestream pressure. Furthermore, the negative pressure is larger in magnitude than the positive pressure. In general, a wider negative-pressure zone is observed in the flow around an object. This is because, as the object occupies a certain space, the airflow needs to accelerate and depressurize to bypass it.

Of course, we can also create a high-pressure zone below the lower airfoil surface to produce lift. Figure 9.51(c) shows the high-pressure zone generated by the obstruction of the lower surface to the airflow. Except for a small low-pressure zone near the trailing edge, the pressure most of the lower surface experiences is higher than the freestream pressure. Strictly speaking, this shape can generate a small lift force, but accompanied by a large drag force. As in the case of the flat plate at some angle of attack in Figure 9.51(b), the lift of a kite is still predominantly generated by the lower pressure on the upper surface rather than the higher pressure on the lower one.

9.24 Principle of Heat Engines: Working Substance Must be Compressible

A heat engine is a system that converts heat or thermal energy into mechanical work, such as steam engines, steam turbines, internal combustion engines, gas turbines, and so on. Theoretically, the higher the temperature of the heat source, the higher the efficiency of a heat engine. Therefore, practical heat engines need a high energy–density heating process as the heat source, mostly provided by chemical or nuclear energy.

The heat absorbed is used to increase the internal energy of a working substance, which is subsequently converted into useful work – that is, mechanical energy. The internal energy is defined as the microscopic disordered energy. Consider, for example, an ideal gas. Its internal energy is associated with the random, disordered motion of molecules. Theoretically, there is no way to make all the molecules move and do work in a particular direction. However, for an air cylinder, as the gas inside expands, it will push and do work on the piston. Since the molecules moving in other directions have been confined inside the cylinder, only those pushing the piston outward can do work. If there is no friction and no transfer of heat, the gas undergoes an isentropic expansion. As the gas does work on the piston, its internal energy will be consumed, causing its temperature to decrease (see Tip 4.3 in Chapter 4 for the principle of temperature drop during gas expansion).

According to the derivation of the energy equation in Chapter 4, we substitute Equation (4.48) into Equation (4.45) and rearrange the internal energy equation as:

$$\rho \frac{\mathrm{d}\hat{u}}{\mathrm{d}t} = -p\left(\nabla \cdot \vec{V}\right) + \Phi_{\mathrm{V}} + \frac{\partial}{\partial x_i}\left(\lambda \frac{\partial T}{\partial x_i}\right) + \rho \dot{q}.$$

where the first two terms on the right-hand side represent the mechanical work, and the last two represent the heat transfer. There are two ways to convert mechanical energy into internal energy in fluids: volume work and dissipation. The dissipation term comes from the shear deformation caused by shear forces, characterized by friction and mixing of adjacent fluid layers. Fluid temperature rise due to compression is reversible, while that owing to friction and mixing is irreversible. Therefore, thermal energy can be converted into mechanical energy only by expansion. A compressible working substance is the most critical factor for a heat engine to work. Since a phase transition leads to considerable change in volume, the first heat engine ever invented was a steam engine. Among the heat engines using air as a working substance, a piston engine works intermittently and compresses the air in the cylinder; an open-type gas turbine requires its compressor to operate at high speeds (high subsonic or even supersonic speed) to compress the gas. This is the reason why piston engines can operate at a very low idle speed but gas turbines cannot.

In theory, a heat engine needs only to heat and allow the working substance to expand and do work on its surroundings. Figure 9.56 shows the first recorded steam engine, the Hero engine, invented by Hero of Alexandria in AD 50. Place a cauldron filled with water over a fire. As the cauldron gets hot, the created steam enters a rotatable hollow sphere via two pipes, and then flows out of two nozzles that are opposite to each other. The combined thrust generated by both nozzles results in torque, causing the sphere to spin around its axis. In this way, the heat is converted into the mechanical energy of the rotating sphere. Of course, this device is not very useful, because it is highly inefficient and cannot work continuously. When the water boils away the rotation stops, so the engine requires a constant supply of water into the cauldron. Because the steam pressure in the cauldron is higher than the atmospheric pressure, the added water needs to be pressurized, corresponding to a compression process. In other words, in order to construct a continuous heat engine, the heating and expansion process alone is insufficient, and a compression process is also necessary.

Figure 9.56 Hero's aeolipile.

If the device is filled with air instead of water, it will be difficult for the sphere to rotate as the cauldron gets hot. When the heated water turns into steam at atmospheric pressure, it expands to nearly 1,600 times its original volume, and can do more expansion work on its surroundings. However, the air expands by a much smaller factor at the same heat input. Steam engines are typical examples of external combustion engines with phase change of working substance. The major role of phase change is to allow the heated working substance to expand and do work on its surroundings.

In a typical heat engine, heating (or combustion) components can be classified into two categories, namely closed system and open system. For heating in a closed system, the total amount of working substance is fixed, and its energy equation can be written as:

$$Q = \Delta \hat{U} + W,$$

where heat Q is added to the working substance, thus increasing its internal energy $\Delta \hat{U}$ and performing work W on its surroundings. The work W must be realized by the mechanical movement of external parts.

Figure 9.57 shows the working principle of a typical four-stroke gasoline engine. In these four strokes, the mechanical work done by the gas only occurs in the third "combustion + doing work" stroke, while the movement of the external parts in the other three strokes rely on the rotational energy stored by a flywheel.

Gas turbine engines, as opposed to four-stroke internal combustion engines, perform the intake and exhaust processes simultaneously. In other words, combustion occurs continuously in open cycle gas turbines. For these two types of engines with completely different structures, the fundamental principles are, however, similar and consistent. Both of them go through four processes: intake → compression → ignition and combustion → expansion. The ideal cycle for four-stroke internal combustion engines is the Otto cycle, and that for gas turbine engines is the Brayton cycle. Figure 9.58 shows the two operating cycles on a p–v diagram. The four strokes to complete one Otto cycle are: isentropic compression (1 → 2), isochoric heat addition (2 → 3), isentropic expansion (3 → 4), and isochoric heat release (4 → 1). The four strokes

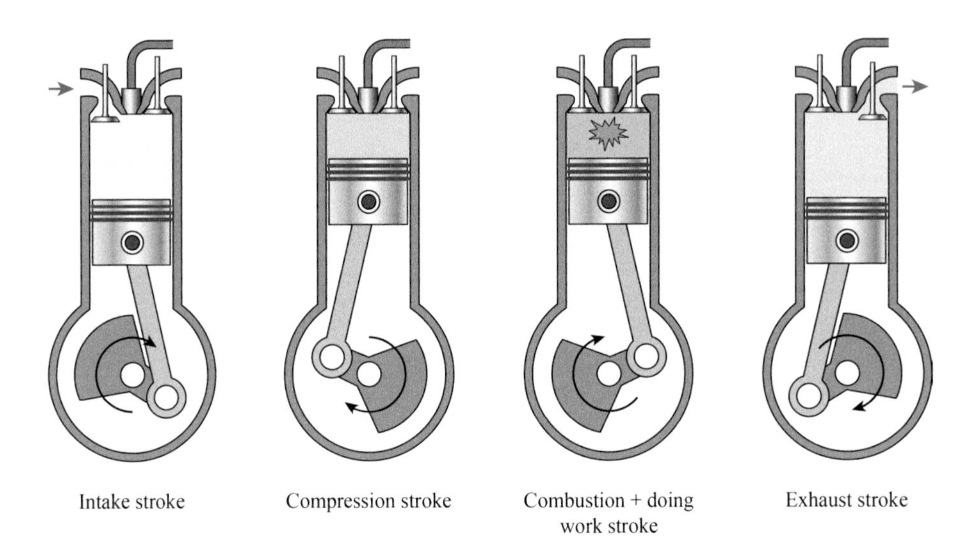

Figure 9.57 A typical four-stroke gasoline engine.

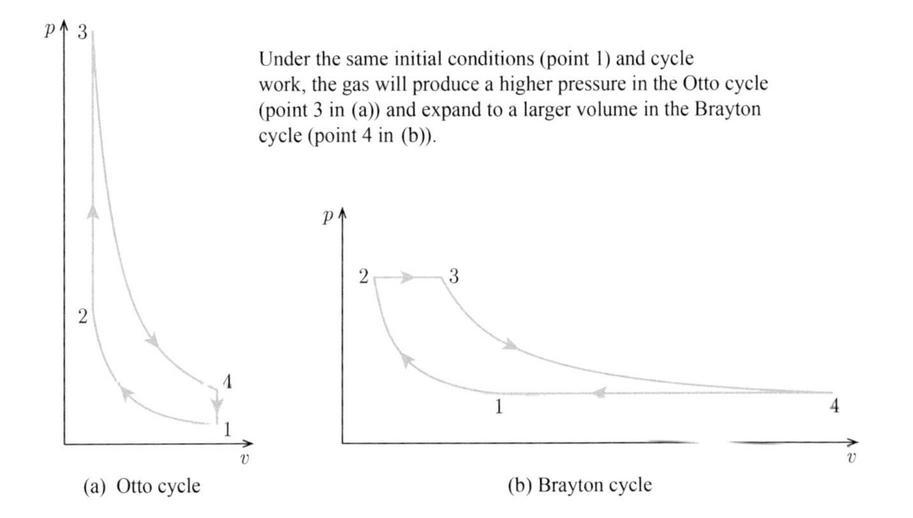

Figure 9.58 p–v diagram of the Otto cycle and the Brayton cycle.

to complete one Brayton cycle are: isentropic compression ($1 \rightarrow 2$), isobaric heat addition ($2 \rightarrow 3$), isentropic expansion ($3 \rightarrow 4$), and isobaric heat release ($4 \rightarrow 1$). The principal difference between the two cycles is that the heat addition takes place in an isochoric process in one cycle, and in an isobaric one in the other.

The p–v diagram reflects the exchange of mechanical work between the working substance and its surroundings. The amount of net work involved in a cyclic process is the enclosed area on the diagram. Obviously, heat engines require both changes in pressure and volume in a cyclic process to function. From the point of view of work, both force and displacement are indispensable, and in heat engines the displacement is

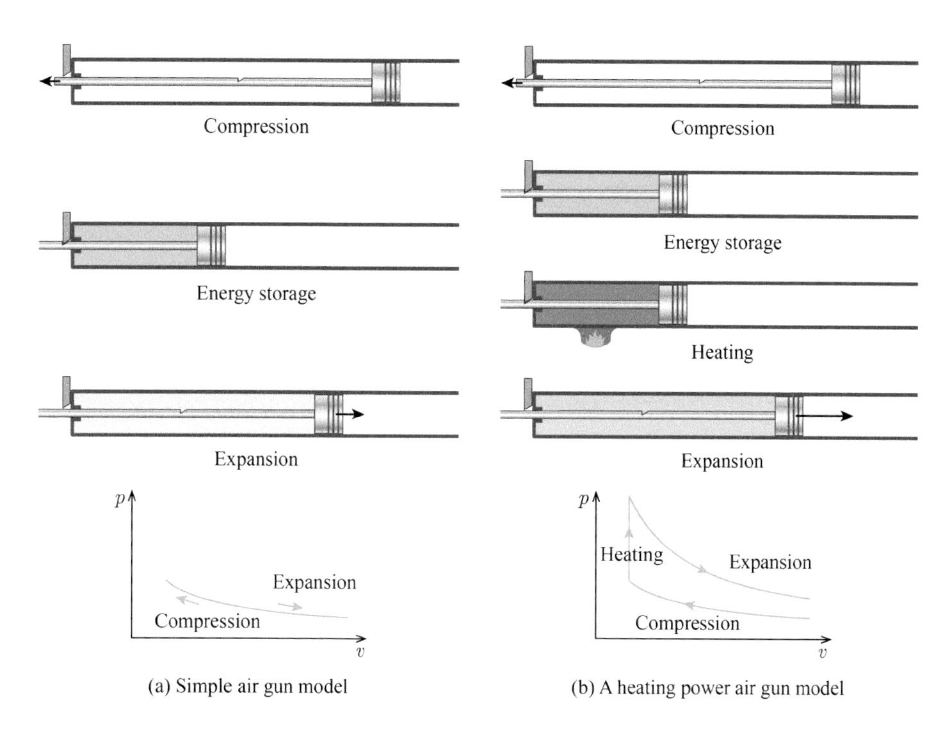

Figure 9.59 An air gun model using heat to increase its power.

characterized by the compression and expansion of the working substance. An incompressible working substance that only moves as a whole to do work does not conform with the principle of a heat engine. There are also machines that directly use the overall movement of the working substance to do work, such as water turbines and wind turbines. These machines are not considered as heat engines since they do not utilize the compressibility of the fluid to do work.

Here's a vivid example to illustrate the relationship between compression and heating. Figure 9.59(a) shows a simple air gun model. Its working principle is as follows: The air trapped in the cylinder is originally compressed by an external force. When the trigger is pulled to release the piston, the air expands and drives the piston forward, so the bullet is fired. If we want to increase the muzzle velocity of the bullet, the air pressure in the cylinder needs to be higher. Consequently, a greater external force is required to compress the air. Figure 9.59(b) shows an alternative method to increase pressure by heating. The compressed air is heated to increase its pressure, and then the piston is released to do more work. In this case, heat is converted into mechanical work. As seen in the p–v diagram, air can do positive work only when it is compressed and then heated. If the air is heated and then compressed, mechanical energy is converted into internal energy. An intuitive explanation goes like this: Compress the unheated air with a small modulus of elasticity, and then heat it to increase its modulus of elasticity. After the heated air is released, a greater force can be obtained.

9.25 Principle of Compressors: Work Done by Unsteady Pressure Forces

Compressors are the major part in a gas turbine engine that uses rotating blades to increase the pressure of the incoming air. Figure 9.60 schematically shows the pressure and velocity profiles through a multistage axial compressor. Many stages of rotors/stators gradually increase the static pressure from inlet to outlet. The air velocity increases in the rotors and decreases in the stators. The velocity at the outlet of the compressor is lower than that at the inlet.

In a broad sense, any mechanical device that increases the pressure of a gas may be called a compressor. There are various types of industrial air compressors, such as reciprocating piston air compressors, rotary vane compressors, rotary screw compressors, and rotary cam compressors, as shown in Figure 9.61.

If we only discuss the air compressor working stationarily, total pressure rise is a good measure to evaluate its performance. When will the total pressure increase? The answer is: when there is input work.

The energy equation (4.60) is rewritten here:

$$\frac{\mathrm{d}h_{\mathrm{t}}}{\mathrm{d}t} = \frac{1}{\rho}\frac{\partial p}{\partial t} + u_j \frac{\partial \tau_{\mathrm{v},ij}}{\partial x_i} + T\frac{\mathrm{d}s}{\mathrm{d}t},$$

where the third term on the right-hand side represents the increase in entropy due to heat transfer, friction, and mixing, which should be kept as small as possible.

The second term on the right-hand side represents the reversible work done by viscous forces dragging the airflow to move as a rigid body. As shown in Figure 4.14, the hub drags the gas particles to move along streamline 2 by viscous shear forces, thus

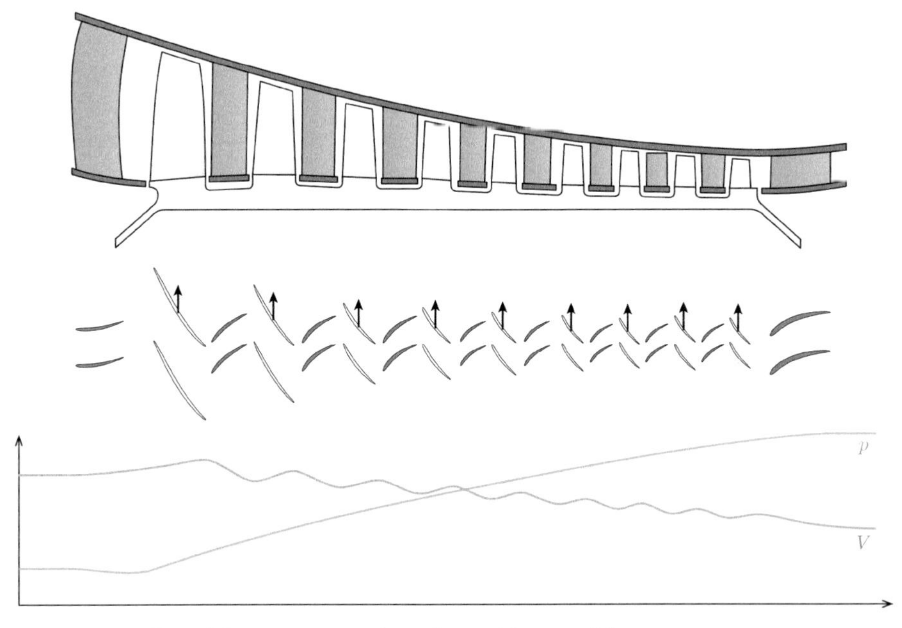

Figure 9.60 Profiles of flow parameters through a multistage axial compressor.

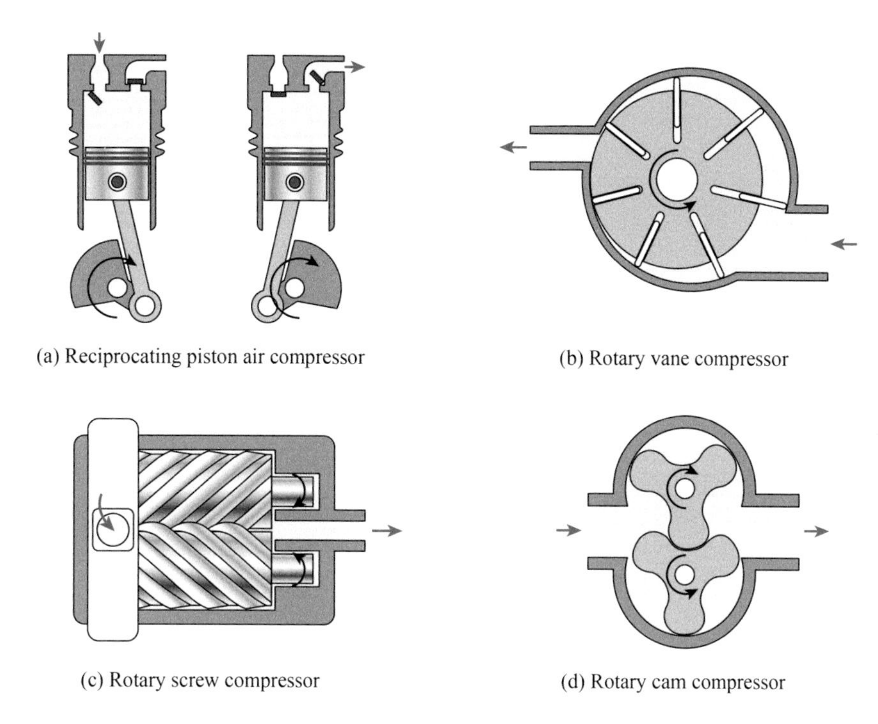

(a) Reciprocating piston air compressor

(b) Rotary vane compressor

(c) Rotary screw compressor

(d) Rotary cam compressor

Figure 9.61 Several types of air compressors.

providing a local increase in total pressure. However, this amount of work in the compressor is very limited, and inevitably accompanied by the loss due to shear deformation.

The first term on the right-hand side represents the reversible work done by unsteady pressure forces. In other words, the work done by unsteady pressure forces is the main way to isentropically increase the air total pressure, which is the basic principle for various types of air compressors. All compressors – reciprocating piston, rotary screw, blade, and cam – can generate unsteady pressure forces that do work on the air. In layman's terms, external forces are required to push, flap, or stir the incoming air to increase its total pressure. The external forces are usually exerted by solid parts such as blades.

The external forces exerted on the airflow can be directed in the streamwise, lateral, or any direction. Since the direction of the fluid flow can change through channels or blades, the work done on it can eventually be converted into increase in pressure. For instance, two common ways of rowing a boat are by paddles or by yuloh. Propelling a boat with paddles on the sides pushes the water backward; while sculling a boat with yulohs over the stern pushes the water horizontally. The earliest invented paddle steamers use the wheels on both sides to push the water backward, just as in paddling. Modern ships use the propellers behind the stern to push the water along the circumferential direction, in a manner similar to sculling with a yuloh. Both rotary vane and rotary cam air compressors push the air along the streamwise direction, while the axial flow compressor shown in Figure 9.60, as well as the common types of fan blades and propellers, push the air in the circumferential direction.

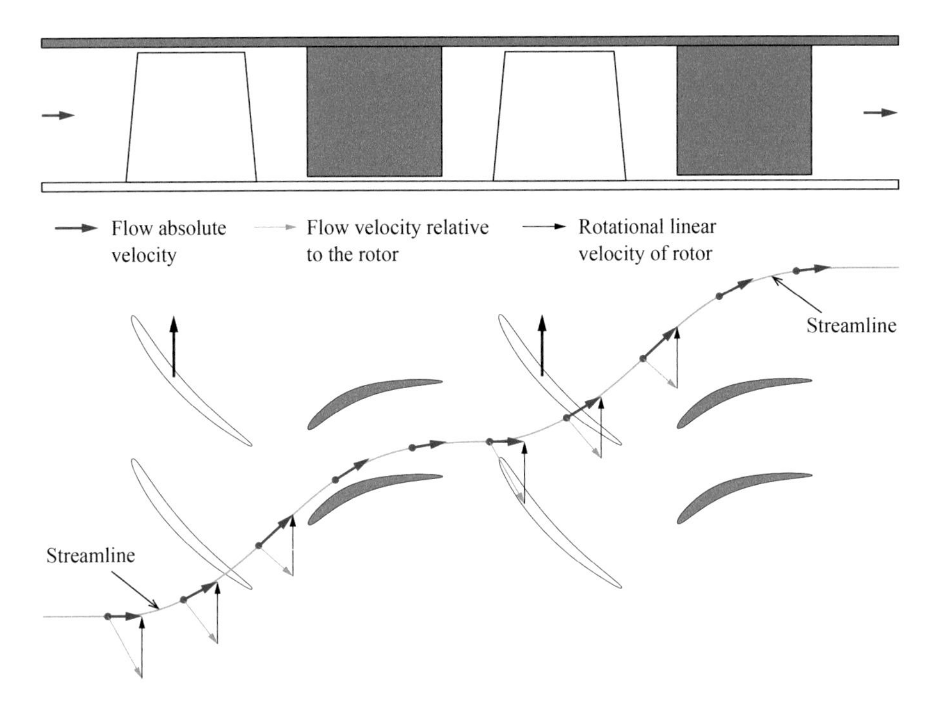

Figure 9.62 A streamline and velocity distribution in the flow through a low-speed, two-stage axial compressor.

The energy added to the airflow is completely caused by the circumferential variation of velocity. Figure 9.62 shows the airflow through a low-speed (incompressible flow), two-stage axial compressor. At the compressor inlet and outlet, the air flows in parallel to the axis. As the air passes through the rotors, its circumferential velocity increases, while its axial velocity remains the same, thus increasing the magnitude of its total velocity. In the rotor blade–fixed reference frame, the air through the diverging blade passage is decelerated with a consequent increase in static pressure. The diffusion in the stator blades also decelerates the air and converts the kinetic energy gained by the rotors into pressure energy, thus causing a further increase in static pressure. The stators, meanwhile, play a role in straightening or aligning the nonaxial airflow ready for the next set of rotor blades. Generally speaking, the air particles under the successive action of pairs of rotor–stator stages move forward along spiral trajectories.

Another compressor type is the contra-rotating compressor. Instead of using stators, this type of compressor lets the sets of rotor blades rotate opposite each other. Theoretically, the desired pressure ratio can be achieved with fewer stages, and the compressor has less weight and higher efficiency. However, due to structural limitations, in a multistage compressor it is significantly difficult to make all the sets of rotor blades rotate opposite each other.

Bibliography

Books

Anderson JD. *Modern Compressible Flow*. McGraw-Hill, 1982.

Anderson JD. *Fundamentals of Aerodynamics*. McGraw-Hill, 1984.

Anderson JD. *Computational Fluid Dynamics: The Basic with Applications*. Photocopy ed. Tsinghua University Press, Beijing, 2002.

Maozhang Chen. *Fundamentals of Viscous Fluid Dynamics*. Higher Education Press, Beijing, 2002.

Cumpsty NA. *Compressor Aerodynamics*. Krieger, Malabar, FL, 2004.

Currie IG. *Fundamental Mechanics of Fluids*. 3rd ed. Marcel Dekker Inc., New York, 2003.

Zurong Ding. *Fluid Mechanics*. 2nd edition. Higher Education Press, Beijing, 2013.

Zengnan Dong, Zixiong Zhang. *Inviscid Hydrodynamics*. Tsinghua University Press, Beijing, 2003.

Dyke MV. *An Album of Fluid Motion*. Parabolic Press, Stanford, CA, 1982.

Elger DF. *Engineering Fluid Mechanics*. 10th ed. Wiley, New York, 2012.

Qing Feng, Shiwu Li, Li Zhang et al. *Engineering Thermodynamics*. Northwest University of Technology Press, Xi'an, 2012.

Fenn JB. *Short History of Heat*. Translated by Naixin Li. Oriental Press, Beijing, 2009. (Chinese translation.)

Greitzer EM, Tan CS, Graf MB. *Internal Flow: Concepts and Applications*, Cambridge University Press, Cambridge, 2004.

Hewitt PG. *Conceptual Physics*. 11th ed., photocopy ed. Machinery Industry Press, Beijing, 2012.

Hughes WF, Brighton JA. *Fluid Dynamics*. Translated by Yanhou Xu et al. Science Press, Beijing, 2002. (Chinese translation.)

Incropera FP, DeWitt DP, Bergman T, et al. *Fundamentals of Heat and Mass Transfer*. Chemical Industry Press, Beijing, 2014.

Krause E. *Fluid Mechanics with Problems and Solutions, and an Aerodynamic Laboratory*. Springer, New York, 2005.

Kuethe AM. *Foundations of Aerodynamics*. 5th ed. Wiley, New York, 1998.

Kundu PK, Cohen I, Dowling DR. *Fluid Mechanics*. 5th ed. Academic Press, New York, 2012.

Landau LD, Lifshitz EM. *Fluid Mechanics*. 2nd ed., photocopy ed. World Book Publishing Company, Beijing, 1989.

Suxun Li. *Complex Flow Dominated by Shock Wave and Boundary Layer*. Science Press, Beijing, 2007.

Peiqing Liu. *The Theory of Free Turbulent Jets*. Beihang University Press, Beijing, 2008.

Yuxin Liu. *Thermology*. Peking University Press, Beijing, 2004.

Genhai Mao. *Wonderful Fluid Sports Science*. Zhejiang University Press, Hangzhou, 2012.

Takei Masahiro. *Fluid Dynamics Cartoons*. Translated by Pijuan Gao. Science Press, Beijing, 2010. (Chinese translation.)

Munson BR, Rothmayer AP, Okiishi TH, et al. *Fundamentals of Fluid Mechanics*. 6th ed. Wiley, New York, 2009.

Oertel H. *Prandtl's Essentials of Fluid Mechanics*. Translated by Ziqiang Zhu et al. Science Press, Beijing, 2008. (Chinese translation.)

Jinshan Pan. *Fundamentals of Aerodynamics*. Rev. ed. Northwest University of Technology Press, Xi'an, 1995.

Jinshan Pan, Peng Shan, Huoxing Liu, et al. *Fundamentals of Aerodynamics*. Rev. ed. National Defense Industry Press, Beijing, 1989.

Zeyan Peng, Gang Liu, Xingmin Gui, et al. *Principles of Aviation Gas Turbine*. National Defense Industry Press, Beijing, 2008.

Perelman YI. *Physics for Entertainment*. Harbin Press, Harbin, 2012.

Perelman YI. *Mechanics for Entertainment*. Harbin Press, Harbin, 2012.

Pope SB. *Turbulent Flows*. Cambridge University Press, Cambridge, 2000.

Yiji Qian. *Aerodynamics*. Beihang University Press, Beijing, 2004.

Schetz JA. *Boundary Layer Analysis*. Prentice-Hall, Hoboken, NJ, 1993.

Tennekes H, Lumley JL. *A First Course in Turbulence*. MIT Press, Cambridge, MA, 1972

Walker J. *The Flying Circus of Physics: Thermodynamics and Fluid problems*. Translated by Na Luo et al. Electronic Industry Press, Beijing, 2012. (Chinese translation.)

Baoguo Wang, Shuyan Liu, Weiguang Huang, et al. *Gas Dynamics*. Beijing Institute of Technology Press, Beijing, 2005.

Xianfu Wang, Aokui Xiong. *Advanced Fluid Mechanics*. Huazhong University of Science and Technology Press, Wuhan, 2003.

Xinyue Wang. *Fundamentals of Gas Dynamics*. Northwest University of Technology Press, Xi'an, 2006.

White FM. *Fluid Mechanics*. 4th ed. McGraw-Hill, New York, 1998.

White FM. *Viscous Fluid Flow*. 3rd ed. McGraw-Hill, New York, 2005.

Ziniu Wu. *Aerodynamics*, vol. 1. Tsinghua University Press, Beijing, 2007.

Wyngaard JC. *Turbulence in the Atmosphere*. Cambridge University Press, Cambridge, 2010.

Xueyu Xia, Xueying Deng. *Engineering Separation Dynamics*. Beihang University Press, Beijing, 1991.

Komine Yoshio. *Graphic Fluid Mechanics*. Translated by Pijuan Gao. Science Press, Beijing, 2012. (Chinese translation.)

Zhilun Xu. *A Concise Course in Elasticity*. Higher Education Press, Beijing, 1980.

Sanhui Zhang. *Thermology*. Tsinghua University Press, Beijing, 2004.

Zhaoshun Zhang. *Turbulence*. National Defense Industry Press, Beijing, 2002.

Zhaoshun Zhang, Guixiang Cui, Chunxiao Xu, et al. *Theory and Modeling of Turbulence*. Tsinghua University Press, Beijing, 2005.

Zixiong Zhang, Zengnan Dong. *Viscous Fluid Mechanics*. Tsinghua University Press, Beijing, 1999.

Guangdi Zhou. *The Prehistoric with Today's Fluid Mechanics Problems*. Peking University Press, Beijing, 2002.

Mingshan Zhu, Ying Liu, Zhaozhuang Lin, et al. *Engineering Thermodynamics*. Tsinghua University Press, Beijing, 2001.

Articles

Bernard SF Laplace and the speed of sound. *ISIS* 1964, Vol. 55, No. 179.

Blasius H. Grenzschichten in flüssigkeiten mit kleiner reibung. *Journal of Math and Physics*, 1908, Vol. 56. (English translation.)

Fowler M. Viscosity. University of Virginia, 2006.

Gandemer J. Discomfort due to wind near buildings: aerodynamic concepts. NBS, 1978.

Hodson HP, Hynes TP, Greitzer EM, et al. A physical interpretation of stagnation pressure and enthalpy changes in unsteady flow. *Journal of Turbomachinery*, 2012, Vol. 134, No. 6.

Hunt JCR. Lewis Fry Richardson and his contributions to mathematics, meteorology, and models of conflict. *Annual Review of Fluid Mechanics*, 1998, Vol. 30, No. 1.

Hunt JCR, Abell CJ, Peterka JA, et al. Kinematical studies of the flows around free or surface-mounted obstacles: applying topology to flow visualization. *Journal of Fluid Mechanics*, 1978, Vol. 86, No. 1.

Jimenez J. The physics of wall turbulence. *Physica A*, 1999, Vol. 263.

Joseph DD. Potential flow of viscous fluids: historical notes. *International Journal of Multiphase Flow*, 2006, Vol. 32.

Jovan J, Frohnapfel B, Jovanovic M, et al. Persistence of the laminar regime in a flat plate boundary layer at very high Reynolds number. *Thermal Science*, 2006, Vol. 10, No. 2.

Oke TR. Street design and urban canopy layer climate. *Energy and Buildings*, 1988, Vol. 11.

Panton RL. Overview of the self-sustaining mechanisms of wall turbulence. *Progress in Aerospace Sciences*, 2001, Vol. 37.

Purcell EM. Life at low Reynolds number. *American Journal of Physics*, 1997, Vol. 45, No. 1.

Reneau LR, Johnston P, Kline SJ. Performance and design of straight two-dimensional diffusers. *Journal of Fluids Engineering: Series D*, 1967, Vol. 89, No. 1.

Simpson RL. Aspects of turbulent boundary-layer separation. *Progress of Aerospace Science*, 1996, Vol. 32.

Wood RM. Aerodynamic drag and drag reduction: energy and energy savings.41st Aerospace Sciences Meeting and Exhibit, Reno, 2003.

Youschkevitch AP. A. N. Kolmogorov: historian and philosopher of mathematics on the occasion of his 80th birthday. *Historia Mathematica*, 1983, Vol. 10.

Keqin Zhu. Magical superfluid. *Mechanics and Engineering*, 2010, Vol. 32, No. 1.

Printed in the United States
by Baker & Taylor Publisher Services